Springer Series in Computational Neuroscience

Volume 3

For other titles published in this series, go to
http://www.springer.com/series/8164

Krešimir Josić • Jonathan Rubin
Manuel A. Matías • Ranulfo Romo
Editors

Coherent Behavior in Neuronal Networks

Springer

Editors
Krešimir Josić
Dept. Mathematics
University of Houston
651 Phillip G. Hoffman Hall
Houston TX 77204-3008
USA
josic@math.uh.edu

Manuel A. Matías
IFISC
CSIC-UIB
07122 Palma de Mallorca
Spain
manuel@ifisc.uib-csic.es

Jonathan Rubin
Dept. Mathematics
University of Pittsburgh
301 Thackeray Hall
Pittsburgh PA 15260
USA
rubin@math.pitt.edu

Ranulfo Romo
Universidad Nacional
Autónoma de México
Instituto de Fisiología Celular
04510 Mexico, D.F.
Mexico
rromo@ifc.unam.mx

Cover illustration: Neuronal Composition. Image by Treina Tai McAlister.

ISBN 978-1-4614-2447-5 e-ISBN 978-1-4419-0389-1
DOI 10.1007/978-1-4419-0389-1
Springer Dordrecht Heidelberg London New York

Printed on acid-free paper

Springer is part of Springer Science+Business Media (www.springer.com)

Preface

New developments in experimental methods are leading to an increasingly detailed description of how networks of interacting neurons process information. These findings strongly suggest that dynamic network behaviors underlie information processing, and that these activity patterns cannot be fully explained by simple concepts such as synchrony and phase locking. These new results raise significant challenges, and at the same time offer exciting opportunities, for experimental and theoretical neuroscientists. Moreover, advances in understanding in this area will require interdisciplinary efforts aimed at developing improved quantitative models that provide new insight into the emergence and function of experimentally observed behaviors and lead to predictions that can guide future experimental investigations.

We have undertaken two major projects to promote the translation of these new developments into scientific progress. First, we organized the workshop *Coherent behavior in neuronal networks*, which took place on October 17–20, 2007, in Mallorca, Spain, funded by the US National Science Foundation, the Spanish Ministerio de Educación y Ciencia, Govern de les Illes Balears, the Office of Naval Research Global, Universitat de les Illes Balears, the Consejo Superior de Investigaciones Científicas, the University of Houston, the University of Pittsburgh, and the Ajuntament de Palma de Mallorca. This unique workshop brought together a highly interdisciplinary and international mixture of 95 researchers with interests in the functional relevance of, and the mechanisms underlying, coherent behavior in neuronal networks. Reflecting the belief that understanding coherent behavior in neuronal networks requires interdisciplinary approaches, a key component of the meeting was the inclusion of linked back-to-back talks by experimental and theoretical collaborators, on their joint research endeavors. Scientifically, the meeting was structured around multiple themes, including the possible roles of globally coherent rhythms in the coordination of distributed processing, the possible roles of coherence in stimulus encoding and decoding, the interplay of coherence of neuronal network activity with Hebbian plasticity, and the mechanisms and functional implications of repeated spiking sequences. Participants responded quite positively to the workshop, expressing a strong desire for further activities to encourage the exchange of ideas and establishment of collaborative efforts in this field.

To address this need, and to reach a wider audience with interests in the broad area of coherent behavior in neuronal networks, our second project has been editing this volume. The chapters collected here include work from some workshop participants as well as some nonparticipants. The goal of the book is not to provide a summary of workshop activities but rather to provide a representative sampling of the diverse recent research activities and perspectives on coherent behavior in neuronal networks, and to serve as a resource to the research community. Nonetheless, we have made sure that the interdisciplinary flavor of the workshop has extended to this volume. Indeed, many of the chapters are coauthored by collaborating theorists and experimentalists. We hope that these chapters will provide useful examples of how theoretical abstractions can be derived from experimental data and used to attain general, mechanistic insights, and how theoretical insights can guide experiments in turn. Several chapters also include reviews or examples of novel methodologies, some experimental and some theoretical, that may be useful in analyzing coherent behavior in neuronal networks.

Scientifically, the book starts with a focus on ongoing or persistent cortical activity, as a baseline upon which sensory processing and faster oscillations must occur. In particular, the first chapters consider spatiotemporal patterning of synaptic inputs during such states, as well as the more abstract question of identifying repeating motifs within these inputs. From there, the book moves to small networks and small-scale interactions, including input-dominated cultured networks, which are particularly well suited for the study of how network dynamics interact with plasticity in an ongoing feedback cycle. Next, we return to larger scale but abstract issues, but with a shift in focus to the spatiotemporal relationships observed in the activity patterns of different cells, such as synchrony or causality. Subsequent chapters offer a broad survey of coherence in encoding and decoding, such as in stimulus discrimination and perception across systems such as motor, olfactory, and visual, with a particular emphasis on the role of noise.

We believe this book is suitable for special topics courses for graduate students, particularly in interdisciplinary neuroscience training programs, and for interdisciplinary journal club discussions. More broadly, we hope this volume will be a valuable resource for the many researchers, across a wide variety of disciplines, who are working on problems relating to neuronal activity patterns. We look forward to following and participating in future developments in the field, as interdisciplinary collaborations become increasingly widespread and continue to generate exciting advances in our understanding of coherent behavior in neuronal networks.

Houston, TX — Krešimir Josić
Pittsburgh, PA — Jonathan Rubin
Palma de Mallorca, Spain — Manuel A. Matías
Mexico, D.F. — Ranulfo Romo

Contents

On the Dynamics of Synaptic Inputs During Ongoing Activity in the Cortex ... 1
Michael Okun, Alik Mokeichev, Yonatan Katz, and Ilan Lampl

Timing Excitation and Inhibition in the Cortical Network ... 17
Albert Compte, Ramon Reig, and Maria V. Sanchez-Vives

Finding Repeating Synaptic Inputs in a Single Neocortical Neuron ... 47
Gloster Aaron

Reverberatory Activity in Neuronal Networks ... 61
Pak-Ming Lau and Guo-Qiang Bi

Gap Junctions and Emergent Rhythms ... 77
S. Coombes and M. Zachariou

The Feed-Forward Chain as a Filter-Amplifier Motif ... 95
Martin Golubitsky, LieJune Shiau, Claire Postlethwaite, and Yanyan Zhang

Gain Modulation as a Mechanism for Switching Reference Frames, Tasks, and Targets ... 121
Emilio Salinas and Nicholas M. Bentley

Far in Space and Yet in Synchrony: Neuronal Mechanisms for Zero-Lag Long-Range Synchronization ... 143
Raul Vicente, Leonardo L. Gollo, Claudio R. Mirasso, Ingo Fischer, and Gordon Pipa

Characterizing Oscillatory Cortical Networks with Granger Causality ... 169
Anil Bollimunta, Yonghong Chen, Charles E. Schroeder, and Mingzhou Ding

Neurophysiology of Interceptive Behavior in the Primate: Encoding and Decoding Target Parameters in the Parietofrontal System 191
Hugo Merchant and Oswaldo Pérez

Noise Correlations and Information Encoding and Decoding 207
Bruno B. Averbeck

Stochastic Synchrony in the Olfactory Bulb 229
Bard Ermentrout, Nathaniel Urban, and Roberto F. Galán

Stochastic Neural Dynamics as a Principle of Perception 247
Gustavo Deco and Ranulfo Romo

Large-Scale Computational Modeling of the Primary Visual Cortex 263
Aaditya V. Rangan, Louis Tao, Gregor Kovačič, and David Cai

Index 297

Contributors

Gloster Aaron Department of Biology, Wesleyan University, Middletown, CT 06459, USA, gaaron@wesleyan.edu

Bruno B. Averbeck Sobell Department of Motor Neuroscience and Movement Disorders, Institute of Neurology, UCL, London WC1N 3BG, UK, b.averbeck@ion.ucl.ac.uk

Nicholas M. Bentley Department of Neurobiology and Anatomy, Wake Forest University School of Medicine, Winston-Salem, NC 27157, USA, nbentley@wfubmc.edu

Guo-Qiang Bi Department of Neurobiology, University of Pittsburgh School of Medicine, Pittsburgh, PA 15261, USA, gqbi@pitt.edu

Anil Bollimunta J. Crayton Pruitt Family Department of Biomedical Engineering, University of Florida, Gainesville, FL 32611, USA, banil@ufl.edu

David Cai Courant Institute of Mathematical Sciences, New York University, 251 Mercer Street, New York, NY 10012, USA, cai@cims.nyu.edu

Yonghong Chen J. Crayton Pruitt Family Department of Biomedical Engineering, University of Florida, Gainesville, FL 32611, USA, ychen@bme.ufl.edu

Albert Compte Institut d'Investigacions Biomèdiques August Pi i Sunyer (IDIBAPS), 08036 Barcelona, Spain, acompte@clinic.ub.es

Stephen Coombes School of Mathematical Sciences, University of Nottingham, Nottingham NG7 2RD, UK, stephen.coombes@nottingham.ac.uk

Gustavo Deco Institució Catalana de Recerca i Estudis Avançats (ICREA), Universitat Pompeu Fabra, Passeig de Circumvallació, 8, 08003 Barcelona, Spain, gustavo.deco@upf.edu

Mingzhou Ding J. Crayton Pruitt Family Department of Biomedical Engineering, University of Florida, Gainesville, FL 32611, USA, mding@bme.ufl.edu

Bard Ermentrout Department of Mathematics, University of Pittsburgh, Pittsburgh, PA 15260, USA, bard@math.pitt.edu

Ingo Fischer School of Engineering and Physical Science, Heriot-Watt University, Edinburgh EH14 4AS, UK, I.Fischer@hw.ac.uk

Roberto F. Galán Department of Neurosciences, Case Western Reserve University School of Medicine, Cleveland, OH 44106-4975, USA, rfgalan@case.edu

Leonardo L. Gollo Instituto de Física Interdisciplinar y Sistemas Complejos (IFISC), Universitat de les Illes Balears-CSIC, Crta. de Valldemossa km 7.5, 07122 Palma de Mallorca, Spain, leonardo@ifisc.uib-csic.es

Martin Golubitsky Mathematical Biosciences Institute, Ohio State University, 1735 Neil Avenue, Columbus, OH 43210, USA, mg@mbi.ohio-state.edu

Yonatan Katz Department of Neurobiology, Weizmann Institute of Science, Rehovot 76100, Israel, yonatan.katz@weizmann.ac.il

Gregor Kovačič Mathematical Sciences Department, Rensselaer Polytechnic Institute, 110 8th Street, Troy, NY 12180, USA, kovacg@rpi.edu

Ilan Lampl Department of Neurobiology, Weizmann Institute of Science, Rehovot 76100, Israel, ilan.lampl@weizmann.ac.il

Pak-Ming Lau Department of Neurobiology, University of Pittsburgh School of Medicine, Pittsburgh, PA 15261, USA

Hugo Merchant Instituto de Neurobiología, UNAM, Campus Juriquilla, Querétaro Qro. 76230, México, merchant@inb.unam.mx

Claudio R. Mirasso Instituto de Física Interdisciplinar y Sistemas Complejos (IFISC), Universitat de les Illes Balears-CSIC, Crta. de Valldemossa km 7.5, 07122 Palma de Mallorca, Spain, claudio@ifisc.uib-csic.es

Alik Mokeichev Department of Neurobiology, Weizmann Institute of Science, Rehovot 76100, Israel, alik.mokeichev@weizmann.ac.il

Michael Okun Department of Neurobiology, Weizmann Institute of Science, Rehovot 76100, Israel, michael.okun@weizmann.ac.il

Oswaldo Pérez Instituto de Neurobiología, UNAM, Campus Juriquilla, Querétaro Qro. 76230, México

Gordon Pipa Department of Neurophysiology, Max-Planck Institute for Brain Research, Deutschordenstrasse 46, 60528 Frankfurt, Germany, pipa@mpih-frankfurt.mpg.de

Claire Postlethwaite Department of Mathematics, University of Auckland, Private Bag 92019, Auckland, New Zealand, c.postlethwaite@math.auckland.ac.nz

Aaditya V. Rangan Courant Institute of Mathematical Sciences, New York University, 251 Mercer Street, New York, NY 10012, USA, rangan@cims.nyu.edu

Ramon Reig Institut d'Investigacions Biomèdiques August Pi i Sunyer (IDIBAPS), 08036 Barcelona, Spain, rreig@clinic.ub.es

Ranulfo Romo Instituto de Fisiología Celular, Universidad Nacional Autónoma de México, 04510 México, D.F., México, rromo@ifc.unam.mx

Emilio Salinas Department of Neurobiology and Anatomy, Wake Forest University School of Medicine, Winston-Salem, NC 27157, USA, esalinas@wfubmc.edu

Maria V. Sanchez-Vives Institut d'Investigacions Biomèdiques August Pi i Sunyer (IDIBAPS), 08036 Barcelona, Spain
Institució Catalana de Recerca i Estudis Avançats (ICREA), 08010 Barcelona, Spain, msanche3@clinic.ub.es

Charles E. Schroeder Nathan Kline Institute for Psychiatric Research, Orangeburg, NY 10962, USA
Columbia University College of Physicians and Surgeons, New York, NY 10027, USA, schrod@nki.rfmh.org

LieJune Shiau Department of Mathematics, University of Houston, Clear Lake, Houston, TX 77058, USA, shiau@uhcl.edu

Louis Tao Center for Bioinformatics, National Laboratory of Protein Engineering and Plant Genetics Engineering, College of Life Sciences, Peking University, Beijing 100871, People's Republic of China, taolt@mail.cbi.pku.edu.cn

Nathaniel Urban Department of Biology, Carnegie Mellon University, Pittsburgh, PA, USA, nurban@cmu.edu

Raul Vicente Department of Neurophysiology, Max-Planck Institute for Brain Research, Deutschordenstrasse 46, 60528 Frankfurt, Germany, raulvicente@mpih-frankfurt.mpg.de

Margarita Zachariou School of Mathematical Sciences, University of Nottingham, Nottingham NG7 2RD, UK, margarita.zachariou@nottingham.ac.uk

Yanyan Zhang Department of Mathematics, Ohio State University, Columbus, OH 43210, USA, yzhang@math.ohio-state.edu

On the Dynamics of Synaptic Inputs During Ongoing Activity in the Cortex

Michael Okun[1], Alik Mokeichev[1], Yonatan Katz, and Ilan Lampl

Abstract In this chapter, we provide an overview of the dynamical properties of spontaneous activity in the cortex, as represented by the subthreshold membrane potential fluctuations of the cortical neurons. First, we discuss the main findings from various intracellular recording studies performed in anesthetized animals as well as from a handful of studies in awake animals. Then, we focus on two specific questions pertaining to random and deterministic properties of cortical spontaneous activity. One of the questions is the relationship between excitation and inhibition, which is shown to posses a well-defined structure, owing to the spatio-temporal organization of the spontaneous activity in local cortical circuits at the millisecond scale. The other question regards the spontaneous activity at a scale of seconds and minutes. Here, examination of repeating patterns in subthreshold voltage fluctuations failed to reveal any evidence for deterministic structures.

Introduction

Even in the absence of sensory stimuli, cortical activity is highly prominent. At the single-cell level, spontaneous activity in the cortex is observed using extracellular, intracellular, and calcium imaging recordings, whereas populations of cells can be seen using voltage sensitive dyes. At a larger scale, spontaneous activity can be observed in EEG, MEG, and fMRI recordings. In this chapter, we focus on the ongoing cortical activity as expressed by the dynamics of the subthreshold membrane potential of single neurons. Since synaptic inputs are the main cause of membrane potential fluctuations in cortical neurons [51], this technique is one of the most powerful tools to probe the network activity. The intracellular recording technique

I. Lampl (✉)
Department of Neurobiology, Weizmann Institute of Science, Rehovot 76100, Israel
e-mail: ilan.lampl@weizmann.ac.il

[1] Equal contribution.

K. Josić et al. (eds.), *Coherent Behavior in Neuronal Networks*, Springer Series in Computational Neuroscience 3, DOI 10.1007/978-1-4419-0389-1_1,

provides the most accurate data in terms of spatial and temporal precision, which comes at the expense of low yield of recorded cells and limited recording duration, because of the mechanical sensitivity of the technique. Nevertheless, an increasing number of studies have used this method to unveil the dynamics of spontaneous activity in the cortex.

A particularly distinctive feature of the subthreshold dynamics in cortical neurons is the appearance of Up-Down states of membrane potential, originally described in anesthetized cats [52] and rats [17]. The Up-Down dynamics is characterized by large (10–20 mV) depolarizations relative to the baseline potential, lasting for several hundreds of milliseconds (the Up state), resulting in bimodal membrane potential distribution (Fig. 1a). This activity pattern was also observed in other species, including mice [42] and ferrets [23, 25]. Indirect EEG evidence for the presence of Up-Down states is also available for monkeys [39] and humans [4,52]. In a series of studies in drug-free cats, it was found that Up-Down dynamics occurs during slow wave sleep (SWS) [53,54]. Similar behavior during SWS and periods of drowsiness was observed in rats and mice as well [33,42]. On the other end of scale, Up-Down dynamics was also reproduced in slices [49].

While Up-Down dynamics is readily observed under some conditions of anesthesia (urethane, ketamine-xylazine), quite a different activity pattern, characterized by rather short (10–50 ms) depolarizations and membrane potential distribution that is not bimodal, emerges with other anesthetics (most distinctively the inhaled ones, such as isoflurane and halothane). This kind of activity appears to be a manifestation of lighter anesthesia when compared with the Up-Down dynamics, since the bimodal distribution of the membrane potential tends to appear when the concentration of the inhaled anesthetic is increased (unpublished results). Furthermore, under light gas anesthesia membrane dynamics is more similar to the activity observed in awake animals (see below).

Since it is plausible that the spontaneous dynamics in awake animals differs substantially from the anesthetized condition, intracellular recordings of cortical neurons in awake animals have been performed as well. Rather unfortunately these data are also most experimentally demanding to obtain, since intracellular recordings are extremely sensitive to mechanical instabilities, which are almost inevitable in awake, drug-free animals. At present time only a handful of such studies were performed, mostly in head fixed animals: monkeys [14, 35], cats [8, 53, 54], rats [12, 20, 22, 34, 40], mice [18, 43], and bats [16]. A methodology for whole-cell recording in behaving rodents is being developed as well [32].

Perhaps somewhat surprisingly, there exist large discrepancies between these studies. Two recent investigations reported diametrically opposing results: one group recorded neurons from the primary auditory cortex (A1) of rats [20, 27] and the other recorded from the parietal association cortex in cats [46]. According to Zador and his colleagues, the spontaneous subthreshold activity in the rat A1 is characterized by infrequent large positive excursions ("bumps"), resulting in membrane potential distribution with sharp peak and heavy tail at its positive side (average kurtosis of $\sim$15), quite distinct from the Gaussian distribution. On the contrary, in [46] the membrane potential exhibits activity resembling a continuous Up state,

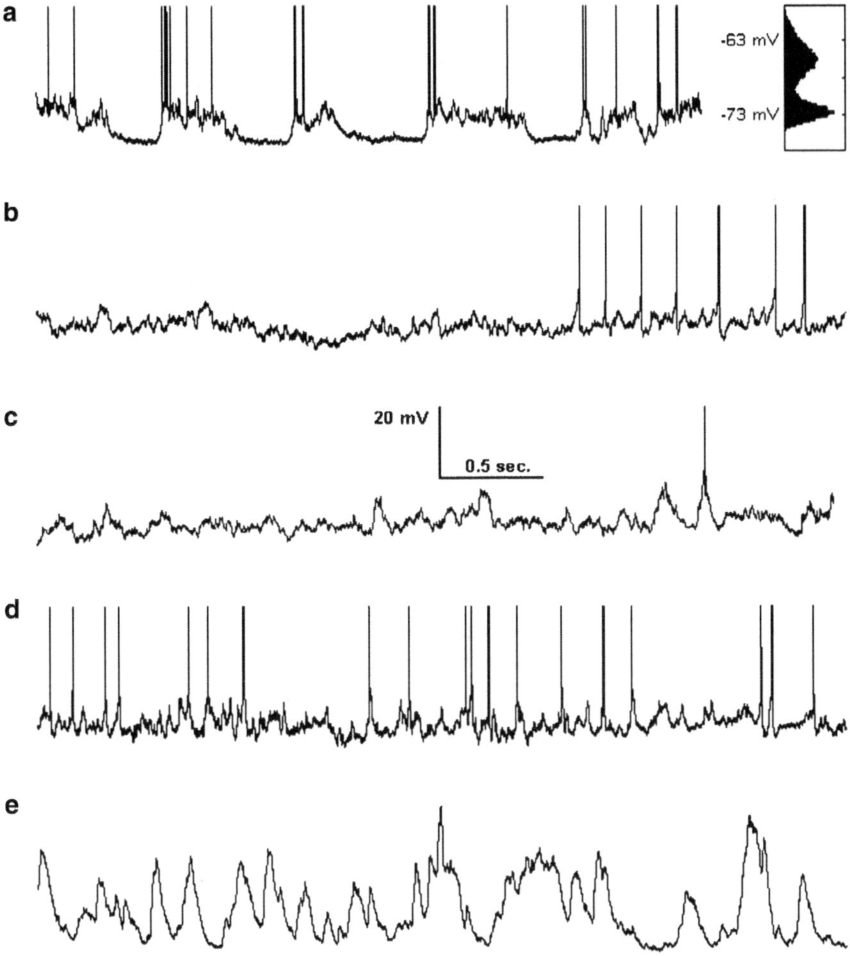

Fig. 1 Examples of subthreshold spontaneous activity. (**a**) Up-Down dynamics in a neuron in the primary somatosensory cortex (S1) of ketamine-anesthetized rat, and the resulting membrane potential distribution. (**b**) Spontaneous activity in parietal association cortex of an awake cat, data from [9]. (**c**) Spontaneous activity in A1 of an awake rat, data from [27]. (**d**) Spontaneous activity in S1 of an awake rat, data from [40]. (**e**) Spontaneous activity in S1 of an awake mouse, data from [18] (To have a uniform scale, in all panels the data is reprinted in modified form from the original publication, and spikes are cut.)

characterized by frequent, small fluctuations and membrane potential distribution which is close to normal. In particular, independent excitatory and inhibitory synaptic inputs that follow the Ornstein–Uhlenbeck stochastic differential equation are shown to provide a rather accurate approximation of the observed activity.

Intracellular recordings in awake animals were carried out in several additional works, but unlike the two papers above, in these studies the investigation of spontaneous dynamics was not the primary goal. Nevertheless, they provide an additional opportunity to examine the ongoing activity in the cortex. In most of these studies recordings were conducted in the rodent barrel cortex: [12, 18, 22, 34, 43]. When inspecting these recordings, as well as our own (Fig. 1d), the dynamics appears to be somewhere in between the two extremes of [20, 27] and [46]. On the one hand, the potential fluctuations do not seem to be produced by entirely uncorrelated synaptic inputs, as suggested in [46] while at the same time the bumps are smaller and more frequent than in [20]. In particular, in our recordings voltage distribution is approximately normal (kurtosis ~0). However, we note that the presently available experimental data on the patterns of spontaneous activity in the barrel cortex of awake rodents are not fully consistent on their own, since recordings in mice [18, 43], see Fig. 1e, show a very bumpy activity. In these mice studies, bump amplitude appears to be several times larger than in rats and more importantly their durations are substantially longer than in the rat traces.

At the present stage we are not aware of any persuasive explanation for the discrepancies just described. Possible factors that might contribute to the observed differences are the animal species used, the cortical areas, layers, and specific neuron types from which the recordings were made, as well as the state of the animal (its level of stress, alertness, etc.). At the first sight the discrepancies between the handful of currently available datasets seem to be of a highly significant nature. Because ongoing activity can have substantial effect on the neural response to sensory stimuli, e.g., see [24, 42], cortical spontaneous activity may play a significant role in sensory processing. However, it is not clear whether the differences in spontaneous cortical dynamics are manifested during behavioral states, such as sensory processing, memory tasks, attention, and awareness. Though it is unlikely, it might be the case that these large differences from the point of view of the researcher are of no major importance for the processing of information in the cortex.

Synchrony in Spontaneous Activity

A significant difference in the amount of synchrony at the network level exists between the model of spontaneous activity proposed in [46] (Fig. 1b) and that of [20, 27] (Fig. 1c). In the first case, the dynamics is suggested to be asynchronous, with each presynaptic neuron firing independently of the others. However, the distinctive short bumps, as in Fig. 1c, indicate that firing of hundreds of presynaptic neurons is synchronized, since unitary synaptic potentials (uPSPs) are of the order of 1 mV or less. Owing to the enormous connectivity in the cortex, even if the presynaptic neurons do fire in synchrony, it is possible that nearby neurons receive inputs from independent pools of inputs. Simultaneous dual intracellular recordings, however, indicate that neurons in the local cortical network receive synaptic inputs with highly similar pattern and magnitude. In anesthetized animals, dual recordings reveal a very high correlation between the subthreshold activities in pairs of cells (Fig. 2a) [25, 31, 40, 56]. Since there is good evidence that the probability of nearby cells to receive inputs from the same presynaptic neuron is low, e.g., [50], and since a lag of several milliseconds may exist between their activities [31,33,40,56], this synchrony indicates that most of the cortical spontaneous activity consists of *waves*, with large portions of local cortical circuits participating in their propagation [29,33,42].

It was shown that synchrony in the cortex increases with the level of anesthesia [15]. For example, the Up-Down activity (Fig. 1a) is synchronous across large areas of the cortex (several mm apart) [56]. The high synchrony of large neuronal populations during Up-Down activity is further evidenced by high correlation between membrane potential and EEG, e.g., see [52]. Moreover, even the small synaptic events that appear *within* individual Up-states are synchronized across cells [25].

Under lighter anesthesia conditions that do not exhibit Up-Down dynamics, coherent subthreshold activity in pairs of nearby neurons is still observed even though global brain synchrony is very low, which is evident from the low correlation of EEG and membrane potential [31,40]. Possibly this kind of local synchrony is similar to the synchrony existing within single Up states. The degree of synchrony in an awake animal is probably even lower than in the lightly anesthetized one; however, recent imaging studies and dual intracellular recordings in awake animals [43] show that it does not disappear altogether.

Excitation and Inhibition During Spontaneous Activity

The interplay between the excitatory and inhibitory synaptic inputs is a long studied topic in the computational neuroscience community. A careful examination of the spiking statistics of single cortical neurons has suggested that they are constantly bombarded by excitatory and inhibitory inputs that on average balance each other [47,48]. Furthermore, theoretical studies showed that networks which exhibit sustained activity with an approximate balance between the excitatory and inhibitory neurons indeed exist, e.g., see the review in [55].

Intracellular recordings allow direct measurement of the excitatory and the inhibitory synaptic inputs associated with some reproducible stereotypical event (such as a sensory stimulus), by a method introduced in [11] (for an up-to-date review see [38]). Using this methodology, a balanced excitatory and inhibitory activity was indeed discovered during Up states in the ferret cortex, both in vitro [49] and in vivo [23]. Specifically, it was found that in the beginning of an Up state, both synaptic conductances are high and they tend to progressively decrease, but their ratio remains constant and approximately equal to 1. It should be noted, however, that questions relating to the balance between excitation and inhibition during the Up state are still not settled. A study of Up-Down activity in association cortex of awake cats [46] reports inhibitory conductance that is several times higher than the excitatory one. On the contrary, it was also argued that during Up states in the rat somatosensory cortex the inhibitory conductance is only about 10% of the excitatory conductance and that the duration of an Up state is at least partially determined by intrinsic mechanisms [58].

It is important to observe that a particular membrane potential value can be produced by different combinations of excitatory and inhibitory synaptic inputs. For example, a positive change in membrane potential can be caused both by an increased excitation without any significant change in inhibition, and by withdrawal of inhibitory input (disinhibition). Hence, the single-cell conductance measurement methodology, whether in voltage or current clamp mode, can only provide the mean relation between these inputs, calculated from the average event recorded at different holding potentials. Because of this important limitation, the relationship between the excitatory and inhibitory inputs during spontaneous activity which does not exhibit stereotypic Up states remained unknown.

As we already discussed at length in the Introduction, different studies report very distinct dynamics of ongoing activity. However, the presence of synchrony in inputs of nearby neurons appears to be common to all types of activity, since it was observed in all studies involving dual intracellular recordings or single-electrode intracellular with nearby imaging or LFP recordings [19, 31, 40, 42, 56], though to a different degree. This synchrony provides a method for measuring at the same time both the excitatory and the inhibitory synaptic inputs to the local circuit, by means of simultaneous dual intracellular recording [40]. In each pair of neurons, one cell is recorded near the reversal potential of inhibition so that positive excursions of its membrane potential reflect excitatory currents, at the same time a positive current is injected into the other cell to reveal inhibitory potentials. In fact, this is the only presently available experimental methodology that provides an adequate single-trial picture of the magnitude and timing of both excitatory and inhibitory inputs and that is suitable for elucidating the excitatory–inhibitory dynamics during ongoing activity and evoked responses. An example of such a recording is presented in Fig. 2. The two cells receive synchronized excitatory (Fig. 2a) and inhibitory (Fig. 2b) inputs. Furthermore, when one cell was depolarized (by positive current injection) to reveal inhibitory potentials, while the second cell was recorded near its resting level (Fig. 2c and vice versa in Fig. 2d), a high synchrony between the excitatory and inhibitory potentials was revealed.

The shape and amplitude of the synaptic events are highly variable when recorded at the neuron's resting potential (e.g., as in Fig. 2a). A priori, the variability in the amplitude of these events could reflect considerable changes in the contribution of the excitatory and inhibitory synaptic inputs, ranging between the following two diametrically opposing possibilities. It might be the case that large bumps occur when the inhibitory activity, which can shunt the excitation, is weak, whereas small bumps reflect shunted excitatory inputs. Alternatively, both types of synaptic inputs might reflect the overall level of activity in the local network and go hand in hand. We have used the same experimental approach to resolve this issue. We found that the amplitudes of spontaneous events were significantly correlated, both for the depolarizing potentials, when the two cells were held near their resting potential, and when one cell was depolarized to reveal inhibitory potentials (Fig. 2e–f). Hence, the latter alternative is correct; that is, the larger the excitatory drive in the local circuit, the larger the inhibitory one. In addition, we used one of the cells as a reference to measure the relative timing of excitatory and inhibitory bumps in the

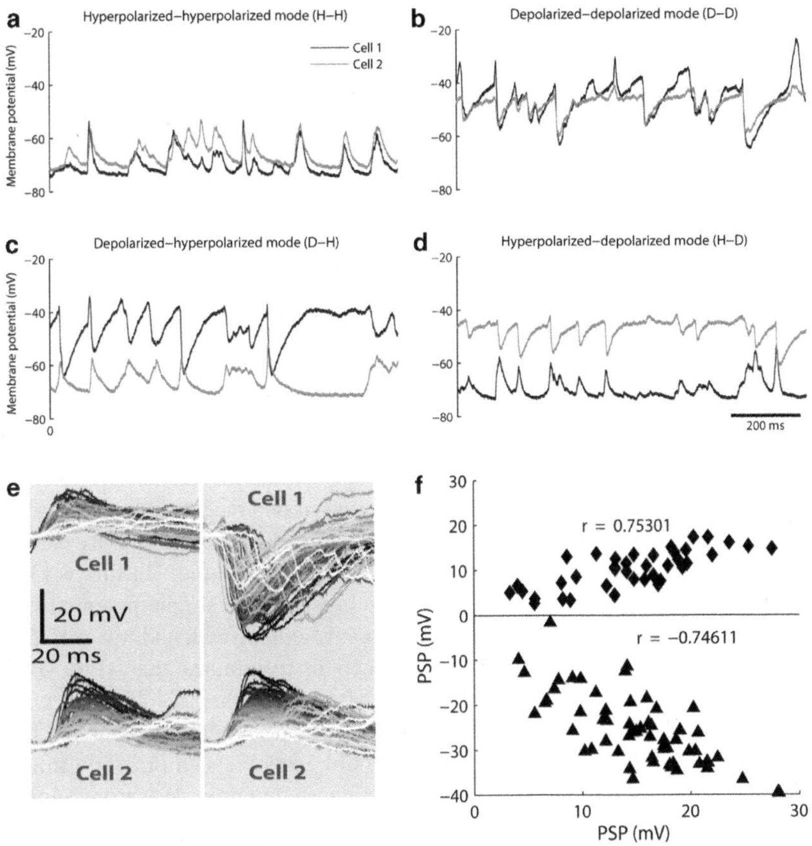

Fig. 2 Correlation of excitatory and inhibitory synaptic inputs during spontaneous activity. (**a–d**) Examples of simultaneous dual intracellular recordings, when both cells are hyperpolarized (**a**), depolarized (**b**) and intermixed (**c, d**). (**e**) Synaptic events in the second cell (Cell 2) of a pair are shown at the bottom, sorted by their amplitude (indicated by the color intensity). The corresponding events in Cell 1 are shown above with the same color. At first, both neurons were in the hyperpolarized mode. Then, the first cell was recorded in the depolarized mode. (**f**) Amplitudes of the events presented in (**e**). A significant correlation between the amplitudes is clearly visible in each case.

other cell. We found that during spontaneous activity the onset of inhibition lags by few milliseconds behind the excitatory input (Fig. 3).

The same experimental method was used more recently in an in vitro study of gamma oscillations in the hippocampus. This study also found a tight amplitude correlation between the excitatory and inhibitory inputs when analyzed on a cycle by cycle basis [6]. The coordinated activity of excitation and inhibition across two neurons strongly suggests that variations in excitation and inhibition reflect mostly changes in the local network activity rather than "private" variability of the inputs of individual cells.

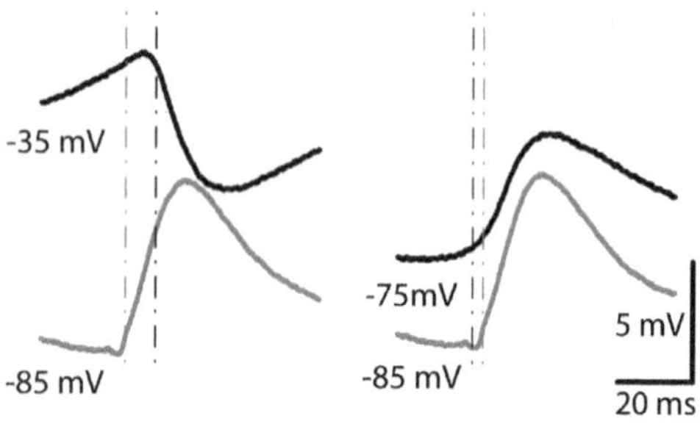

Fig. 3 Lag analysis based on onsets of the average events. Averages of the membrane potentials for the two recording conditions, triggered by events during ~100 s of recording. *Dashed lines* mark the onsets (10% from the peak).

Repeating Patterns in the Spontaneous Subthreshold Membrane Potential Fluctuations of Cortical Neurons

The notion that distinct and large depolarizing excursions of membrane potential reflect synchronized activity in the network was used in a recent study which reported on the surprising finding that patterns of spontaneous synaptic activity can repeat with high precision [28] (see also Chapter 3 in this volume). In this work, long and continuous records of spontaneous subthreshold membrane potential fluctuations obtained both in vivo and in brain slices were analyzed, and it was found that specific patterns of activity can reappear seconds to minutes apart (Fig. 4). These repeats of membrane potential fluctuations, also termed "motifs" [28, 37], typically span 1–2 s and include several large bumpy synaptic potentials separated by quiescent periods. What cortical mechanism could generate such precisely timed and long activity patterns? In [28], motifs were suggested to provide a strong supporting evidence for the existence of special cortical mechanisms, such as the synfire chain operation mode of the cortical network, which generate exact firing patterns with a millisecond precision.

According to the synfire chain model, cortical cells are organized into pools of cells [1]. Each pool is connected to the next by a set of diverging and converging connections, forming together a chain of pools. Despite the low transmission reliability between pairs of cells, a secure transmission of information is suggested to be accomplished by synchronous firing within each pool and its propagation from a pool of neurons to the following one. That is, after one pool of neurons was synchronously activated, a synchronous activation of the next pool of neurons follows with a typical synaptic delay. Processing of information may include multiple feedbacks, so that a single neuron might be a member in several pools along the chain. For such a neuron, activation of the chain would result in a sequence of large synaptic potentials, where each one is generated by a different group of presynaptic neurons. Because of the high reliability of the chain, such sequence of synaptic potentials is expected to repeat with high temporal precision once the first pool of neurons in the chain is reactivated, generating a repeated firing and synaptic pattern (motif).

The synfire chain model proposes an efficient mechanism for propagating a signal with a low number of spikes, in addition to a compact way of information encoding. The amount of information that might be encoded by precise temporal

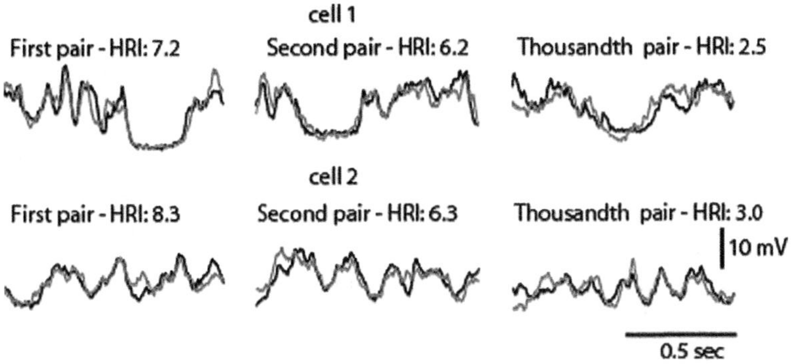

Fig. 4 Examples of repeating motifs from two different cells. The similarity between motif repetitions was quantified by the High Resolution Index (HRI, for further details see [28, 37] and Chapter 3 in this volume). Each *row* presents examples from the same cell, first the two motifs with the highest HRI values, then a motif which has the 1,000th highest HRI rank (still showing a marked similarity). Such highly similar motifs were found in all the recorded cells in spite of the differences in the statistical properties of their subthreshold activities.

structures is far larger when compared with encoding by spike rate or spike count alone [36]. If motifs do not result from stimulation locking [41], then they may support higher level processes such as binding together different features of an object [1, 21, 57]. Few synaptic connections between reverberating synfire chains may facilitate the binding of several smaller chains, representing specific features, into a larger super-assembly [2, 3, 10] which could be a substrate of higher level perceptual processes. For example, it has been recently suggested that large-scale spatio-temporal activation patterns spreading over superior-temporal and inferior-frontal cortices, observed during processing of speech stimuli, are best explained by interarea synfire chain propagation [45].

The synfire chain model was supported by an analysis of spike sequences recorded simultaneously from several neurons of the frontal cortex of behaving monkeys [3, 44], which showed that the number of spike motifs exceeded what could be expected by chance in surrogate data [3]. While the propagation of the signal might be purely feedforward, so that a motif includes only one spike per neuron, most of the motifs that were described in the above studies were composed of patterns of spikes originating from a single unit. Therefore, the authors concluded that synfire chains typically contain feedback connections [3,44].

Subsequent studies questioned the significance of temporal patterns by comparing the number of repeating patterns of spikes in the recordings and in surrogate data generated by shuffling the original spike times using different stochastic models [7,41]. Their analysis of spike times recorded from the lateral geniculate nucleus (LGN) and primary visual cortex of behaving monkeys showed that adding more constraints to the surrogate data brings the number of repeating patterns closer to the recorded data. Hence, choosing the right stochastic model has critical consequences regarding the conclusions. For example, in surrogates that preserved the number of spikes per trial and the firing probability distribution, the number of motifs was much closer to the original than in surrogates generated with more simple assumptions such as a Poisson process. Baker and Lemon and Oram et al. [7, 41] therefore suggested that the appearance of spike motifs reflects the coarse dynamics of firing rate modulations rather than the existence of special network mechanisms for their generation. A lack of evidence for the occurrence of precise spike patterns beyond what is expected by chance was also reported in [7], where spike patterns recorded from multiple units of the primary motor cortex (M1) and the supplementary motor area (SMA) in monkeys were analyzed.

Research aimed to find in the subthreshold traces the synaptic inputs that create the precise spike patterns observed in some of the above described studies was reported by Ikegaya et al. in [28]. In contrast to the above studies, which searched for precise spike patterns in awake behaving monkeys, the repeating patterns reported in [28] were found in subthreshold membrane potential fluctuations recorded *in-vivo* in anesthetized cats and were generated spontaneously, in the absence of external sensory stimulation. Recently, we have reexamined the statistical significance of such spontaneously repeating patterns in intracellular recordings from the rat barrel cortex (S1) and the cat primary visual cortex (V1) of anesthetized animals [37]. In most of the recordings, the dynamics of spontaneous activity was similar to those reported in [28]. Using a search algorithm similar to the one described by Ikegaya and his colleagues, we found a large number of motifs. To test their statistical significance, we used three different methods to generate surrogate data, each corresponding to a different model of randomness. In the first method, the surrogate data were constructed by a time domain shuffling of the original trace (Fig. 5a). In the second method, the data were randomized in the frequency domain. In the third method, the parameters of Poisson distributed synaptic inputs of a simulated passive cell were optimized to elicit an activity with dynamics similar to that of the recorded cell. Perhaps surprisingly, a large number of motifs were found in all types of surrogate data.

The close resemblance between the distributions of similarity scores of motifs found in physiological spontaneous activity and in the different types of surrogate

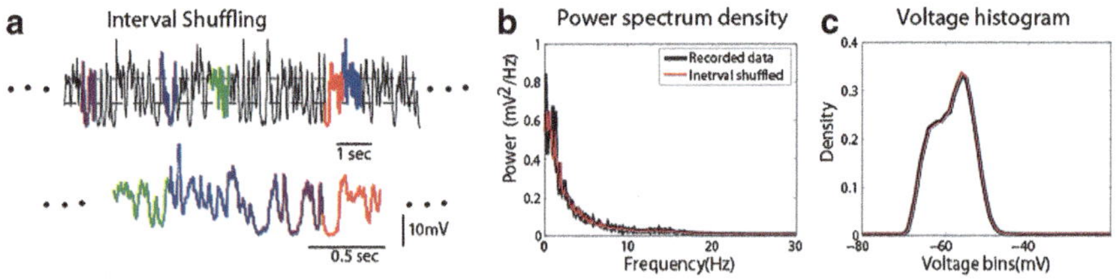

Fig. 5 Generation of surrogate data by time domain interval shuffling. (**a**) Time domain interval shuffling: Using two levels of potentials (determined from 1/3 and 2/3 of the density distribution), the data were fragmented into short segments. Each segment starts and ends at one of the two levels and its duration was the longest possible below 500 ms. Five different segments are marked. The fragments were then randomly assembled to generate a new continuous voltage trace. (**b**) Membrane potential distribution and (**c**) power spectrum of the recorded data and its time domain shuffled surrogate.

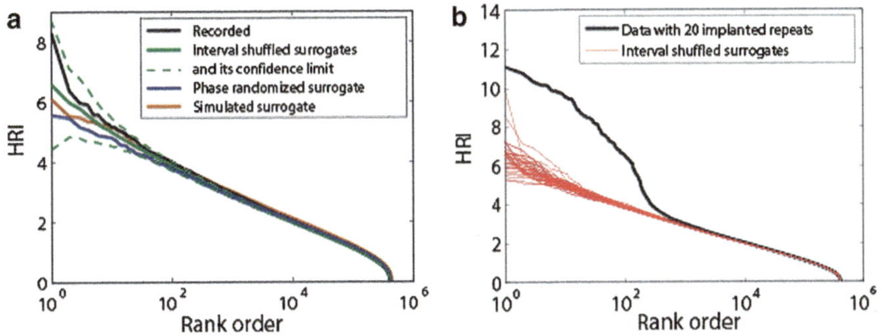

Fig. 6 (**a**) Example of HRI scores distributions in one of the cells analyzed in [37], sorted in decreasing order. Recorded data and the three types of surrogate data have very similar distributions of HRI scores. The interval shuffled curve is the average of 40 independent shuffles of the original recording, and the dashed green curves are its confidence intervals ($p = 0.99$). (**b**) Rank ordered HRI scores of all motifs found in surrogate data of 20 min in duration with 20 implanted motif repeats, and its 40 interval shuffled surrogates. To test whether motifs of high similarity that occur beyond chance level could be detected, we implanted into a surrogate data of 20 min in duration a single motif of 1-s duration that repeated every minute on average. The surrogate was produced from physiological recordings by shuffling it in the time domain. It is evident that the HRI scores of the top ranked motifs that were found in data with implanted motifs are much higher than those of all 40 shuffles. These results demonstrate that even a single motif that repeats several times with high similarity is identified by our methods.

data (Fig. 6a) suggests that the motifs in physiological data emerge by chance. Of the three methods for generating surrogate data, the time domain interval shuffling preserved both the power spectrum and the voltage distribution of recorded data most accurately (Fig. 5b–c). Surrogates produced with this method also had motif statistics that were closest to the original. These results suggest that physiological motifs could simply arise as a result of the coarse constraints on the subthreshold fluctuations dynamics, imposed by a wide range of properties of the cortical neuronal networks. An important issue of concern with any method that produces surrogate data from the original recordings is its effect on genuine motifs, if they do exist. To test this, we have implanted several highly similar repeats into a long continuous recording. Such artificial motif is easily detected by comparing the similarity scores distribution of the synthetic data to its shuffled versions (Fig. 6b). These results further support the conclusion that physiological records did not contain repeating motifs above what is expected at chance level.

Additional statistic that compared between physiological and surrogate data sets was the number of times motifs reappear. This particular comparison is of a particular interest since a theoretical analysis of synfire chains [26] demonstrates that typical cortical columns are not large enough to support many synfire chains of a duration as long as 1 s. Therefore, if the recorded 1 s repeats are generated at the column level, one would expect a small number of motifs that repeat numerous times rather than a large number of different motifs repeating small number of times. The analysis performed in [37] found that the statistics of the number of motif repetitions in the original traces and the surrogates was the same.

The tight continuous synchrony of spontaneous membrane potential fluctuations of cortical neurons that is observed across extended cortical regions [5, 56] also stands in contrast to the idea that the observed long repeating patterns reflect a propagation of synfire chain within a single cortical column. This discrepancy is further demonstrated by our experiment in which intracellular voltages were recorded simultaneously from pairs of neurons (in the barrel cortex of an anesthetized rat). Some of the synchronized cells were laterally separated by about $\sim$500 μm, thus typically belonged to two distinct columns and to different layers. The same measure that was used to quantify the degree of similarity between repeats of a motif may also be used to measure similarity of simultaneously (or almost simultaneously[2]) recorded epochs in a pair of cells. The inter-neuron similarity of simultaneous intervals was very high, much higher than the similarity between repeats within the recording of individual neuron (Fig. 7). This indicates that the vast majority of reoccurring temporal patterns in the spontaneous activity of the cortex do not reflect a column specific processing, rather they are a consequence of waves propagating across wider cortical regions. The dual recordings also provide an answer to an issue of concern not fully addressed in [37], regarding the possibility that the stochastic nature of motifs is due to intrinsic noise, unrelated to network activity. The results in Fig. 7 indicate that this is not the case. Finally, we note that [28] described in vitro motifs of firing sequences within a population of neurons whose size is of the order

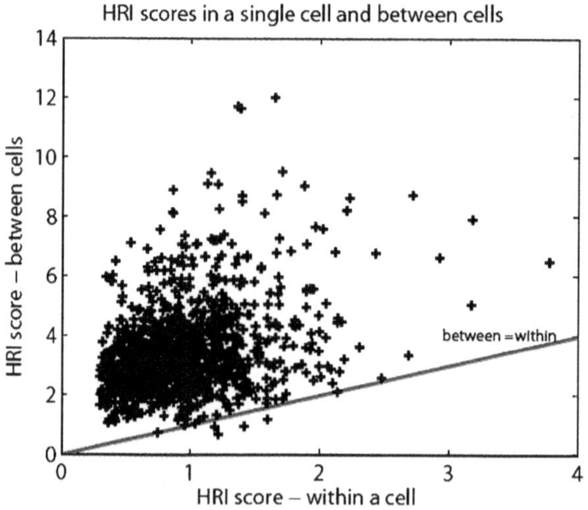

Fig. 7 Comparison between the similarity of motif repeats that were found within a continuous recording of a single cell, and the similarity of the first repeat to the corresponding epoch in another simultaneously recorded cell. Most points lie above the diagonal line. This implies that nearby cells have similar patterns of synaptic inputs and that these patterns do not reoccur with a comparable similarity at different times.

[2] In most of the cases, the temporal pattern of synaptic inputs recorded in one cell appeared in the other cell after a short delay of a few milliseconds, an indication of a traveling wave in the cortex.

of a single cortical column, an indication that the mechanisms that are involved in their generation are different than those that may support the generation of motifs in the much larger scales of in vivo cortical activity.

The above results provide no evidence for the existence of cortical mechanisms to support precisely timed, long-lasting spike patterns that propagate recurrently in the local cortical network. These results are consistent with other electrophysiological studies in vivo [30, 33, 48] that do not support a millisecond precision in cortical function. Moreover, [13] showed that the jitter of firing (measured by the standard deviation of the latency across multiple trials following local stimulation) is directly proportional to the latency of the propagating wave that activates them, suggesting that precise propagation of spike patterns cannot be maintained for long durations. However, we cannot exclude the existence of statistically significant motifs of much shorter duration. A study of Luczak et al. [33] has shown that different cortical cells exhibit unique patterns of modulation in firing rate that are associated with the onset of Up states and last around 100 ms. The typical patterns emerge regardless of the direction from which the propagating wave arrives, and therefore it has been hypothesized that they reflect the local functional organization of the network [33]. In summary, our data provide indications that long motifs of spontaneous activity are generated stochastically as a result of the coarse dynamics of cortical activity.

Conclusions

Spontaneous activity in the cortex was extensively investigated in the recent years, from the level of individual cells all the way to activity in whole cortical areas. The intracellular recording technique provides a powerful tool for probing cortical dynamics and synaptic inputs during spontaneous and evoked activities. In particular, our own works examined different and even opposing views regarding the organization of neuronal activity in the local cortical network, ranging from approximately independent firing to highly structured spike patterns, generated by synfire chains. On one hand, using in-vivo dual intracellular recordings of nearby cortical cells, we have shown that a highly coordinated activity in the local network exists at the millisecond time scale. This coordinated activity, in which both excitatory and inhibitory inputs participate, is manifested as brief bursts of network activity, interleaved with lower activity levels. On the contrary, on the time scale of seconds, spontaneous network activity appears to be governed by stochastic mechanisms, with no strong evidence for temporal patterns repeating above chance levels.

Acknowledgments We would like to thank Profs. Carl Petersen, Igor Timofeev, and Tony Zador for providing examples of intracellular recordings in awake animals from experiments conducted in their laboratories (Fig. 1). We thank all the members of Lampl lab for their contribution to this work. This work was supported by grants from the Israel Science Foundation (1037/03, 326/07), the National Institute for Psychobiology in Israel, by the Henry S. and Anne Reich Research Fund for Mental Health, the Asher and Jeanette Alhadeff Research Award, and Sir Charles Clore fellowship.

References

1. Abeles M (1991) Corticonics. Cambridge: Cambridge University press.
2. Abeles M (2004) Neuroscience. Time is precious. Science 304:523–524.
3. Abeles M, Bergman H, Margalit E, Vaadia E (1993) Spatiotemporal firing patterns in the frontal cortex of behaving monkeys. J Neurophysiol 70:1629–1638.
4. Achermann P, Borbely AA (1997) Low-frequency (<1 Hz) oscillations in the human sleep electroencephalogram. Neuroscience 81:213–222.
5. Arieli A, Shoham D, Hildesheim R, Grinvald A (1995) Coherent spatiotemporal patterns of ongoing activity revealed by real-time optical imaging coupled with single-unit recording in the cat visual cortex. J Neurophysiol 73:2072–2093.
6. Atallah BV, Scanziani M (2008) Proportional excitatory and inhibitory conductances are maintained during gamma oscillations. In: COSYNE. Salt Lake City, UT.
7. Baker SN, Lemon RN (2000) Precise spatiotemporal repeating patterns in monkey primary and supplementary motor areas occur at chance levels. J Neurophysiol 84:1770–1780.
8. Baranyi A, Szente MB, Woody CD (1993) Electrophysiological characterization of different types of neurons recorded in vivo in the motor cortex of the cat. I. Patterns of firing activity and synaptic responses. J Neurophysiol 69:1850–1864.
9. Bazhenov M, Timofeev I, Steriade M, Sejnowski TJ (2002) Model of thalamocortical slow-wave sleep oscillations and transitions to activated states. J Neurosci 22:8691–8704.
10. Bienenstock E (1995) A model of neocortex. Network-Computation in Neural Systems 6:179–224.
11. Borg-Graham LJ, Monier C, Fregnac Y (1998) Visual input evokes transient and strong shunting inhibition in visual cortical neurons. Nature 393:369–373.
12. Bruno RM, Sakmann B (2006) Cortex is driven by weak but synchronously active thalamocortical synapses. Science 312:1622–1627.
13. Buonomano DV (2003) Timing of neural responses in cortical organotypic slices. Proc Natl Acad Sci U S A 100:4897–4902.
14. Chen D, Fetz EE (2005) Characteristic membrane potential trajectories in primate sensorimotor cortex neurons recorded in vivo. J Neurophysiol 94:2713–2725.
15. Contreras D, Steriade M (1997) State-dependent fluctuations of low-frequency rhythms in corticothalamic networks. Neuroscience 76:25–38.
16. Covey E, Kauer JA, Casseday JH (1996) Whole-cell patch-clamp recording reveals subthreshold sound-evoked postsynaptic currents in the inferior colliculus of awake bats. J Neurosci 16:3009–3018.
17. Cowan RL, Wilson CJ (1994) Spontaneous firing patterns and axonal projections of single corticostriatal neurons in the rat medial agranular cortex. J Neurophysiol 71:17–32.
18. Crochet S, Petersen CC (2006) Correlating whisker behavior with membrane potential in barrel cortex of awake mice. Nat Neurosci 9:608–610.
19. DeWeese MR, Zador AM (2004) Shared and private variability in the auditory cortex. J Neurophysiol 92:1840–1855.
20. DeWeese MR, Zador AM (2006) Non-Gaussian membrane potential dynamics imply sparse, synchronous activity in auditory cortex. J Neurosci 26:12206–12218.
21. Engel AK, Konig P, Kreiter AK, Schillen TB, Singer W (1992) Temporal coding in the visual cortex: new vistas on integration in the nervous system. Trends Neurosci 15:218–226.
22. Fee MS (2000) Active stabilization of electrodes for intracellular recording in awake behaving animals. Neuron 27:461–468.
23. Haider B, Duque A, Hasenstaub AR, McCormick DA (2006) Neocortical network activity in vivo is generated through a dynamic balance of excitation and inhibition. J Neurosci 26:4535–4545.
24. Hasenstaub A, Sachdev RN, McCormick DA (2007) State changes rapidly modulate cortical neuronal responsiveness. J Neurosci 27:9607–9622.
25. Hasenstaub A, Shu Y, Haider B, Kraushaar U, Duque A, McCormick DA (2005) Inhibitory postsynaptic potentials carry synchronized frequency information in active cortical networks. Neuron 47:423–435.

26. Herrmann M, Hertz JA, Prugelbennett A (1995) Analysis of synfire chains. Network: Comput Neural Syst 6:403–414.
27. Hromádka T (2007) Representation of Sounds in Auditory Cortex of Awake Rats, PhD thesis.
28. Ikegaya Y, Aaron G, Cossart R, Aronov D, Lampl I, Ferster D, Yuste R (2004) Synfire chains and cortical songs: temporal modules of cortical activity. Science 304:559–564.
29. Kenet T, Bibitchkov D, Tsodyks M, Grinvald A, Arieli A (2003) Spontaneously emerging cortical representations of visual attributes. Nature 425:954–956.
30. Kerr JN, Greenberg D, Helmchen F (2005) Imaging input and output of neocortical networks in vivo. Proc Natl Acad Sci U S A 102:14063–14068.
31. Lampl I, Reichova I, Ferster D (1999) Synchronous membrane potential fluctuations in neurons of the cat visual cortex. Neuron 22:361–374.
32. Lee AK, Manns ID, Sakmann B, Brecht M (2006) Whole-cell recordings in freely moving rats. Neuron 51:399–407.
33. Luczak A, Bartho P, Marguet SL, Buzsaki G, Harris KD (2007) Sequential structure of neocortical spontaneous activity in vivo. Proc Natl Acad Sci U S A 104:347–352.
34. Margrie TW, Brecht M, Sakmann B (2002) In vivo, low-resistance, whole-cell recordings from neurons in the anaesthetized and awake mammalian brain. Pflugers Arch 444:491–498.
35. Matsumura M, Chen D, Sawaguchi T, Kubota K, Fetz EE (1996) Synaptic interactions between primate precentral cortex neurons revealed by spike-triggered averaging of intracellular membrane potentials in vivo. J Neurosci 16:7757–7767.
36. McClurkin JW, Optican LM, Richmond BJ, Gawne TJ (1991) Concurrent processing and complexity of temporally encoded neuronal messages in visual perception. Science 253:675–677.
37. Mokeichev A, Okun M, Barak O, Katz Y, Ben-Shahar O, Lampl I (2007) Stochastic emergence of repeating cortical motifs in spontaneous membrane potential fluctuations in vivo. Neuron 53:413–425.
38. Monier C, Fournier J, Fregnac Y (2008) In vitro and in vivo measures of evoked excitatory and inhibitory conductance dynamics in sensory cortices. J Neurosci Methods 169:323–365.
39. Okamoto H, Isomura Y, Takada M, Fukai T (2006) Combined Modeling and Extracellular Recording Studies of Up and Down Transitions of Neurons in Awake or Behaving Monkeys. In: The Basal Ganglia VIII, pp. 555–562.
40. Okun M, Lampl I (2008) Instantaneous correlation of excitation and inhibition during ongoing and sensory-evoked activities. Nat Neurosci 11:535–537.
41. Oram MW, Wiener MC, Lestienne R, Richmond BJ (1999) Stochastic nature of precisely timed spike patterns in visual system neuronal responses. J Neurophysiol 81:3021–3033.
42. Petersen CC, Hahn TT, Mehta M, Grinvald A, Sakmann B (2003) Interaction of sensory responses with spontaneous depolarization in layer 2/3 barrel cortex. Proc Natl Acad Sci U S A 100:13638–13643.
43. Poulet JF, Petersen CC (2008) Internal brain state regulates membrane potential synchrony in barrel cortex of behaving mice. Nature 454:881–885.
44. Prut Y, Vaadia E, Bergman H, Haalman I, Slovin H, Abeles M (1998) Spatiotemporal structure of cortical activity: properties and behavioral relevance. J Neurophysiol 79:2857–2874.
45. Pulvermuller F, Shtyrov Y (2009) Spatiotemporal signatures of large-scale synfire chains for speech processing as revealed by MEG. Cereb Cortex 19:79–88.
46. Rudolph M, Pospischil M, Timofeev I, Destexhe A (2007) Inhibition determines membrane potential dynamics and controls action potential generation in awake and sleeping cat cortex. J Neurosci 27:5280–5290.
47. Shadlen MN, Newsome WT (1994) Noise, neural codes and cortical organization. Curr Opin Neurobiol 4:569–579.
48. Shadlen MN, Newsome WT (1998) The variable discharge of cortical neurons: implications for connectivity, computation, and information coding. J Neurosci 18:3870–3896.
49. Shu Y, Hasenstaub A, McCormick DA (2003) Turning on and off recurrent balanced cortical activity. Nature 423:288–293.
50. Song S, Sjostrom PJ, Reigl M, Nelson S, Chklovskii DB (2005) Highly nonrandom features of synaptic connectivity in local cortical circuits. PLoS Biol 3:e68.

51. Steriade M (2001) Impact of network activities on neuronal properties in corticothalamic systems. J Neurophysiol 86:1–39.
52. Steriade M, Nunez A, Amzica F (1993) A novel slow (<1 Hz) oscillation of neocortical neurons in vivo: depolarizing and hyperpolarizing components. J Neurosci 13:3252–3265.
53. Steriade M, Timofeev I, Grenier F (2001) Natural waking and sleep states: a view from inside neocortical neurons. J Neurophysiol 85:1969–1985.
54. Timofeev I, Grenier F, Steriade M (2001) Disfacilitation and active inhibition in the neocortex during the natural sleep-wake cycle: an intracellular study. Proc Natl Acad Sci U S A 98:1924–1929.
55. Vogels TP, Rajan K, Abbott LF (2005) Neural network dynamics. Annu Rev Neurosci 28:357–376.
56. Volgushev M, Chauvette S, Mukovski M, Timofeev I (2006) Precise long-range synchronization of activity and silence in neocortical neurons during slow-wave oscillations. J Neurosci 26:5665–5672.
57. von der Malsburg C (1995) Binding in models of perception and brain function. Curr Opin Neurobiol 5:520–526.
58. Waters J, Helmchen F (2006) Background synaptic activity is sparse in neocortex. J Neurosci 26:8267–8277.

Timing Excitation and Inhibition in the Cortical Network

Albert Compte, Ramon Reig, and Maria V. Sanchez-Vives

Abstract The interaction between excitation and inhibition in the cerebral cortex network determines the emergent patterns of activity. Here we analyze the specific engagement of excitation and inhibition during a physiological network function such as slow oscillatory activity (<1 Hz), during which up and down cortical states alternate. This slow rhythm represents a well-characterized physiological activity with a range of experimental models from in vitro maintained cortical slices to sleeping animals. Excitatory and inhibitory events impinging on individual neurons were identified during up and down network states, which were recognized by the population activity. The accumulation of excitatory and inhibitory events at the beginning of up states was remarkably synchronized in the cortex both in vitro and in vivo. The same synchronization prevailed during the transition from up to down states. The absolute number of detected synaptic events pointed as well towards a delicate balance between excitation and inhibition in the network. The mechanistic and connectivity rules that can support these experimental findings are explored using a biologically inspired computer model of the cortical network.

Excitation and Inhibition During Cortical Up and Down States

Basal excitability and recurrent connectivity in the cerebral cortical network [18, 22] induce neuronal firing that reverberates in the circuit, resulting in an emergent network activity. During slow-wave sleep and anesthesia, this activity is organized in the cerebral cortex network in a slow (<1 Hz) rhythmic pattern consisting of interspersed up (or activated) and down (or silent) states (Fig. 1) [21, 41, 44]. This rhythm is recorded in the thalamocortical loop, but persists in the cortex following thalamectomy [42]. Furthermore, it can be generated in cortical slices maintained

M.V. Sanchez-Vives (✉)
Institut d'Investigacions Biomèdiques August Pi i Sunyer (IDIBAPS), 08036 Barcelona, Spain
Institució Catalana de Recerca i Estudis Avançats (ICREA), 08036 Barcelona, Spain
e-mail: msanche3@clinic.ub.es

K. Josić et al. (eds.), *Coherent Behavior in Neuronal Networks*, Springer Series in Computational Neuroscience 3, DOI 10.1007/978-1-4419-0389-1_2,

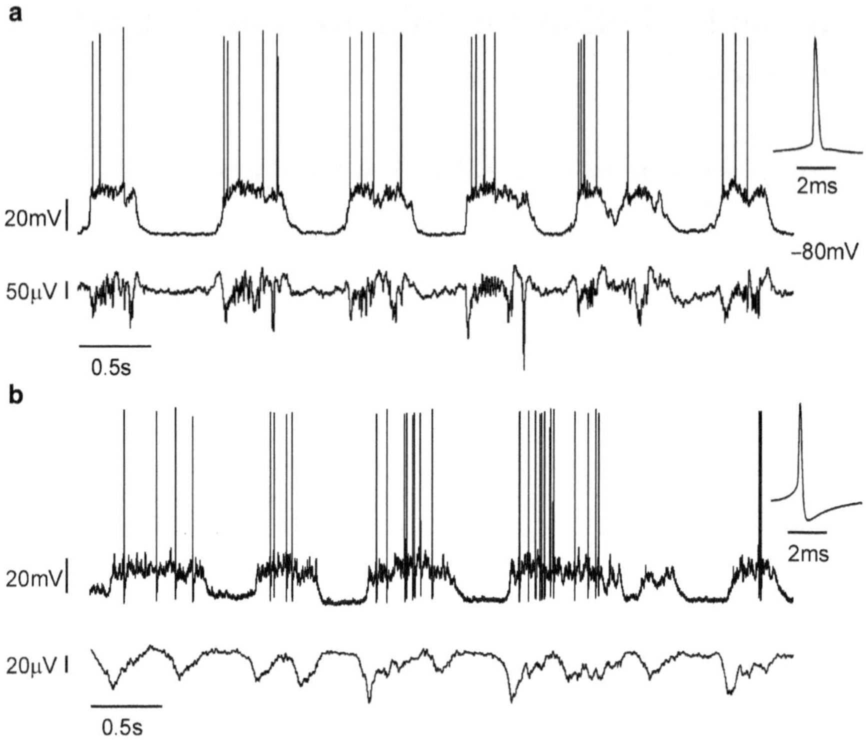

Fig. 1 Slow rhythmic activity in excitatory and inhibitory neurons in vivo. (**a**) Successive up states recorded intracellularly from a regular spiking neuron in vivo (*top trace*). Inset illustrates an averaged action potential. Local field potential in the close vicinity (ca 100 μm) reflects network activity (*bottom trace*). (**b**) Successive up states recorded intracellularly from a fast spiking neuron in vivo (*top trace*). Note the higher firing frequency during up states displayed by the fast spiking neuron. Averaged action potential is represented in the inset. Local field potential from the vicinity in the *bottom trace*

in vitro [35], bearing a remarkable similarity to cortical activity during slow-wave sleep or anesthesia. Spontaneous slow rhythmic activity can also be recorded from disconnected cortical slabs in vivo [47]. Therefore, the slow rhythm is generated in the local cortical network, although the thalamic network can generate a similar rhythm if activated by metabotropic glutamate receptors [52].

The understanding of the detailed cellular and network mechanisms that regulate the aforementioned emergent activity provides a valuable insight into cortical function, and more generally into properties and regulation in neuronal networks. A key element in the balance and control of either spontaneous emergent or evoked cortical activity is the relation between excitation and inhibition. Slow oscillatory activity represents a well-characterized physiological activity with a range of experimental models, from in vitro to sleeping animals, where the specific engagement of excitation and inhibition in physiological network function can be studied. This chapter will be devoted to excitatory and inhibitory activation during the occurrence of up and down spontaneous cortical states both in the real and in a modeled cortical network. The purpose is to understand how the network properties are tuned to achieve functional equilibrium and how this equilibrium can be eventually lost, as for instance in epilepsy. The approach we will present is both experimental and theoretical. In the experiments, we measure the time of occurrence of excitatory and inhibitory synaptic potentials during network activity. In the computational model, the relationship between structural parameters of network connectivity and the timing of excitatory and inhibitory inputs is explored.

The activated periods during rhythmic activity, or up states, are periods of intense synaptic activity that generate neuronal firing by pushing neuronal membrane potential above firing threshold. Both excitatory and inhibitory neurons fire during up states, while they remain relatively silent during down states. Several lines of evidence confirm that both types of neurons fire during up states (Fig. 1). From the first studies oscillations [41] it was already reported that not only excitatory electrophysiological types but also inhibitory also inhibitory ones (fast spiking neurons) fired during up states. membrane potential to different values by means of current injection further illustrated the coexistence of both excitatory and inhibitory potentials during up states in vivo and in vitro (Fig. 2) [35, 41]. Indeed, practically every recorded neuron participated in the rhythm with enhanced firing during the up state [7, 8, 35, 41, 43]. Quantification in striatal neurons also confirmed the participation of both excitatory and inhibitory events during participation of both excitatory and inhibitory events during Although all this evidence supports the simultaneous activation of Although all this evidence supports the simultaneous activation of oscillation, the issue of the timing of both types of events remains unsettled. A computational model of propagating slow oscillations predicted that inhibitory neurons should activate to their maximal rate slightly ahead in time than neighboring pyramidal neurons at the beginning of the up states pyramidal neurons at the beginning of the up states that this could be supported experimentally, although the trend did not reach statistical significance [17]. At the end of the up state, instead, experiments in vivo indicate that excitatory firing outlasts inhibitory firing [17].

So far, most studies have analyzed the relative contribution of excitation and inhibition to the conductance changes that neurons experience in the course of network activity [1,4,17,29,30,34,38]. A related aspect that has received much less attention is how the timing of excitatory and inhibitory events contributes to the excitation-inhibition balance [6]. Conductance measurements during up states reveal that the weight of excitation and inhibition is well balanced in vivo [17], and similarly in vitro [38], as also argued theoretically [5]. Still, there are some contradictory findings reported in the literature. Conductance measurements suggest that excitatory conductance dominates slightly at the beginning and the end of the up states but is otherwise comparable to inhibitory conductance [17]. Other studies, instead, report that the inhibitory conductance is much larger during up states [34].

The general understanding achieved by different methods is that excitation and inhibition balance each other, and this has been reported both during spontaneous or sensory activated cortical activity [1, 17, 27, 29, 37–39, 51]. However, we do not

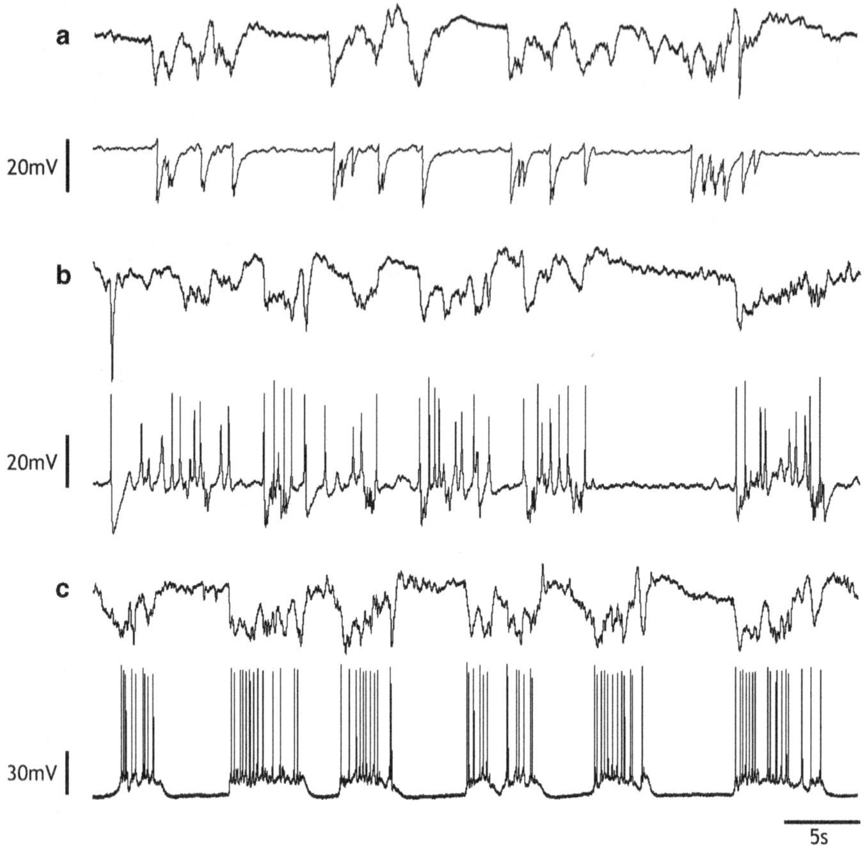

Fig. 2 Excitatory and inhibitory synaptic potentials during slow oscillations in the auditory cortex *in vivo*. In the three panels (A, B and C) the unfiltered local field potential on top and intracellular recordings at the bottom. (**a**) Intracellular membrane potential at −15 mV to illustrate the IPSPs occurring during the up states. Sodium action potentials have been inactivated by depolarization. (**b**) Intracellular membrane potential at −45 mV illustrates a mix of IPSPs and EPSPs, while sodium action potentials are partially inactivated by depolarization. (**c**) Intracellular membrane potential at −75 mV illustrating suprathreshold up states. All intracellular recordings are from the same neuron

know exactly how this balance is achieved in terms of the contrasting proportions of inhibitory and excitatory mechanisms in the cortex. Indeed, changes in conductance during synaptic activity are determined by the combination of a number of factors: the firing rate of presynaptic neurons, the number of presynaptic neurons, the number of synaptic contacts from each presynaptic neuron, or the conductance change (excitatory or inhibitory) induced at a single contact by a presynaptic action potential, among others. It is estimated that a single pyramidal cell in the cortex receives its input from as many as 1,000 other excitatory neurons (that would make some 5,000 contacts) and as many as 75 inhibitory neurons (that would make some 750 contacts) [31], and the proportion is of 30,000 excitatory against 1,700 inhibitory for CA1 pyramidal neurons [26]. This anatomical disproportion contrasts with the functional balance between excitation and inhibition. Different factors seem to contribute to the counter-balance of inhibition. Lesser failures of inhibitory transmission achieved by multiple presynaptic contacts from the same inhibitory neuron [40,45] is one of them, as well as the larger synchronization between inhibitory neurons due to electric coupling [15, 16]. Even more critical is the segregation of inputs onto pyramidal neurons, where inhibitory contacts are restricted to the soma and proximal dendrites [20, 26], while excitatory inputs only innervate further than 50 μm away from the soma [13, 14]. Not only this results in a larger weight at the soma for inhibitory inputs, but also in a control over the excitatory inputs that reach the soma [2, 36, 48]. There is an additional element and that is the firing rate of inhibitory neurons with respect to excitatory neurons. Fast spiking neurons are known to fire at higher frequencies (Fig. 1b) [25, 28], as a consequence of the presence of K^+ channels (Kv3) that allow for a fast repolarization [12] and the lack of fast spike frequency adaptation [10,25], which can contribute to the excitation/inhibition balance by imposing larger numbers of presynaptic events.

Resolving the contribution of all these mechanisms in achieving the physiological excitation-inhibition balance in the cortex remains a challenge. Current estimations from slow oscillatory activity in the cortex indicate that firing rates differing by around a mere factor of two between regular spiking and fast spiking neurons result in excitatory and inhibitory synaptic conductances that are in balance [17]. This is surprising given the difference in orders of magnitude of the connectivity parameters for excitation and inhibition in the cortex (see above). In order to dissect further the mechanisms that link spiking activity and synaptic current for excitation and inhibition in the cortex, we look here at the relative timing of excitation and inhibition by detecting the times of occurrence of excitatory and inhibitory synaptic events impacting on pyramidal cortical neurons. We then analyze how synaptic event timing and neuronal spiking are related through some connectivity parameters in a computer model of slow oscillatory activity in the cortex [5]. Finally, we discuss the implications that these computational results have in interpreting our experimental findings and their relation to functional structure and dynamics of excitation and inhibition in the cortical network.

Experimental Procedures and Detection of Synaptic Events

Intracellular and Extracellular Recordings In Vitro and In Vivo

In vitro recordings were obtained as previously described [6, 35] and detailed in the Appendix. In brief, cortical slices from ferret prefrontal or visual cortex were prepared and bathed in an ACSF solution containing ionic concentrations that closely mimic the conditions in situ. In these conditions, spontaneous rhythmic activity (<1 Hz) is generated in the circuit [35]. Recordings were also obtained from anesthetized rat neocortex (auditory and barrel cortex) [32] and the recorded activity showed the characteristic slow oscillations of this state, which is closely related to slow-wave sleep [41]. Thus, both the in vitro and in vivo preparations reflect a very similar rhythm and are presumably engaging similar mechanisms of the cortical circuit [35].

We investigated the properties of this rhythmic activity in vitro and in vivo by recording intracellularly with sharp electrodes and extracellularly with tungsten electrodes. In all cases, intracellular recordings were recorded in close vicinity of the extracellular recording, in order to relate single-neuron activity to the surrounding population dynamics. For a more detailed methodological description, see the Appendix.

Data Analysis

We used an analysis protocol described elsewhere [6] to identify the timing of excitatory and inhibitory synaptic events recorded intracellularly at different membrane potentials (Fig. 2) and relate them to the ongoing population dynamics (up and down states). The extracellular recording was used to detect the times of transitions between up and down states as illustrated in Fig. 3a and described in the Appendix. From intracellular recordings at different holding voltages, the times of synaptic events were identified as sharp upward or downward deflections in the membrane potential (Fig. 3b, c) and were aligned to the beginning or end of the up state by using the transitions detected from the extracellular recording. This alignment allows to compare the timing of events recorded nonsimultaneously at different holding voltages, because the extracellular recording remains unchanged as the conditions of the intracellular recording are modified. A more detailed description of these methods can be found in the Appendix and in [6].

A Short Discussion on the Method

In order to detect excitatory and inhibitory events in this study we recorded intracellularly from a neuron at membrane potentials that are the reversal potentials of glutamatergic excitation (0 mV) and $GABA_A$ inhibition (−70 mV). In this way, the postsynaptic events that correspond to excitation and inhibition respectively can be isolated. Extremes of the first derivative provide timings for transitions of either EPSPs (if at −70 mV) or IPSPs (if at 0 mV). A threshold is set based on statistical criteria and those events that surpass the threshold separating them from the noise are taken into account as valid synaptic events. An envelope of the local field potential trace determines the times of transition and separates the periods of up and down states. This study focuses on the timing of excitatory and inhibitory events with respect to up and down transitions.

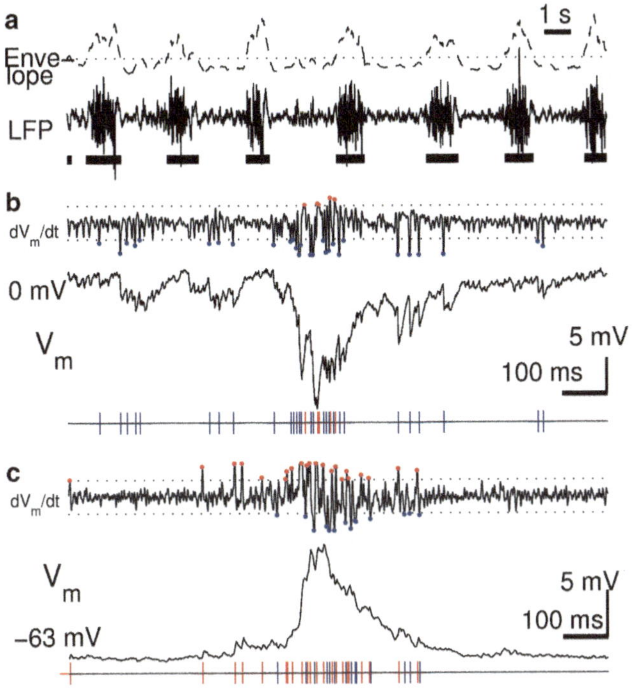

Fig. 3 Detection of IPSPs and EPSPs during slow oscillations in the cortex. (**a**) Detection of up and down states from the extracellular recording was performed by filtering it between 2 and 150 Hz to obtain a local field potential (LFP) signal. An Envelope was then computed (see the Appendix), from which a simple thresholding allowed us to detect up states (*thick black lines below* LFP). (**b, c**) We detected the timing of synaptic events from each intracellular recording (V_m). To this end, we computed its derivative (dV_m/dt), and then thresholded it at 2–4 interquartile ranges to detect excitatory events (*red dots*) and inhibitory events (*blue dots*). We did this for an intracellular recording at a depolarized membrane potential (**b**) and at a hyperpolarized membrane potential (**c**) for each neuron, so we could have a more reliable identification of synaptic events of each kind. Because the extracellular recording remained unchanged while we modified V_m, aligning event timing to the up state beginning and end (detected from the extracellular record), allowed us to compare the timing of excitatory and inhibitory events. Data shown here correspond to an in vitro recording, but identical methods were applied to in vivo data.

There are some caveats associated to this method. The absolute number of events may be underestimated since those events below threshold are not considered. They may also remain undetected if their rise time is not sharp enough to appear as an independent event, e.g., because they are embedded in a group of events or because they occurred far out in the dendritic arbor. Regarding this, synchronous events may be underestimated by being considered under the same detected event. Because of the higher synchronization of inhibitory neurons [15, 16], this may affect especially inhibitory events. Similarly, slower post-synaptic voltage dynamics will blur post-synaptic responses and induce more false negatives in our detection method. Thus, there may be limitations derived from the different excitatory and inhibitory kinetics and from the particular distribution of inhibitory connections (soma and proximal dendrites) vs. excitatory connections (distal dendrites). This could bias the detection towards inhibition, since it is going to generate faster events and therefore easier to detect with the first derivative method. Different excitatory and inhibitory potential kinetics could thus bias the detection towards faster, sharper events. Still, the system has been carefully validated in [6], where the influence of threshold on event detection was explored.

Another aspect of this method to consider (and indeed of all conductance detection methods, for a review see [27] is that given that the V_m is held at different values (0, −70 mV) for the detection of IPSPs/EPSPs respectively, the up states that are studied are never the same for both types of events. Still, each quantification of synaptic events (Figs. 5–10) is the result of averaging 17–187 up states, and therefore individual variations between up states are not taken into account. Finally, our derivative method included a low-pass filter with cut-off at 200 Hz. This could also limit the detection of closely spaced events (<5 ms). However, we tested this by repeating the analysis using a cut-off at 500 Hz and we did not find any significant increase in the number of synaptic events detected. Despite all these caveats, this is to our knowledge the only method so far to have an approximation to the timing of the individual synaptic inputs being received by a single pyramidal neuron during physiological network activity. Apart from other possible sources of error, both EPSPs and IPSPs are being recorded ≅70 mV apart from their reversal potential and therefore their driving force should be the same. Our main interest is on the relative timing of both types of events. We consider the method particularly valid on that regard, given that for timing considerations the absolute number of events has been normalized. Still, we dare to have a look into the absolute number of events (see Figs. 5, 8), assuming a comparable error in the detection of excitatory and inhibitory events and considering that we can still learn from their proportions.

Experimental Results

We applied our synaptic event detection method to $n = 10$ neurons recorded in vitro and $n = 5$ neurons recorded in vivo. For each of these neurons, intracellular recordings of variable duration (range 60–729 s) were obtained, one at a depolarized potential (around 0 mV) and one at a hyperpolarized potential (around −70 mV). A closely adjacent extracellular recording was simultaneously registered to determine the transition times between up states and down states (Fig. 3). We were thus able to obtain putative excitatory events (from the −70 mV recording), and putative inhibitory events (from the 0 mV recording) for each neuron, and attribute them to the up state or down state (as identified from the extracellular signal).

We first extracted general statistics from this analysis, concerning the characteristics of individual EPSPs and IPSPs (Fig. 4, Table 1) and the comparative quantification of synaptic events during up and down states (Table 2). We found that, in vitro, the amplitudes of putative excitatory postsynaptic potentials were

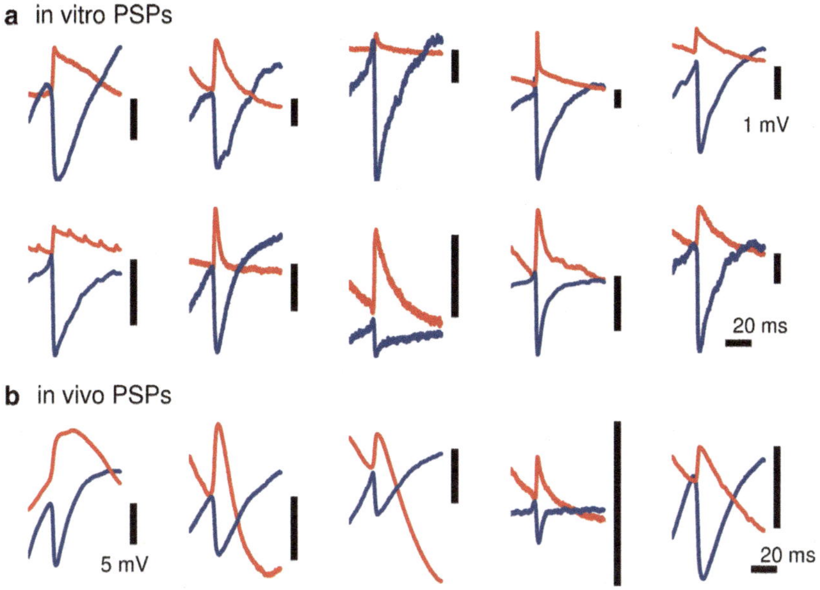

Fig. 4 Amplitudes and time-course of average post-synaptic potentials for neurons in our database, in vitro (**a**, $n = 10$) and in vivo (**b**, $n = 5$). Inhibitory (excitatory) events were detected from intracellular recordings in neurons held at ~0 mV (~−70 mV) as illustrated in Fig. 3. Events that did not occur within 100 ms of other events were used to align pieces of the intracellular signal and average them to obtain the average inhibitory (*blue*) and excitatory (*red*) post-synaptic potentials. Different number of events were used for averaging each trace, ranging from 17 to 187 in vitro, and 70 to 846 in vivo. Each panel shows averages for a given neuron in our database. Vertical calibration bars indicate 1 mV in (**a**) and 5 mV in (**b**). The time base is the same for all panels, as indicated in the last set of traces in each panel.

Table 1 Amplitude and decay time of V_m deflections for isolated synaptic events (no other event occurring in a 100-ms window) detected at either depolarized ($V_m \sim 0$ mV) or hyperpolarized ($V_m \sim -70$ mV) voltages (Fig. 4)

	In vitro recordings ($n = 10$)			In vivo recordings ($n = 5$)	
	$V_m \sim 0$ mV		$V_m \sim -70$ mV	$V_m \sim 0$ mV	$V_m \sim -70$ mV
Amplitude of isolated events	2.58 ± 0.45 mV (0.44–5.23 mV)	>	1.17 ± 0.21 mV (0.41–2.67 mV)	4.6 ± 1.1 mV (1.2–7.4 mV)	3.6 ± 0.96 mV (1.2–6.3 mV)
Decay time of isolated events	27.9 ± 10.1 ms (0.66–13.8 s)		28 ± 13 ms (3–127 ms)	31 ± 10 ms (4–52 ms)	33 ± 13 ms (7–62 ms)

Population data is reported as mean ± s.e.m. and ranges are indicated in parenthesis. Significant differences (paired t-test, $p < 0.05$) are indicated with the symbol>

significantly smaller than those of putative inhibitory postsynaptic potentials (paired t-test, $p = 0.007$). Such difference was not detected in vivo (Table 1). We did not detect differences in the decay dynamics of synaptic events either in vivo or in vitro. Regarding up and down states (Table 2), we found that up states were significantly

Table 2 Comparative statistics of excitatory and inhibitory events during up and down states in vitro and in vivo

	In vitro recordings ($n = 10$)			In vivo recordings ($n = 5$)		
	Up states		Down states	Up states		Down states
Duration	878 ± 66 ms† (531–1,607 ms)	<	3.21 ± 0.77 s (0.66–13.8 s)	792 ± 296 ms† (342–2,172 ms)		580 ± 86 ms† (329–886 ms)
No. detected excitatory events	27 ± 5 (14–65)		15.7 ± 8.9 (0–19)	25.9 ± 12.2 (4–72)		16.1 ± 6.4 (1–36)
No. detected inhibitory events	24 ± 4 (13–53)		18.5 ± 9.3 (0–81)	19.3 ± 5.5 (10–39)		10.4 ± 4.3 (1–26)
Rate detected excitatory events	36.6 ± 5.8 s$^{-1\ddagger}$ (10–73 s^{-1})	>	3.2 ± 0.8 s^{-1} (0.2–7.2 s^{-1})	33.5 ± 10 s$^{-1\ddagger}$ (7–67 s^{-1})	>	25.2 ± 8.9 s$^{-1\ddagger}$ (1.5–56 s^{-1})
Rate detected inhibitory events	26.7 ± 2.5 s$^{-1\ddagger}$ (15–43 s^{-1})	>	5.9 ± 2.1 s^{-1} (0.1–23 s^{-1})	31.4 ± 7.7 s$^{-1\ddagger}$ (16–60 s^{-1})	>	21.1 ± 9.5 s$^{-1\ddagger}$ (2–57 s^{-1})

Population data is reported as mean ± s.e.m. and ranges are indicated in parenthesis. Significant differences (one-tailed paired t-test, $p < 0.05$) are indicated with the symbols > and <. † This is an over-estimate, as states shorter than 250 ms were discarded from the analysis. ‡ This is an under-estimate, because of the over-estimation in †

shorter than down states in vitro (one-tailed paired t-test, $p = 0.0037$, $n = 20$), but not so in vivo (paired t-test $p = 0.2$, $n = 5$). Synaptic event rates were always significantly higher in the up states than in the down states, both in vitro and in vivo, although the difference was more accentuated in vitro.

Comparing the statistics numbers between in vivo and in vitro, we found that up state durations were comparable between the two conditions (two-sample t-test, $p = 0.64$), but down state durations were significantly shorter in vivo than in vitro (two-sample t-test, $p = 0.023$). The amplitudes of excitatory postsynaptic events were significantly smaller in vitro than in vivo (two-sample t-test, $p = 0.006$), but their kinetics were comparable. Regarding detected synaptic events, only the rate of events during the down state differed significantly between these two conditions (two-sample t-test, $p < 0.05$). Inhibitory events occurred significantly more frequently in the down state in vivo than in vitro ($p = 0.004$), and excitatory events showed a similar, marginally significant trend ($p = 0.054$). This result indicates the presence of more basal synaptic activity in vivo than in vitro, and also shows that network activations in the two conditions do not differ significantly.

Excitatory and Inhibitory Events During Risetime of Up States In Vitro

For each neuron, excitatory and inhibitory events detected intracellularly were aligned at the time of up state initiation, as detected in the neighboring simultaneous extracellular recording (Fig. 3). Synaptic event histograms (in bins of 5 ms) were then averaged across neurons after realigning them to the steepest increase in excitatory synaptic events (arbitrarily considered time zero, see Fig. 7). In terms of detected event rate, we found no consistent difference between the event rate of

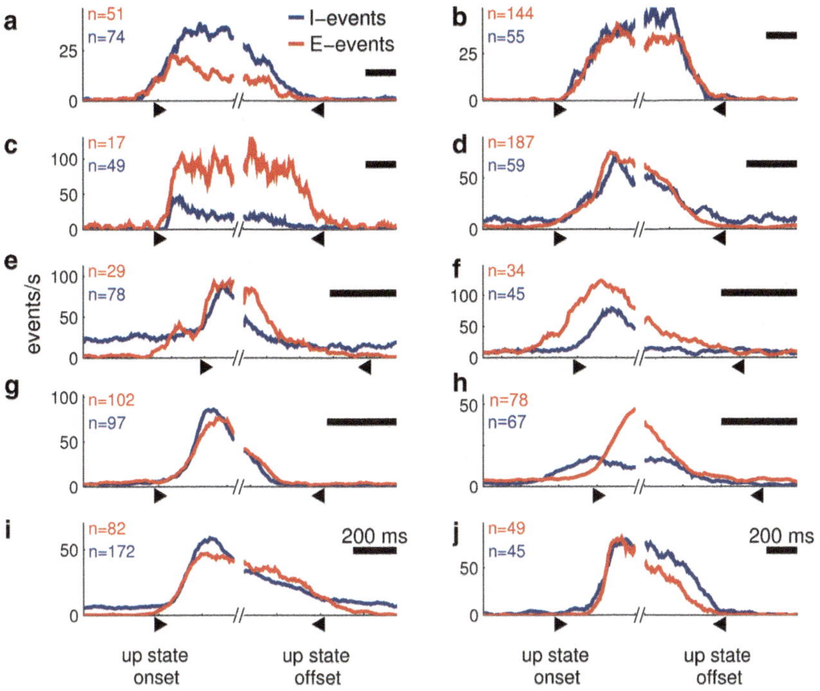

Fig. 5 Synaptic event rates through the duration of the up state in each of 10 neurons in vitro. Time-histograms of synaptic events detected from intracellular records (Fig. 3) and aligned to up state onset (►) or up state offset (◄), as detected from the extracellular record (Fig. 3). Excitatory events (*red*) arrive at a higher peak rate than inhibitory events (*blue*) in (**c**), (**f**) and (**h**), whereas the opposite is true in (**a**). For the rest of panels, peak event rate is approximately similar for excitatory and inhibitory events. For each trace, the number *n* of up states from which spike events were gathered is indicated. Horizontal calibration bars indicate 200 ms.

excitation and inhibition during the up state in vitro (Fig. 5). While some neurons showed a higher peak rate of excitatory events (Fig. 5c, f, h), others showed a higher peak rate for inhibitory events (Fig. 5a), and most ($n = 6/10$) presented approximately equal rates for both types of events. Such a delicate balance of event rate between excitation and inhibition is remarkable. Even when the method used may have a number of limitations (see above), it seems unlikely that this almost identical number could be reached by chance. When the normalized risetime of EPSPs/IPSPs to their maximal event rate was evaluated, individual neurons showed in most cases ($n = 8/10$) a matching time course for excitation and inhibition (Fig. 6). From the other two cells, one shows an early rise of excitation (by ~50 ms, Fig. 6f) and the other one of inhibition (by ~20 ms, Fig. 6h). The occurrence of excitatory and inhibitory events during the risetime of up states averaged across cells revealed that the increase in both types of events is synchronous in our population of in vitro

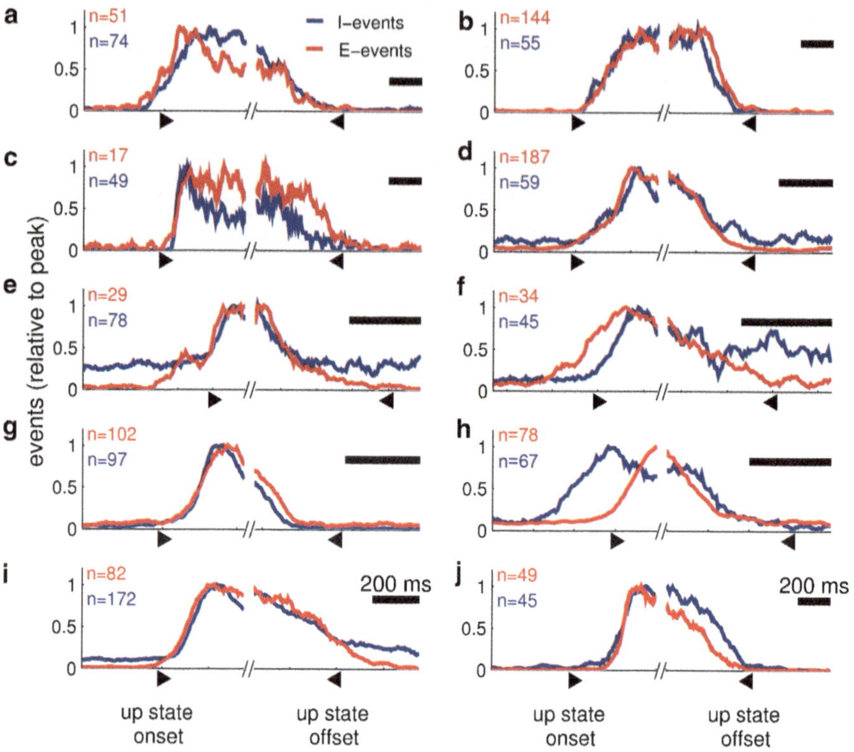

Fig. 6 Normalized synaptic events through the duration of the up state in each of 10 neurons in vitro. Time-histograms of synaptic events from Fig. 6 were normalized to peak event rate to compare the dynamics of excitation and inhibition at up state onset (▶) and at up state offset (◀). During up state onset, excitatory events (*red*) increase ahead than inhibitory events (*blue*) in (**f**), whereas the opposite is true in (**h**). For the rest of panels, excitatory and inhibitory events increase to their maximal rate at approximately the same point in time. Also extinction at up state offset occurs concomitantly for excitation and inhibition, except in (**b**), (**c**), and (**g**) (excitation outlasts inhibition) and in (**h**) and (**j**) (inhibition outlasts excitation). For each trace, the number *n* of up states from which spike events were gathered is indicated. Horizontal calibration bars indicate 200 ms.

recordings. The take off from the down state occurs simultaneously, with no significant deviation between excitation and inhibition until well after the peak in PSPs is reached (Fig. 7a, b).

Excitatory and Inhibitory Events During the End of Up States In Vitro

In order to analyze the occurrence of excitatory and inhibitory events during the termination of the up states, synaptic events in each cell were aligned at the offset of the up state, as detected from extracellular recordings, and then realigned to

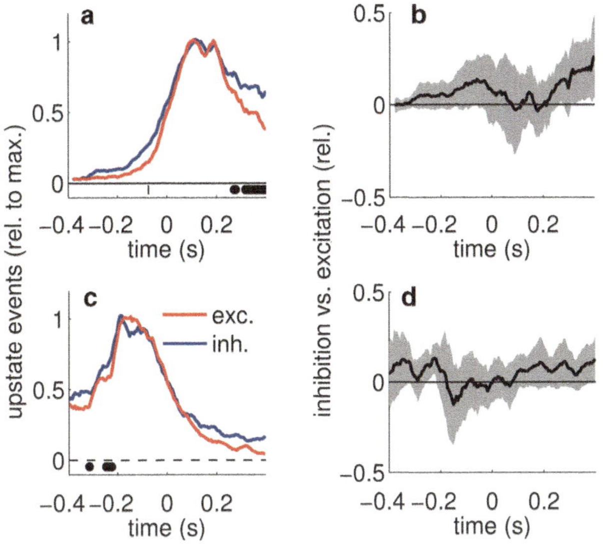

Fig. 7 Population analysis of EPSPs/ IPSPs timing at the beginning and end of the up states in vitro. (**a**) Excitatory (*red*) and inhibitory (*blue*) synaptic events detected from intracellular recordings in vitro ($n = 10$) accumulated at comparable rate at the onset of the up state. Synaptic event histograms from Fig. 7 were averaged across neurons after aligning them to the steepest increase in excitatory synaptic events (corresponding to time zero in **a**, **b**). (**b**) Difference in event accumulation to peak event rate between inhibition and excitation (*black line*). Positive (negative) values indicate excess of inhibition (excitation). *Gray shadow* is the 95% confidence interval calculated with a jackknife procedure over neurons ($n = 10$). During up state onset, there was no significant difference in the time of fastest accumulation of excitatory and inhibitory events. Periods with significant difference between excitation and inhibition are marked on **a** with a thick black line along the x-axis. (**c**) Same as **a**, but synaptic events into each cell were aligned at the offset of the up state (◀ in Fig. 6), and then realigned to the steepest decrease in the excitatory histogram before computing the average over neurons. (**d**) Same as **b** for the data in **c**. Synaptic event extinction at the end of the up state did not differ for excitation and inhibition at the 95% confidence level.

the steepest decrease in the excitatory histogram before computing the average over neurons (Fig. 7c, d). When the average obtained in this way is observed, we can see that the peak in both EPSPs and IPSPs before the transition to the down state is initiated is reached at the same time. From that moment, the decrease in excitation and inhibition is in average mostly synchronous, with no significant difference between both within a 95% confidence interval (Fig. 7c, d). When we look at individual neurons, the relation between excitation and inhibition during up state termination is more heterogeneous than during its initiation (Fig. 6). While in 5 out of 10 neurons, there is a synchronous decrease in EPSPs and IPSPs, in 4 out of 10 the decrease in inhibition precedes that of excitation in time, although with a similar time course (Fig. 6b, c, g, h). In just one case, it is the excitation the one that decreases first, followed by inhibition (Fig. 6j). Therefore, even when the result of the average suggests an equilibrium between the timing of decrease of EPSPs and IPSPs at the end of the up states, the EPSPs and IPSPs at the end of the up states, the individual cases heterogeneity also exists regarding the absolute number of excitatory and inhibitory events. Even when the general trend gravitates towards the comparable number in both cases, individual neurons display either larger numbers of EPSPs or IPSPs (Fig. 5).

In spite of individual variations, we conclude that when the timing of excitatory and inhibitory events is analyzed during the risetime and the repolarization of the up states in vitro it is noteworthy that both events increase to start an up state, and decrease to finish it up with a remarkable synchrony. Furthermore, the total number of events, even within certain individual variability, could be considered to be quite similar, in at least half the neurons virtually identical.

Excitatory and Inhibitory Events During Risetime of Up States In Vivo

For each neuron, events detected intracellularly were aligned at the time of up state initiation, as detected in the neighboring simultaneous extracellular recording. Synaptic event histograms were then averaged across neurons after realigning them to the steepest increase in excitatory synaptic events (corresponding to time zero in Fig. 10a, b). During up state onset, there was no significant difference between excitation and inhibition rate of increase in our population of in vivo recordings (Fig. 10a, b, $n = 5$), quite similarly to what was observed in vitro (Fig. 7). Different from in vitro was, though, a faster rate in the accumulation of synaptic events. Excitatory events accumulated at a rate of 1.307%/ms (range 0.309–3.93%/ms) in vitro and 1.815%/ms (range 0.312–4.97%/ms) in vivo. Inhibitory events accumulated at a rate of 1.380%/ms (range 0.533–3.32%/ms) in vitro and 1.704%/ms (range 1.396–2.076%/ms) in vivo. Although nonsignificant (two-sample t-test $p > 0.5$; Wilcoxon rank sum test $p > 0.3$, $n = 10, 5$), the trend in difference between in vitro and in vivo measurements agrees with what is observed at the membrane level (Figs. 1 and 2), where transitions between up and down states are often faster in vivo than in vitro.

Another interesting difference between in vitro and in vivo conditions revealed by the average of synaptic events is the decay in accumulated EPSPs as soon as the up state is reached (Fig. 10a), what could be the result of spike frequency adaptation in pyramidal neurons, and/or synaptic depression in excitatory synapses to pyramidals. This time course observed for excitatory events is not followed by inhibitory events, that remained in a plateau once the up state was reached. Note that this analysis is normalized and provides information about timing of occurrence, but not about the absolute number of excitatory/inhibitory events.

Focusing on individual neurons, the normalized events (Fig. 9) are indicative of a remarkable analogous timing of accumulation of excitatory and inhibitory events in vivo as well as in vitro (see above). Even when there is a slight variation in 2 out of 5 cases, the predominant trend is a well synchronized accumulation of events. If the

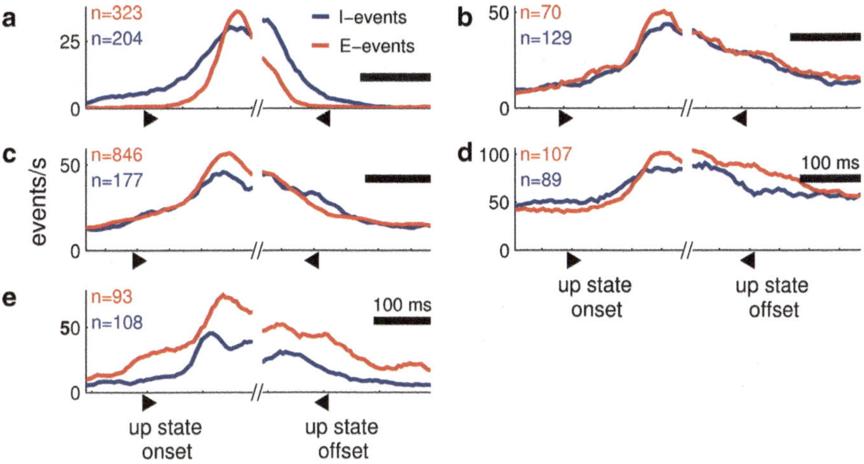

Fig. 8 Synaptic event rates through the duration of the up state in each of 5 neurons in vivo. Time-histograms of synaptic events detected from intracellular records (Fig. 3) and aligned to up state onset (▶) or up state offset (◀), as detected from the extracellular record (Fig. 3). Excitatory events (*red*) arrive at a higher peak rate than inhibitory events (*blue*) in all panels, but only in (**e**) the difference was sizeable. For each trace, the number *n* of up states from which spike events were gathered is indicated. *Horizontal calibration bars* indicate 100 ms.

absolute – and not the normalized – number of events is considered, the same trend is maintained, although it allows to evaluate the relative number of events (Fig. 7). In all five neurons studied here the absolute number of excitatory synaptic events that lead to the up state is larger than that of inhibitory events, but for a proportion not larger than 25% (except in Fig. 8e, where the excess of excitatory events is around 50%).

Excitatory and Inhibitory Events During the End of Up States In Vivo

Synaptic events in each cell were aligned at the offset of the up state, as detected from extracellular recordings, and then realigned to the steepest decrease in the excitatory histogram before computing the average over neurons (Fig. 10c, d). The average across five neurons recorded in vivo revealed a simultaneous decay in the rate of occurrence of EPSPs and IPSPs, although for a brief time (few tens of ms) excitatory events extinguished earlier than inhibitory ones, as assessed at the 95% confidence level (Fig. 10c, d). A larger sample would be necessary to confirm this trend.

The timing of synaptic events at the end of the up states for individual neurons is illustrated in Fig. 9. Again, the simultaneous decrease of both excitatory and inhibitory events is striking in these plots. When each individual case is explored in detail, we can see that both excitatory or inhibitory events can lead the extinction

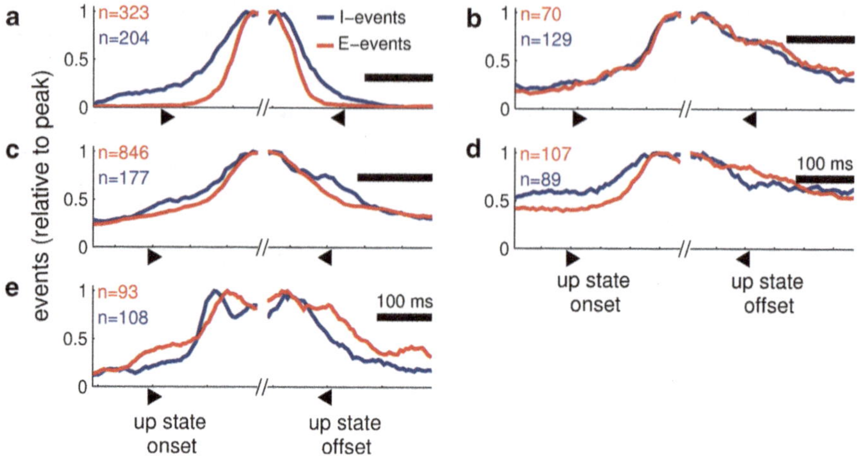

Fig. 9 Normalized synaptic events through the duration of the up state in each of 5 neurons in vivo. Time-histograms of synaptic events from Fig. 9 were normalized to peak event rate to compare the dynamics of excitation and inhibition at up state onset (►) and at up state offset (◄). During up state onset, inhibitory events (*blue*) increase ahead than excitatory events (*red*) in (**a**), (**c**), and (**d**). For the rest of panels, excitatory and inhibitory events increase to their maximal rate at approximately the same point in time. Extinction at up state offset occurs first for excitation in (**a**) and (**c**), first for inhibition in (**e**), and concomitantly for excitation and inhibition in (**b**) and (**d**). For each trace, the number n of up states from which spike events were gathered is indicated. Horizontal calibration bars indicate 100 ms.

of synaptic events, but that the time course of the decay is invariably similar. If the absolute number of events are evaluated (Fig. 8), then it can be observed that the similarity of time course can indeed conceal a remarkable difference in the number of synaptic events.

Excitation and Inhibition in Up and Down states Generated in a Cortical Model

The results of our experimental study of excitation and inhibition are difficult to reconcile with the predictions of the computer model of slow oscillatory activity ([5]; Fig. 11a). This computer model can reproduce intracellular and extracellular data of slow oscillatory activity in cortical slices [35], with interneurons and pyramidal neurons firing practically in phase through the slow oscillation (Fig. 11b). Excitatory and inhibitory conductances were found to maintain a proportionality in this model, as found experimentally [38]. Model inhibitory neurons display a higher firing rate during up states (ca. 30 Hz) while excitatory neurons have a firing frequency which is lower (ca. 15 Hz) (for comparison with experimental results see Fig. 1). However, one feature of the model seems at odds with the experimental results reported here. The presynaptic firing of model inhibitory neurons leads the beginning of the

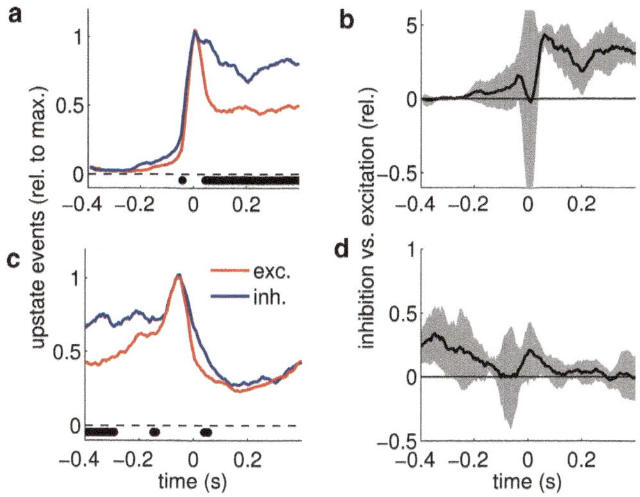

Fig. 10 Population analysis of EPSPs/IPSPs timing during the beginning and end of the up states in vivo. (**a**) Excitatory (*red*) and inhibitory (*blue*) synaptic events detected from intracellular recordings in vivo ($n = 5$) accumulated at comparable rate at the onset of the up state. Normalized synaptic event histograms (Fig. 10) were averaged across neurons after aligning them to the steepest increase in excitatory synaptic events (corresponding to time zero in **a**). (**b**) Difference in event accumulation to peak event rate between inhibition and excitation (*black line*). Positive (negative) values indicate excess of inhibition (excitation). Gray shadow is the 95% confidence interval calculated with a jackknife procedure over neurons ($n = 5$). During up state onset, there was no significant difference between excitation and inhibition rate of increase. However, a significant fraction of excitatory events remained confined to a short time window after up state initiation, possibly indicating adaptation dynamics. Instead, inhibitory events remained constant over up state duration. (**c**) Same as (**a**), but synaptic events into each cell were aligned at the offset of the up state, and then realigned to the steepest decrease in the excitatory histogram before computing the average over neurons. (**d**) Same as **b** for the data in **c**. Inhibitory synaptic events extinguished later than excitatory synaptic events during the transition from the up state to the down state, as assessed at the 95% confidence level.

up states by tens of milliseconds and persists at their ending (Fig. 11c). In contrast, in our experiments we found that synaptic events detected intracellularly, both in vitro and in vivo, showed a remarkable matching of both event rate and timing of onset for excitatory and inhibitory events. We therefore turned back to our computer model to explore mechanistically the compatibility between the model and the experimental results regarding the timing and event rate magnitude of excitation and inhibition during the slow oscillation.

Modeling the Cortex

We used the network model of [5], with exactly the same parameters as in their control condition. Briefly, the network model consists of a population of 1024 pyramidal

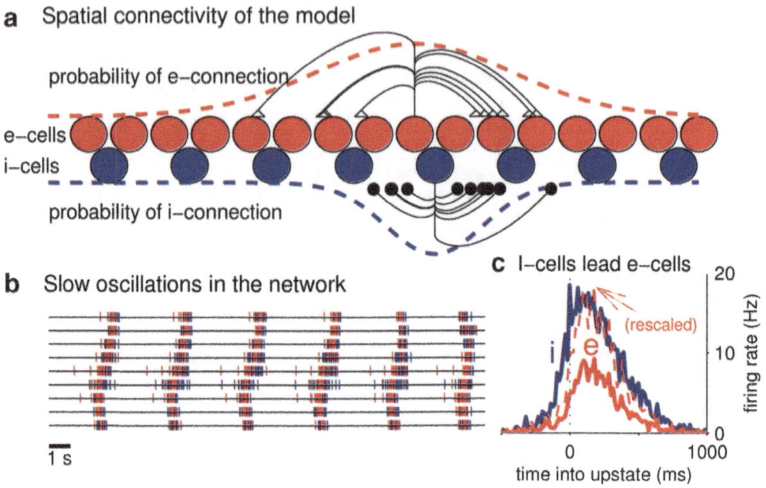

0

Fig. 11 Model architecture and function: i-cells lead e-cells during up state initiation. (**a**) The model consisted of excitatory (*red*) and inhibitory (*blue*) neurons (in a relation 4:1) connected through conductance-based synapses. The existence of a functional synapse between any two neurons was decided at the beginning of the simulation based on a Gaussian probability distribution. The footprint σ of this connectivity distribution could differ for excitatory and inhibitory connections. In the control case in [5], $\sigma_E = 2\sigma_I$. Each neuron only had a limited number of postsynaptic partners. In the control network in [5], both pyramidals and interneurons connected to 20 pyramidal neurons and 20 interneurons, respectively. (**b**) Sample network activity, shown as an array of multiunit spike trains, reflects slow oscillatory activity with interneurons (*blue*) and pyramidal neurons (*red*) firing in phase during the slow oscillation. (**c**) A closer look at neuronal activity around the time of up state initiation, shows that interneurons rise to their maximal firing rate ahead of closely adjacent pyramidal neurons (Adapted with permission from Figs. 2 and 3 in [5].).

cells and 256 interneurons equidistantly distributed on a line and interconnected through biologically plausible synaptic dynamics (Fig. 11a). Some of the intrinsic parameters of the cells are randomly distributed, so that the populations are heterogeneous. This and the random connectivity are the only sources of noise in the network.

Our model pyramidal cells have a somatic and a dendritic compartment. The spiking currents, I_{Na} and I_K, are located in the soma, together with a leak current I_L, a fast A-type K^+-current I_A, a noninactivating slow K^+-current I_{KS} and a Na^+-dependent K^+-current I_{KNa}. The dendrite contains a high threshold Ca^{2+} current I_{Ca}, a Ca^{2+}-dependent K^+-current I_{KCa}, a noninactivating (persistent) Na^+ current I_{NaP} and an inward rectifier (activated by hyperpolarization) noninactivating K^+ current I_{AR}. Explicit equations and parameters for these Hodgkin–Huxley-type currents can be found in [5]. In our simulations, all excitatory synapses target the dendritic compartment and all inhibitory synapses are localized on the somatic compartment of postsynaptic pyramidal neurons. Interneurons are modeled with just Hodgkin–Huxley spiking currents, I_{Na} and I_K, and a leak current I_L in their single compartment [50]. Model pyramidal neurons set according to these parameters fire at an average of 22 Hz when they are injected a depolarizing current of 0.25 nA for 0.5 s. The firing pattern corresponds to a regular spiking neuron with some adaptation. In contrast, a model interneuron fires at about 75 Hz when equally stimulated and has the firing pattern of a fast spiking neuron.

Synaptic currents are conductance-based and their kinetics are modeled to mimic AMPAR-, NMDAR-, and $GABA_A$ R-mediated synaptic transmission [5,49]. All parameters for synaptic transmission are taken from the control network in [5]. These values were chosen so that the network would show stable periodic propagating discharges with characteristics compatible with experimental observations.

The neurons in the network are sparsely connected to each other through a fixed number of connections that are set at the beginning of the simulation. In our control network, neurons make 20 ± 5 contacts (mean $\pm$ standard deviation) to their postsynaptic partners (multiple contacts onto the same target, but no autapses, are allowed). For each pair of neurons, the probability that they are connected in each direction is decided by a Gaussian probability distribution centered at 0 and with a prescribed standard deviation.

The model was implemented in C++ and simulated using a fourth-order Runge–Kutta method with a time-step of 0.06 ms.

Excitatory and Inhibitory Events During Up States In Computo

We analyzed spiking activity in inhibitory and excitatory neurons, and the timing of excitatory and inhibitory synaptic events into excitatory neurons, averaging data from five different network simulations (with different noise realizations to define the connectivity and neuron properties). We confirmed that the average firing rates of neurons followed the results reported in [5], namely that inhibitory neurons fired at higher rates (Fig. 12b), and increased earlier to their maximal rate (Fig. 13b) than excitatory neurons in our control network model. However, when the rates of incoming synaptic events into excitatory neurons were analyzed we found that the peak rate of excitatory events exceeded that of inhibitory events (Fig. 12a), while both excitatory and inhibitory events raised to their maximal rate in synchrony (Fig. 13a). These results may appear paradoxical: although interneurons fired more and ahead in time, inhibitory event rate was lower and did not show appreciable advance with respect to excitatory event rate. This reflects the multiple parameters that link firing rate to incoming synaptic rates, to the point of being able to distort significantly the relative values for excitation and inhibition, both in magnitude and timing. We examined this point in the model by testing two specific parameters that define the connectivity in our model network.

The relative length of excitatory and inhibitory horizontal connections in our network controlled the relative timing of arrival of excitatory and inhibitory events into model neurons. In the control network, excitatory neurons connected to other neurons in the network with a symmetric Gaussian probability distribution of standard deviation σ_E twice as large as the standard deviation σ_I of inhibitory projections

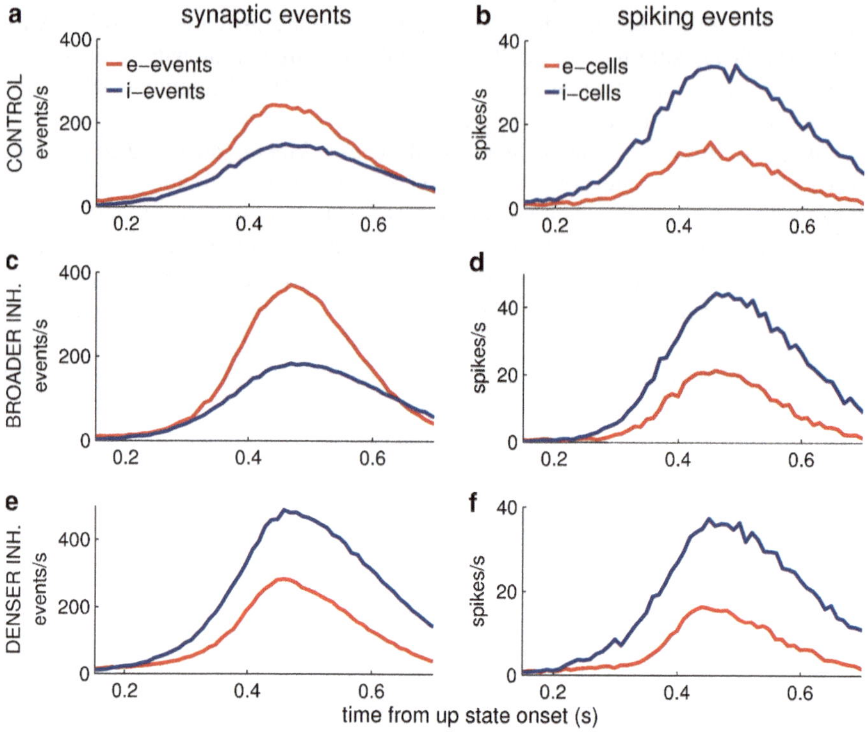

Fig. 12 Rate of excitatory and inhibitory synaptic events and spiking of excitatory and inhibitory neurons during the beginning and end of the up states in computo. (**a**) Rate of synaptic events (*red* = excitatory, *blue* = inhibitory) into pyramidal neurons and (**b**) Firing rate of adjacent excitatory (*red*) and inhibitory (*blue*) neurons during the up state in the computational network model of slow oscillatory activity [5]. In the model, although inhibitory neurons fire at more than double the rate than excitatory neurons, the rate of synaptic events coming into an excitatory neuron is higher for excitation than inhibition. This is due to the larger fraction of excitatory neurons in the network and their approximately equal connectivity (all neurons have 20 postsynaptic partners of each kind, excitatory or inhibitory). One-minute-long simulation data from 128 neurons equidistantly spaced along the network were used for the analysis. When the divergence of inhibitory connections was increased (by a factor four), firing rates increased slightly (**d**) and so did synaptic event rates (**c**). When inhibition was made denser than excitation by increasing twofold the number of synaptic contacts that each interneuron makes, inhibitory synaptic event rates increased markedly (**e**) whereas firing rates remained unaffected (**f**).

(Fig. 11a). Excitatory neurons had longer horizontal connections than inhibitory ones. Thus, when a front of activity propagated along the network, changes in excitatory rates were projected to neurons further away than changes in inhibitory rates. This compensated for the delayed firing of excitatory neurons at up state onset (Fig. 13b), and neurons received synchronous increases of excitatory and inhibitory synaptic events (Fig. 13a). These synaptic events caused firing first in the inhibitory neurons, possibly because of their lower firing threshold and their faster time constant. To test this mechanistic interpretation, we modified the relative footprint of

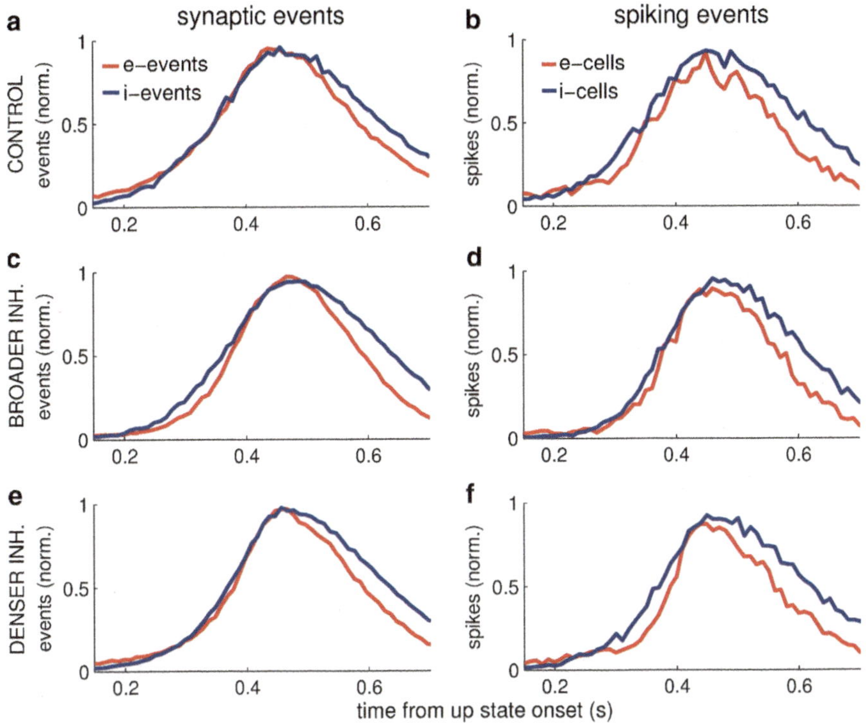

Fig. 13 Timing of excitatory and inhibitory synaptic events and spiking of excitatory and inhibitory neurons during the beginning and end of the up states in computo. (**a**) Rate of synaptic events normalized to peak event rate (*red* = excitatory, *blue* = inhibitory) into pyramidal neurons and (**b**) Normalized firing rate of adjacent excitatory (*red*) and inhibitory (*blue*) neurons during the up state in the computational network model of slow oscillatory activity [5]. In the model, although inhibitory neurons increase to their maximal rate ahead than excitatory neurons during up state onset (**b**), the accumulation of synaptic events coming into an excitatory neuron is equal for excitation than inhibition (**a**). This is due to the broader connectivity footprint for excitation than for inhibition. When the divergence of inhibitory connections was increased (by a factor four), inhibitory rates still accumulated slightly ahead than pyramidal neurons (**d**) and so did now synaptic event rates, too (**c**). When inhibition was made denser than excitation by increasing twofold the number of synaptic contacts that each interneuron makes, the timing relations of the control network (**a**, **b**) were not affected: synaptic event rates varied concomitantly for excitation and inhibition (**e**) and interneurons increased their firing ahead than pyramidals (**f**). In all cases, inhibitory events (neurons) outlasted excitatory events (neurons) at the end of the up state.

excitatory and inhibitory connections to make inhibitory projections more divergent ($\sigma_I = 2\sigma_E$). We found that the slight advance in inhibitory neuron firing rate increase at up state onset (Fig. 13d) was mimicked by an advanced arrival of inhibitory events to excitatory neurons in the network (Fig. 13c). In this case, because excitatory projections did not exceed the inhibitory footprint, they could not compensate interneuron advanced firing at up state onset. This manipulation also increased slightly the event rates and firing rates of neurons during the self-sustained slow oscillation, but did not modify the relative magnitudes between excitation and inhibition (Fig. 12c, d).

The average number of connections that each cell type made with postsynaptic neurons of either kind in the network were parameters that controlled the relative magnitude of inhibitory and excitatory synaptic event rates. In the control network, although inhibitory neurons fired at a higher rate (Fig. 12b), because there were four times more excitatory neurons in the network, and excitatory and inhibitory neurons made the same average number of contacts on postsynaptic neurons, the rate of excitatory events received by postsynaptic neurons exceeded by a significant factor the rate of inhibitory events (Fig. 12a). Instead, if we manipulated the connectivity of the network and had interneurons make more postsynaptic contacts on average than excitatory neurons (per i-cell, 80 ± 5 inhibitory contacts to e-cells, same to i-cells; per e-cell, 20 ± 5 excitatory contacts to e-cells, same to i-cells; mean $\pm$ s.d. Concomitantly to this increase in number of inhibitory synapses, we diminished the conductance of an individual inhibitory synapse by a factor $^1/_4$, so that overall inhibitory currents remained unchanged), we found that the number of inhibitory events received by pyramidal neurons now exceeded that of excitatory events (Fig. 12e) by approximately the same ratio as in neuronal firing rates (Fig. 12f). This is consistent with the fact that the neuronal ratio of 4:1 in cell number (excitatory to inhibitory) was now compensated by a connectivity contact ratio of 1:4, so spiking events translated by a common factor to synaptic events and maintained their relative relationship. Notice that neuronal firing rates during the up states of the slow oscillations did not change appreciably with respect to the control network (Fig. 12f compared with Fig. 12b), because of the rescaling of inhibitory conductances to compensate the increase in inhibitory connectivity. In relation to the timing of excitation and inhibition at up state onset, this manipulation did not induce any appreciable change relative to the control case (Fig. 13a, b): Interneurons kept firing ahead than pyramidal neurons at up state onset (Fig. 13f), but inhibitory and excitatory events arrived in synchrony to their postsynaptic targets (Fig. 13e).

We found that in our model network, inhibitory firing and inhibitory synaptic events outlasted in all cases excitation at the end of the up state (Fig. 13). This persistence of inhibitory events after excitatory event extinction was accentuated when the footprint of inhibitory projections was increased (Fig. 13c), as would be expected. This effect was seen in some of the experimental recordings (Fig. 6h, j and Fig. 9a), but was not generally true in our population of neurons in vitro (Fig. 7c), although it appeared significant in our small sample of in vivo neurons (Fig. 10c).

Timing of Excitation and Inhibition in Cortical Activity

Here we have analyzed the timing of inhibitory and excitatory events during the up and down states occurring in the cortex in vitro, in vivo, and in a computer model.

An equilibrium between excitation and inhibition in the recurrent network of the cerebral cortex has been proposed to be critical to maintain the stability of its function. Changes in excitatory and inhibitory conductances in vitro reveal that both increase and decrease at the beginning/end of up states in close association with each other [38]. Not only in time, but also the amplitude of both were related, with a slope of 0.68 (G_i/G_e) in the aforementioned study. Our approach is different, and provides information regarding timing of both types of events as well as an estimation of the number of events. In agreement to what was reported in [38], we find a remarkable coincidence in the accumulation of both excitatory and inhibitory events during the rise of an up state, suggesting reverberation of activity in the local cortical microcircuits. In six out of ten cases the absolute number of excitatory and inhibitory synaptic events recorded from neurons in vitro is very similar as well.

Individual pyramidal neurons receive on average inputs from 1,000 excitatory neurons vs. 75 inhibitory ones, resulting in a number of contacts of 5,000 vs. 750, respectively [31]. Besides, most of cortical neurons participate in up states [41]. The open question is then, how can the number of excitatory and inhibitory events received by a pyramidal neuron be similar? A simple answer to it is the higher firing rate of inhibitory neurons, that would compensate for the lesser number of inhibitory synaptic connections. Cortical fast spiking neurons, known to be gabaergic [19,45], have steeper input–output (intensity-frequency) relationships [25,28] as a result of their intrinsic properties [12]. Furthermore, fast spiking neurons respond with much longer trains of action potentials when activated synaptically during up states [38]. We and others [43] have also observed that the firing of fast spiking neurons activated synaptically during up states [38]. We and (Fig. 1) although we have not carried out a systematic fast spiking neurons during up states is higher than that of rate could compensate, at least in part, for the lesser number of inhibitory presynaptic contacts. In spite of the disproportion between anatomical excitatory and inhibitory contacts onto pyramidal cells, not only did we find that there are similar numbers of excitatory and inhibitory events but also that the inhibitory ones are of significantly larger average amplitude (2.78 mV) than the excitatory ones (0.8 mV) in vitro. Somatic and proximal innervation of gabaergic inputs is probably a main factor on this difference, although synchrony of inputs due to presynaptic electrical coupling could also contribute [15, 16]. Even when we consider that the caveats of this method (see section "A Short Discussion on the Method") would equally affect both IPSPs and EPSPs, the possibility remains that one of them was consistently underestimated. The method used here could result in an overestimation of inhibitory synaptic events with respect to the excitatory ones. Inhibition occurs in the soma or proximal dendrites [13, 14] while excitation takes place further away from the soma. Therefore, excitatory events would have smaller amplitudes and remain below threshold, or because occurring further away from the soma, their kinetics are slower and they are more difficult to detect. We cannot rule out that possiblity. However, our detection procedure has been tested in detail and the number of excitatory (inhibitory) events decreases (increases) as expected with depolarizing (hyperpolarizing) membrane potential values [6]. Moreover, even if EPSPs underestimation happens, only the absolute EPSPs/IPSPs measurements would be affected, but not the normalized comparisons, and thus the relative times of occurrence of both types of events that would remain valid.

In few cases, synaptic activity is detected during down states, predominantly inhibitory activity (Fig. 5d, e, i). In spite of down states being periods of hyperpolarization [7, 35] and excitatory disfacilitation [46], there is some neuronal firing during down states, mostly reported in layer 5 neurons, where up states start. This activity is illustrated in ([35], Fig. 2b), or in ([5], Fig. 1). In [38], it is reported that 43% of the recorded layer 5 neurons have some firing during down states, of an average rate of 3.6 Hz vs. 17.1 Hz during up states. This firing is, according to our model, implicated in the generation of the subsequent up state [5]. The synaptic events reported in Fig. 5d, e, i were obtained from supragranular layers, and thus the inhibitory activity during the down states, which was of 5–20 Hz can be the result of excitatory innervation from layer 5 to inhibitory interneurons in layers 2/3 [9]. Still, such continuous rate of IPSPs was rather unusual. The average rate for down states was 3 and 6 events/s while 37 and 27 events/s (excitatory and inhibitory, respectively) during up states. Up states in vivo revealed an almost identical average rate of events during up states (33 and 31 events/s excitatory and inhibitory, respectively). These numbers are remarkably lower than the ones reported for up states in striatum-cortex-substantia nigra cocultures, which reached a rate of 800 events/s against 10–20 events/s during down states [3].

Activation of excitatory vs. inhibitory neurons during up states in vivo has been reported in [17, 43]. There, the initiation, the initiation, maintenance, and termination of up states in fast to follow network dynamics similar to those in pyramidal cells. Here, we find that the accumulation of excitatory and inhibitory synaptic events is also quite synchronous in vivo as we reported for in vitro. Similar findings were reported in [17], where PSTHs were built with the firing of both excitatory and inhibitory neurons. In our case, we find that the accumulation of synaptic events is 1.4 times faster in vivo than in vitro. We make similar observations if we look at the rise time of the membrane potential, given that the average time for depolarization to the up state is shorter in vivo than in vitro (unpublished observations). The preservation of the thalamocortical loop and of horizontal connections while in vivo can contribute to this faster slope of up states.

Our findings also relate to the debate regarding the relative magnitude of excitatory and inhibitory conductances during the up state. In vitro, we find that inhibitory and synaptic events arrive at a similar rate in the postsynaptic neuron, which could lend support to the balanced conductance observations of Shu et al. [38, 39]. However, we also observed that voltage deflections caused by inhibitory events were almost 4 times larger than those caused by excitatory events, in conditions where driving forces should be approximately equal for both types of events. Then, the overall inhibitory conductance would be larger by a factor 4 than excitatory conductances, as proposed by [33]. In vivo, instead, we did not find a major difference between excitatory and inhibitory synaptic potential amplitudes while we still observed similar rates for inhibitory and excitatory events during up states, in agreement with the results in [17]. Other authors though report larger inhibitory conductances during up states in vivo in average, although approximately half of their recorded neurons showed similar levels of excitatory and inhibitory conductance [34].

Our computational model has allowed us to demonstrate how the precise relationship between excitation and inhibition inputs depends on the structural parameters defining the connectivity in the local cortical circuit. In the light of the large divergence in connectivity parameters for excitatory and inhibitory transmission in the cortex (see "Introduction"), the approximate balance both in timing and magnitude of excitatory and inhibitory synaptic conductances measured experimentally [17, 33, 38, 39] is remarkable and reflects compensation in various of these parameters. By detecting synaptic event timing, rather than synaptic conductance, we can now eliminate one parameter from the equations: the value of excitatory and inhibitory unitary synaptic conductance changes. We find that the number and timing of incoming synaptic events are also approximately matched, so that the unitary synaptic conductances are not the major compensating mechanism for achieving the excitatory–inhibitory balance in the cortex. Instead, our computational model suggests that compensation might be achieved through the tuning of presynaptic firing rate and postsynaptic contacts. Thus, inhibitory interneurons fire at higher rates than pyramidal neurons, and each individual interneuron makes more contacts onto a given postsynaptic neuron [31], so that these factors can balance the fact that pyramidal neurons outnumber inhibitory neurons in the local cortical circuit. However, in light of the caveats of our detection method (section "A Short Discussion on the Method".), we are not in a position of making a strong case in relation with the absolute value of synaptic event rates in the up states.

Instead, the relative timing of excitatory and inhibitory events seems a more robust estimation. Given that pyramidal cortical neurons are known to have a rich local axonal arborization which is typically larger than that of most GABAergic interneurons [23], our experimental finding of a simultaneous arrival of excitatory and inhibitory events is likely to reflect the early firing of inhibitory neurons relative to neighboring excitatory neurons in the transition to the up state, as suggested computationally in [5]. Experimentally, a nonsignificant trend for inhibitory firing leading excitatory firing has been reported in [17,34], but this data generally indicates an activation close to simultaneous for fast spiking and regular spiking neurons. Based on our model simulations, this would indicate an approximately equal horizontal projection length for excitation and inhibition in the cortex, suggesting that intracortical inhibition in the wavefront of the slow oscillation might be principally mediated by the subclass of inhibitory neurons formed by basket cells, which have the longest projection axons among cortical interneurons [24, 40]. This prediction can be tested experimentally in the future.

Acknowledgments Financial support from Ministerio de Ciencia e Innovación (MICINN) to MVSV and AC is acknowledged. RR was partially supported by the FP7 EU (Synthetic Forager FP7- ICT-217148) and by MICINN.

Appendix

Intracellular and Population Recordings In Vitro and In Vivo

In Vitro Recordings

The methods for preparing cortical slices were similar to those described previously [35]. Briefly, 400 μm cortical prefrontal or visual slices were prepared from 3- to 10 month-old ferrets of either sex that were deeply anesthetized. After preparation, slices were placed in an interface-style recording chamber and bathed in ACSF containing (in mM): NaCl, 124; KCl, 2.5; $MgSO_4$, 2; $NaHPO_4$, 1.25; $CaCl_2$, 2; $NaHCO_3$, 26; and dextrose, 10, and was aerated with 95% O_2, 5% CO_2 to a final pH of 7.4. Bath temperature was maintained at 35–36°C. Intracellular recordings were initiated after 2 h of recovery. In order for spontaneous rhythmic activity to be generated, the solution was switched to "in vivo-like" ACSF containing (in mM): NaCl, 124; KCl, 3.5; $MgSO_4$, 1; $NaHPO_4$, 1.25; $CaCl_2$, 1–1.2; $NaHCO_3$, 26; and dextrose, 10.

In Vivo Recordings

Intracellular recordings in vivo were obtained from rat neocortex (auditory and barrel cortex) as in [32]. Anesthesia was induced by intraperitoneal injection of ketamine (100 mg/kg) and xylacine (8–10 mg/kg) and were not paralyzed. The maintenance dose of ketamine was 75 mg/kg/h. Anesthesia levels were monitored by the recording of low-frequency electroencephalogram (EEG) and the absence of reflexes. Through a craniotomy over the desired area the local field potential was recorded with a tungsten electrode. Intracellular recordings (see below) were obtained in close vicinity from the extracellular recording electrode with identical micropipettes to the ones used to record from the cortical slices.

Recordings and Stimulation

Extracellular multiunit recordings were obtained with 2–4 MΩ tungsten electrodes. The signal was recorded unfiltered at a sampling frequency between 1 and 10 kHz. For intracellular recordings (sampling frequency 10–20 kHz), sharp electrodes of 50–100 MΩ filled with 2 M potassium acetate were used. Sodium channel blocker QX314 (100 μM) was often included in the electrode solution to hold the membrane voltage (Vm) at depolarized potentials while preventing firing.

Data Analysis and Detection of Synaptic Events

Extracellular recordings were used to identify up and down state onsets. To this end, extracellular recordings were high-pass filtered above 1 Hz to remove slow linear trends in the signal. Then, the envelope of the resulting time series was evaluated as the amplitude of its analytic signal (complex Hilbert transform), high-pass filtered above 0.1 Hz to remove the DC, and smoothed with a running-average square window of 100 ms (Fig. 3). The mean value of this signal was the threshold for the detection of transitions between up state and down state in all recordings. Up states and down states shorter than 250 ms were discarded from the analysis.

Intracellular current clamp recordings were maintained at different membrane voltages by means of current injection. At least two membrane voltages were usually attained: (1) around −70 mV, to achieve chloride reversal potential and isolate EPSPs and (2) around 0 mV, to isolate IPSPs. The timing of presynaptic events of excitatory or inhibitory type were extracted from these intracellular recordings at different membrane voltages (Fig. 3). This was achieved by passing the membrane voltage signal through a differentiator filter with a low-pass cutoff at 200 Hz, thus evaluating a smoothed first time derivative (Fig. 3). This cutoff was not significantly limiting the number of detected synaptic events, as changing it to 500 Hz did not modify our conclusions appreciably. The method has been described in detail in [6]. The timing of synaptic events was detected from sharp voltage deflections in intracellular recordings. Local maxima (minima) are then candidates for excitatory (inhibitory) events, as they represent the fastest voltage upward (downward) deflections in a neighborhood of data points. The central values of these local extremes are typically Gaussian distributed, but extreme values are distributed according to long tails. These long tails presumably contain actual synaptic events, which stick out from noisy membrane voltage fluctuations. To estimate the threshold value that separates these random voltage fluctuations from actual synaptic event voltage deflections, we detected events in the tails of the distribution beyond thresholds set at a fixed number n of interquartile ranges σ from the median of the distribution. We used n in the range $n = 2$–5, and its precise value was chosen independently for each recorded cell so that more inhibitory events were detected in the depolarized than in the hyperpolarized recording, while at the same time more excitatory events were detected in the hyperpolarized relative to the depolarized recording [6]. For all our analyses here, inhibitory events were extracted just from the depolarized membrane voltage recording and excitatory events just from the hyperpolarized membrane voltage recording.

References

1. Anderson JS, Carandini M, Ferster D (2000) Orientation tuning of input conductance, excitation, and inhibition in cat primary visual cortex. J Neurophysiol 84:909–926.
2. Bernander O, Koch C, Douglas RJ (1994) Amplification and linearization of distal synaptic input to cortical pyramidal cells. J Neurophysiol 72:2743–2753.

3. Blackwell K, Czubayko U, Plenz D (2003) Quantitative estimate of synaptic inputs to striatal neurons during up and down states in vitro. J Neurosci 23:9123–9132.
4. Borg-Graham LJ, Monier C, Fregnac Y (1998) Visual input evokes transient and strong shunting inhibition in visual cortical neurons. Nature 393:369–373.
5. Compte A, Sanchez-Vives MV, McCormick DA, Wang XJ (2003) Cellular and network mechanisms of slow oscillatory activity (<1 Hz) and wave propagations in a cortical network model. J Neurophysiol 89:2707–2725.
6. Compte A, Reig R, Descalzo VF, Harvey MA, Puccini GD, Sanchez-Vives MV (2008) Spontaneous high-frequency (10–80 Hz) oscillations during up states in the cerebral cortex in vitro. J Neurosci 28:13828–13844.
7. Contreras D, Timofeev I, Steriade M (1996) Mechanisms of long-lasting hyperpolarizations underlying slow sleep oscillations in cat corticothalamic networks. J Physiol 494 (Pt 1): 251–264.
8. Cowan RL, Wilson CJ (1994) Spontaneous firing patterns and axonal projections of single corticostriatal neurons in the rat medial agranular cortex. J Neurophysiol 71:17–32.
9. Dantzker JL, Callaway EM (1998) The development of local, layer-specific visual cortical axons in the absence of extrinsic influences and intrinsic activity. J Neurosci 18:4145–4154.
10. Descalzo VF, Nowak LG, Brumberg JC, McCormick DA, Sanchez-Vives MV (2005) Slow adaptation in fast-spiking neurons of visual cortex. J Neurophysiol 93:1111–1118.
11. Douglas RJ, Martin KA (2004) Neuronal circuits of the neocortex. Annu Rev Neurosci 27:419–451. Review.
12. Erisir A, Lau D, Rudy B, Leonard CS (1999) Function of specific K(+) channels in sustained high-frequency firing of fast-spiking neocortical interneurons. J Neurophysiol 82:2476–2489.
13. Fairen A, DeFelipe J, Regidor J (1984) Non pyramidal neurons: general account. In: Cerebral Cortex (Peters AaJEG, ed.). London: Plenum Press.
14. Feldman ML (1984) Morphology of the neocortical pyramidal neuron. In: Cerebral Cortex (Peters AaJEG, ed.). London: Plenum Press.
15. Galarreta M, Hestrin S (1999) A network of fast-spiking cells in the neocortex connected by electrical synapses. Nature 402:72–75.
16. Gibson JR, Beierlein M, Connors BW (1999) Two networks of electrically coupled inhibitory neurons in neocortex. Nature 402:75–79.
17. Haider B, Duque A, Hasenstaub AR, McCormick DA (2006) Neocortical network activity in vivo is generated through a dynamic balance of excitation and inhibition. J Neurosci 26:4535–4545.
18. Hebb DO (1949) The organization of behavior. New York:Wiley.
19. Kawaguchi Y, Kubota Y (1993) Correlation of physiological subgroupings of nonpyramidal cells with parvalbumin- and calbindinD28k-immunoreactive neurons in layer V of rat frontal cortex. J Neurophysiol 70:387–396.
20. Kruglikov I, Rudy B (2008) Perisomatic GABA release and thalamocortical integration onto neocortical excitatory cells are regulated by neuromodulators. Neuron 58:911–924.
21. Lampl I, Reichova I, Ferster D (1999) Synchronous membrane potential fluctuations in neurons of the cat visual cortex. Neuron 22:361–374.
22. Lorente de Nó R (1938) Analysis of the activity of the chains of internuncial neurons. J Neurophysiol 1:207–244.
23. Lund, JS and Wu, CQ (1997) Local circuit neurons of macaque monkey striate cortex: IV. Neurons of laminae 1–3A. J Comp Neurol 384:109–126.
24. Markram H, Toledo-Rodriguez M, Wang Y, Gupta A, Silberberg G, Wu C (2004) Interneurons of the neocortical inhibitory system. Nat Rev Neurosci 5:793–807.
25. McCormick DA, Connors BW, Lighthall JW, Prince DA (1985) Comparative electrophysiology of pyramidal and sparsely spiny stellate neurons of the neocortex. J Neurophysiol 54:782–806.
26. Megias M, Emri Z, Freund TF, Gulyas AI (2001) Total number and distribution of inhibitory and excitatory synapses on hippocampal CA1 pyramidal cells. Neuroscience 102:527–540.
27. Monier C, Fournier J, Fregnac Y (2008) In vitro and in vivo measures of evoked excitatory and inhibitory conductance dynamics in sensory cortices. J Neurosci Methods 169:323–365.

28. Nowak LG, Azouz R, Sanchez-Vives MV, Gray CM, McCormick DA (2003) Electrophysiological classes of cat primary visual cortical neurons in vivo as revealed by quantitative analyses. J Neurophysiol 89:1541–1566.
29. Okun M, Lampl I (2008) Instantaneous correlation of excitation and inhibition during ongoing and sensory-evoked activities. Nat Neurosci 11:535–537.
30. Pare D, Shink E, Gaudreau H, Destexhe A, Lang EJ (1998) Impact of spontaneous synaptic activity on the resting properties of cat neocortical pyramidal neurons in vivo. J Neurophysiol 79:1450–1460.
31. Peters A (2002) Examining neocortical circuits: some background and facts. J Neurocytol 31:183–193.
32. Reig R, Sanchez-Vives MV (2007) Synaptic transmission and plasticity in an active cortical network. Plos One 2(7):e670.
33. Rudolph M, Pelletier JG, Paré D, Destexhe A (2005) Characterization of synaptic conductances and integrative properties during electrically induced EEG-activated states in neocortical neurons in vivo. J Neurophysiol. Oct 94(4):2805–2821. Epub 2005 Jul 13.
34. Rudolph M, Pospischil M, Timofeev I, Destexhe A (2007) Inhibition determines membrane potential dynamics and controls action potential generation in awake and sleeping cat cortex. J Neurosci 27:5280–5290.
35. Sanchez-Vives MV, McCormick DA (2000) Cellular and network mechanisms of rhythmic recurrent activity in neocortex. Nat Neurosci 3:1027–1034.
36. Schwindt PC, Crill WE (1997) Modification of current transmitted from apical dendrite to soma by blockade of voltage- and Ca^{2+}-dependent conductances in rat neocortical pyramidal neurons. J Neurophysiol 78:187–198.
37. Shadlen MN, Newsome WT (1998) The variable discharge of cortical neurons: implications for connectivity, computation, and information coding. J Neurosci 18:3870–3896.
38. Shu Y, Hasenstaub A, McCormick DA (2003a) Turning on and off recurrent balanced cortical activity. Nature 423:288–293.
39. Shu Y, Hasenstaub A, Badoual M, Bal T, McCormick DA (2003b) Barrages of synaptic activity control the gain and sensitivity of cortical neurons. J Neurosci 23:10388–10401.
40. Somogyi P, Kisvarday ZF, Martin KA, Whitteridge D (1983) Synaptic connections of morphologically identified and physiologically characterized large basket cells in the striate cortex of cat. Neuroscience 10:261–294.
41. Steriade M, Nunez A, Amzica F (1993a) A novel slow (<1 Hz) oscillation of neocortical neurons in vivo: depolarizing and hyperpolarizing components. J Neurosci 13:3252–3265.
42. Steriade M, Nunez A, Amzica F (1993b) Intracellular analysis of relations between the slow (< 1 Hz) neocortical oscillation and other sleep rhythms of the electroencephalogram. J Neurosci 13:3266–3283.
43. Steriade M, Timofeev I, Grenier F (2001) Natural waking and sleep states: a view from inside neocortical neurons. J Neurophysiol 85:1969–1985.
44. Stern EA, Kincaid AE, Wilson CJ (1997) Spontaneous subthreshold membrane potential fluctuations and action potential variability of rat corticostriatal and striatal neurons in vivo. J Neurophysiol 77:1697–1715.
45. Thomson AM, West DC, Hahn J, Deuchars J (1996) Single axon IPSPs elicited in pyramidal cells by three classes of interneurones in slices of rat neocortex. J Physiol 496 (Pt 1):81–102.
46. Timofeev I, Grenier F, Steriade M (2001) Disfacilitation and active inhibition in the neocortex during the natural sleep-wake cycle: an intracellular study. Proc Natl Acad Sci U S A 98:1924–1929.
47. Timofeev I, Grenier F, Bazhenov M, Sejnowski TJ, Steriade M (2000) Origin of slow cortical oscillations in deafferented cortical slabs. Cereb Cortex 10:1185–1199.
48. Trevelyan AJ, Watkinson O (2005) Does inhibition balance excitation in neocortex? Prog Biophys Mol Biol 87:109–143.
49. Wang XJ (1999) Synaptic basis of cortical persistent activity: the importance of NMDA receptors to working memory. J Neurosci 19:9587–9603.
50. Wang XJ, Buzsaki G (1996) Gamma oscillation by synaptic inhibition in a hippocampal interneuronal network model. J Neurosci 16:6402–6413.

51. Wehr M, Zador AM (2003) Balanced inhibition underlies tuning and sharpens spike timing in auditory cortex. Nature 426:442–446.
52. Zhu L, Blethyn KL, Cope DW, Tsomaia V, Crunelli V, Hughes SW (2006) Nucleus- and species-specific properties of the slow (<1 Hz) sleep oscillation in thalamocortical neurons. Neuroscience 141:621–636.

Finding Repeating Synaptic Inputs in a Single Neocortical Neuron

Gloster Aaron

Abstract A goal in neuroscience is to understand what occurs when large numbers of interconnected neurons actively communicate with each other to create perception. An important part in this goal is to observe large numbers of neurons engaged in such communication. Outlined here is an approach to this challenge. This approach uses a single neuron as a "microphone" of cortical activity. As potentially thousands of neurons may connect with a single neuron in the mammalian sensory neocortex, then it may be possible to record large networks by recording the synaptic inputs to a single neuron. In pursuing this goal, we observed patterns in the recordings that appeared to repeat with remarkable precision. Whether this finding is evidence that the cortex can produce precisely repeating patterns is a matter of contention, and we describe recent investigations of this question.

Introduction

A neuron in the mammalian sensory neocortex can receive thousands of synaptic inputs from other neurons. The great majority of those synaptic inputs originate from other neocortical neurons within the same column of cortex. It is thus not surprising that when a slice of neocortex is observed, bereft of any thalamic inputs, that the remaining neocortical circuits in the slice generate spontaneous activity. This spontaneous activity has been shown to contain structure in the form of upstates [1, 2] – events when many neurons fire action potentials simultaneously – and in the form of oscillations [3]. It has also been shown that the spontaneous upstates in layer 4 are significantly similar to the activation of layer 4 that immediately follows stimulation of the thalamus in a thalamo-cortical slice preparation [4]. Such findings suggest that the neocortex is a pattern generator, a device that can create patterned output in the absence of any patterned input [5]. Other studies suggest that the spatio-temporal

G. Aaron (✉)
Department of Biology, Wesleyan University, Middletown, CT 06459, USA
e-mail: gaaron@wesleyan.edu

K. Josić et al. (eds.), *Coherent Behavior in Neuronal Networks*, Springer Series in Computational Neuroscience 3, DOI 10.1007/978-1-4419-0389-1_3,

patterns of neuronal activity that can be produced in cortical circuits are surprisingly precise in their repeatability, possibly displaying a precision on the scale of a millisecond [6–11], while other studies dispute these findings [12–14].

What kind of patterns of activity are created in cortical circuits, and how can they be observed? This chapter examines a technique that may help answer such questions, and while we do not have a priori knowledge of the characteristics of these patterns, we can still attempt to find patterns. One approach is to search for repeats – that is, sequences of synaptic inputs that repeat later in the recording with significant precision. This is analogous to a study of human language by a completely naive observer: a language has a finite vocabulary, and the words composing this vocabulary can be identified via a search for repeats.

Of course, neuronal activity is harder to read. In "seeing" activity, we wish to have a record of the action potentials produced by single neurons, and ideally we record a large number of those neurons simultaneously. The number of neurons in one cortical column in the rodent sensory neocortex is on the order of 10^4, so the challenge of capturing a significant fraction of that number is great. There are now many ways of capturing the activity of several hundreds of neurons with the use of multiple electrodes and imaging techniques.

Repeat Detection

The technique discussed here uses intracellular recordings from single neurons as a means to "listen" to potentially all of the activity of all neurons that form synapses with that recorded neuron. As a single pyramidal neuron may receive 1,000s of synapses from other neurons, most of them locally, then this technique has the potential to yield information about a large fraction of a cortical column (Fig. 1).

The technique is to record a single neuron intracellularly for several minutes and then examine the recording, looking for repeats [10, 14]. The cross-covariance function is at the heart of this analysis, and this function quantifies the temporal similarities of the recorded waveforms. There are two stages in the analysis: LRI (low resolution index) and HRI (high resolution index). The LRI compares 1-s segments of the recorded waveform, using a nested loop of template matching with cross-covariance as the measurement of the match or mismatch. Equation (1) describes the cross-covariance calculation that is performed every time a potential motif and repeat are lifted from the original recording for comparison:

$$h(\tau) = \frac{\sum_{t=-T}^{T} (x_t - \bar{x})(y_{t+\tau} - \bar{y})}{\sqrt{\sum_{t=-T}^{T} (x_t - \bar{x})^2 \sum_{t=-T}^{T} (y_t - \bar{y})^2}} \tag{1}$$

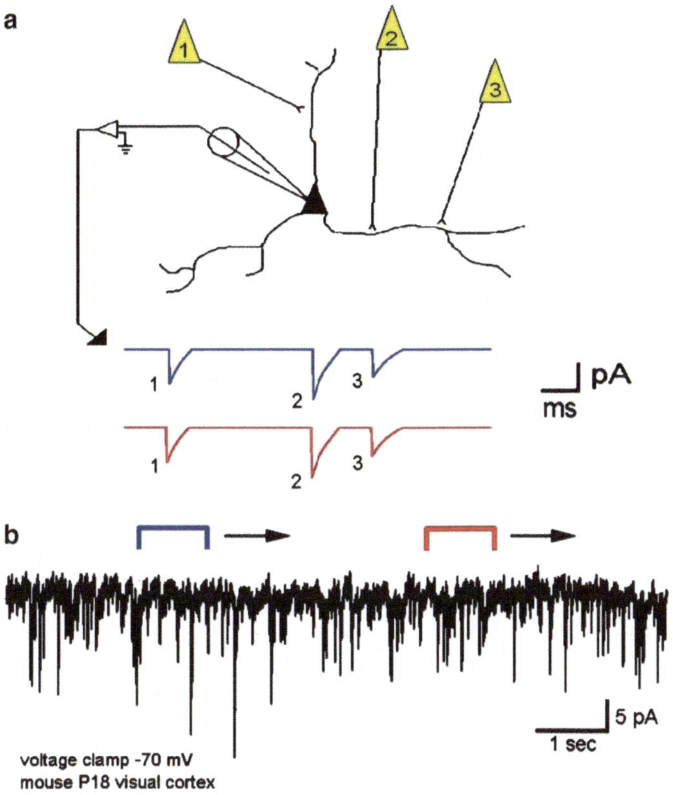

Fig. 1 Searching for and finding putative repeats with the low resolution index (LRI). (**a**) A cartoon illustrating how repeats of action potential sequences in a cortical network can be recorded in a single neuron. The picture depicts a pyramidal neuron being recorded with an intracellular electrode that measures postsynaptic currents (PSCs) during voltage-clamp recordings. A series of action potentials in three neurons forming synapses with the neuron can be recorded. The *blue* trace represents such a sequence that was recorded, and the *red trace* shows the same sequence repeating at some later time. (**b**) A continuous 10-s stretch of intracellularly recording postsynaptic currents (PSCs) is displayed. The program scans this recording, comparing every 1-s interval with every other 1-s interval. Here, the *blue* and *red brackets* represent these 1-s scanning windows. These 1-s segments are compared against each other via a cross-covariance equation (1). If there were a perfect repeat of intracellular activity, then the correlation coefficient at the zeroth lag time would be 1.0.

Here, x and y are amplitudes from the respective motif and its potential repeat, and $2T + 1$ are the number of samples in each at 1 point per ms. The length of x and y is 1 s (1,000 points at 1 point per millisecond), and τ represents the lag time between x and y. The actual time delay between the two traces x and y, however, is not accounted for in this equation. Rather, that time delay is remembered by the program for subsequent retrieval. The denominator normalizes the result so that in all cases the range of answers is from -1 to 1. In Matlab, the software platform that supports the detector, the above equation is represented in the following

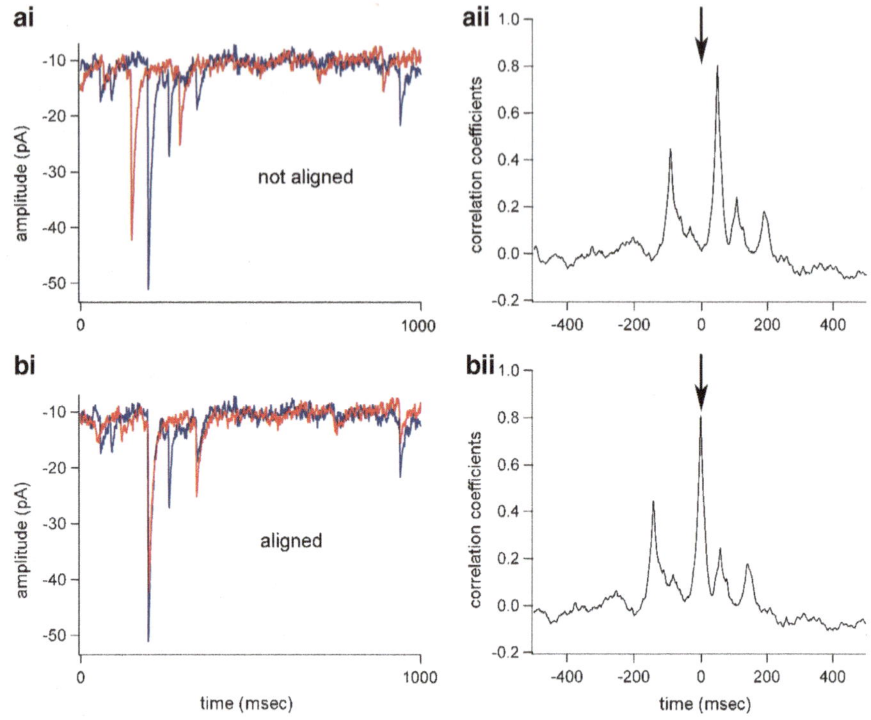

Fig. 2 Aligning the extracted motif-repeat. The peak of the cross-correlogram produced by (1) is used to reposition the 2nd search window (*red*) relative to the 1st (*blue*). (**ai**) Two 1-s segments of the recording have been extracted and aligned according to the 250-ms jump in the extraction windows. (**aii**) Applying (1), a correlogram is produced, indicating a large peak at a 50-ms lag time. *Arrow* indicates the zeroth lag time. (**bi**) The 1-s motif-repeat candidates are realigned according to the peak lag time in **bi**. (**bii**) Using (1), a new correlogram is recalculated from the traces shown in **bi**, producing a correlogram with the maximum peak value at approximately the zeroth lag time (i.e., h (0) produces the largest correlation coefficient).

command: xcov(x,y,'coef'). The motifs and repeats are defined by these lengths and the incremental jump from one potential repeat to another is 250 ms (in Fig. 1 this would represent the incremental movements of the colored brackets). As jumps of 250 ms are unlikely to find the regions of precise overlap, the program realigns the traces according to the difference between the peak value of the covariance function (correlation coefficient) and the zeroth lag of this function (i.e., the value at $\tau = 0$) and then recomputes the function (Fig. 2). This alignment procedure is reiterated up to 4×, or until the peak is within 1 ms of the zeroth lag. The value at the zeroth lag ($h(0)$) is then recorded. The highest values for each 1-s interval and those passing a set threshold were collected for each recording and formed our low resolution similarity index (LRI). The threshold was set according to a level that yielded a reasonable number of putative motif-repeats that could be analyzed with subsequent HRI analysis. "Reasonable" is defined here as taking less than a few

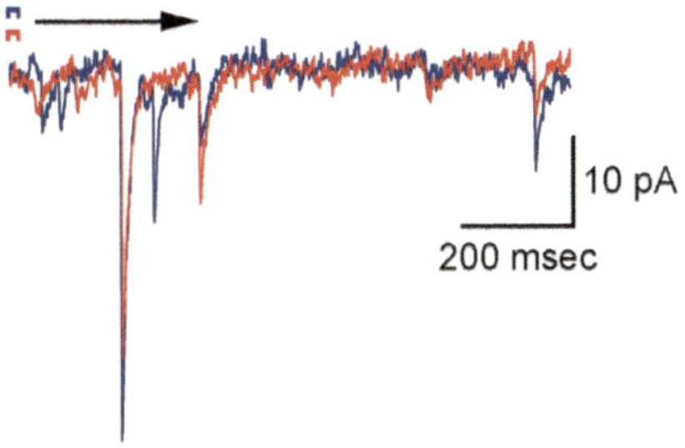

Fig. 3 Motif-repeat segments that yield a minimum threshold h (0) value during LRI calculations are remembered and subsequently analyzed via a high resolution index (HRI). Using voltage-clamp recordings, a 20-ms window is used to extract small segments from the aligned motif and repeat for calculation of T values (2). As shown above, this 20-ms window corresponds to the width of the brackets to the *left* of the *arrow*, indicating multiple extract of 20-ms segments. This 20-ms window corresponds approximately to the average width of a postsynaptic current, shown above as the sharp, downward deflections in the motif and repeat.

days of computation time with HRI analysis, and per recording this would mean on the order of 10,000 putative repeats. For this particular case, the threshold was set to 0.45. In this sense, the thresholds here not in any way considered definitive.

The 1-s length of the potential motif and repeat is also arbitrary, and, as discussed later, problematic. This initial identification is, however, somewhat justified in reducing what would otherwise be an overly burdensome computational task. That is, the LRI is used to identify putative repeats, remember the locations of those putative repeats, and then analyze more carefully those segments in subsequent analyses. Segments that do not pass a minimum threshold are passed over and not analyzed further, saving some time in the subsequent intensive analysis.

HRI examines the 1-s intervals indicated by LRI using 20-ms comparison windows, respectively ((2) and (3), Fig. 3). The 20 ms is roughly matched to the length of the average PSC in the recording. In analyzing current-clamp recordings, a 100-ms window is used, as the longer window better matches the broadening in time of the postsynaptic potential imposed by the RC filtering of the neuronal membrane. In contrast, the 1-s window used in the LRI was chosen arbitrarily and isn't necessarily matched well for putative repeats, a problem discussed later in the manuscript:

$$T = h_{20}(0)\left(1 \quad \frac{\overline{|m-r|}}{|m|+|r|}\right) \tag{2}$$

$$\mathrm{HRI} = \frac{\left(\sum_{i=1}^{n} T_i\right)^2}{n} \frac{\sqrt{\mathrm{std\,(motif)\,std\,(repeat)}}}{\overline{|\mathrm{motif{-}repeat}|}} \tag{3}$$

The calculation of HRI itself is a two-step process, and the goal of this process is to find very similar PSCs recurring in a precise sequence. This procedure begins by retrieving a temporally aligned motif and repeat saved during the LRI procedure

(Fig. 1, Fig. 2, and (1)). This motif and repeat are then also aligned in the amplitude domain by normalizing the entire motif and repeat according to the average pA or mV value in the motif and repeat, respectively. Thereby, the aligned *motif-repeat*, as designated in (3), is then scanned with a 20-ms time window (Fig. 3). This 20-ms window extracts 20 points from the *motif* and *repeat*, with each 20 point segment designated in (2) as $\boldsymbol{m}$ and $\boldsymbol{r}$, respectively. Correlation coefficients are computed according to (1). In this calculation, h_{20} (0) corresponds to (1), but here x and y are replaced with $\boldsymbol{m}$ and $\boldsymbol{r}$, and in this case only a correlation coefficient for $\tau = 0$ is calculated as the segments are already aligned. This 20-ms window selects multiple $\boldsymbol{m}$ and $\boldsymbol{r}$ segments at 1-ms increments, calculating a T value at each of these increments (yielding 981 T values per *motif-repeat*). The largest T values for all nonoverlapping 20-ms increments that pass a set threshold are recorded (threshold is typically 0.55, again, somewhat arbitrary). If there are more than two such threshold-passing $\boldsymbol{m}$ and $\boldsymbol{r}$ segments, then results are passed to (3). Finally, the number (n) of precisely occurring PSCs in a *motif-repeat* sequence and their respective threshold-passing T values (from (2)) are incorporated in the HRI calculation (3), yielding an index of repeatability. As shown, HRI incorporates the average threshold-passing T value, the number of such T values, and is normalized according to the standard deviations and amplitude differences in the *motif-repeat* segments.

We used these search programs in analyzing long voltage-clamp intracellular recordings (8 min) from slices of mouse visual cortex as well as recordings from single neurons in cat visual cortex and mouse somatosensory cortex, both in vivo. These were all spontaneous recordings, meaning no stimulation was applied to the slices or to the unconscious, anesthetized animals. Thus, the currents or potentials identified in the recordings were presumably the result of synaptic activity, created in large part by the action potential activity of synaptically coupled neurons. The search algorithms described here were able to find instances of surprising repeatability, as judged by eye (Fig. 4).

Significance Testing

Given a long enough recording (several minutes) and many synaptic events, it should be expected that some repeating patterns are found. The question is then whether the patterns we find are beyond a level that could be expected to occur by chance. In fact, it is undetermined as to whether the specific examples shown in Fig. 4 are to be expected in a randomly firing network, although there is strong evidence that nonrandom patterns do emerge in these recordings ([10], however, see also [14]). The development of surrogate recordings – artificial data that is based on the real – is one way to deal with this issue. These surrogate data can then be compared with the real to see which produces more putative repeats (repeats being judged by the index methods in Figs. 1–3). If the real data produce more putative repeats than a large number of generated surrogate data sets, then some evidence is given for a nonrandom generation of these repeats.

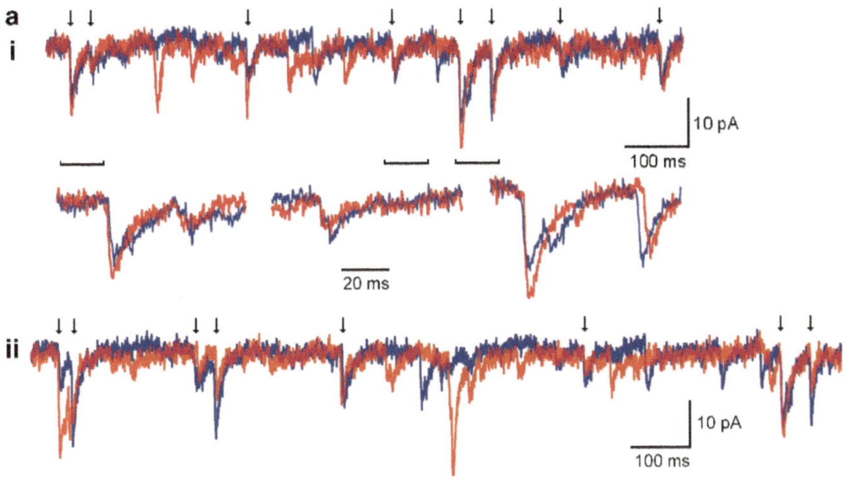

Fig. 4 Apparent repeats of synaptic events found. (**ai**) Two segments from a voltage-clamp recording are displayed, superimposed on each other. The *blue trace* occurred some time before the *red trace*, and yet the sequence of synaptic events appears similar. *Arrows* indicate time points where these synaptic events appear to repeat, and the *brackets* indicate segments that are temporally expanded below. (**aii**) Another example of a repeat from a different recording.

To address this issue, Ikegaya et al. [10] identified the putative PSCs (or PSPs) using a correlation procedure and pulled them out of the original recording, imposing them on a zero baseline. PSCs were detected by computing a covariance function of a mean PSC risetime waveform against the entire spontaneous recording, ms by ms: this produced a correlation trace whose peaks marked the onset of PSCs, and peaks passing a set threshold (typically, 0.9) were taken as the start times of PSCs. PSCs were then extracted from the recording by pulling the start time +20-ms window of the recording out for each identified PSC. This procedure preserves the shape of the individual PSCs as well as the timing of those events, creating an "extracted trace." Surrogate traces were constructed from the extracted trace by shuffling the time intervals between the PSCs, while preserving the temporal order of those events.

A more recent study [14] has argued that our shuffling method may be too lenient in that trivial repeats comprised of just two PSCs, possibly produced by the stereotypical firing pattern of a single presynaptic neuron, would be destroyed by our shuffling method. We agree with this argument. Their solution was to devise a shuffling technique that divided the intracellular recording into segments of approximately 400 ms. Surrogates were constructed by shuffling these segments. Thus, most of the two-event sequences are preserved in this manner.

A potential problem is that this shuffling procedure essentially shuffles the trace less thoroughly, and so the difference between surrogates and the original may not be detectable, even if deterministic repeats do exist. That is, the sensitivity of the detector (i.e., the search program that finds repeats) may not be equipped for the task. A test of the sensitivity is applied by injecting a 1-s artificial repeat (i.e., inject the same 1-s segment multiple times) into the original recording, and then performs the

400-ms shuffling tests on this repeat-injected trace [14]. The detector does indeed distinguish the original with artificial repeat very well from the shuffled surrogates, arguing that the detector is sufficiently sensitive.

However, there is a problem with this sensitivity test: the detector window itself is matched perfectly to the length of the artificial repeat (1 s). The basis of the detector algorithm is cross-covariance, and this function performs poorly if the detector window (set at 1 s in this program) does not match the actual length of the repeat to be detected. As previously stated, the original rationale for this suboptimal detector (i.e., the LRI detection) is that it is merely a first-pass and saves much computation time. The actual values produced in the final analysis from HRI do not suffer from this defect since the detector window is matched to the width of the individual PSCs (20 ms) or PSPs (100 ms). Unfortunately, there can be many false negatives from this 1st pass in the detector algorithm such that many candidates never exceed the threshold for gaining HRI analysis.

Implanted, Artificial Repeats

We investigate this potential problem in the detector program by implanting repeats that are not matched to the LRI detector window: the implanted motif is 850 ms, vs. the 1-s detector window (Fig. 5). The implanted motif consisted of a series of 5 PSPs, and this motif was summed into the original 190-s cat in vivo current-clamp recording every 10 s, yielding 171 motif-repeat pairs. This implanted trace was then shuffled using the 400-ms interval shuffling technique, producing 50 surrogate traces. Using the LRI-HRI detection program, no difference could be found between the implanted trace and its shuffled surrogates (Fig. 5).

An Improved Repeat Detector

In response to these results, we strived to create a detector program that could detect artificial implanted repeats in the face of the 400-ms interval shuffle test [15]. Our first goal was to remove the pitfalls of an arbitrary 1-s LRI detector. Instead, putative repeats were detected by the timings of PSPs. This new detector, PHRI (PSP-based detection, high resolution index), identifies the onsets of PSPs by their stereotypical risetimes, and then uses those timings as the pointers for the subsequent HRI analysis (Fig. 6). That is, every identified PSP is used as a point of alignment for a motif-repeat pair; the two selected PSPs, occurring at disparate times in the recording, are aligned, and the trace that follows each is included as the motif-repeat pair to be examined. The PSPs are identified by their risetimes in a method nearly identical to that from Ikegaya et al. (2004) with regards to the extraction of PSPs in that paper. PSPs were detected by computing a covariance function of a mean PSP risetime waveform (4–6 ms in duration) against the entire spontaneous recording;

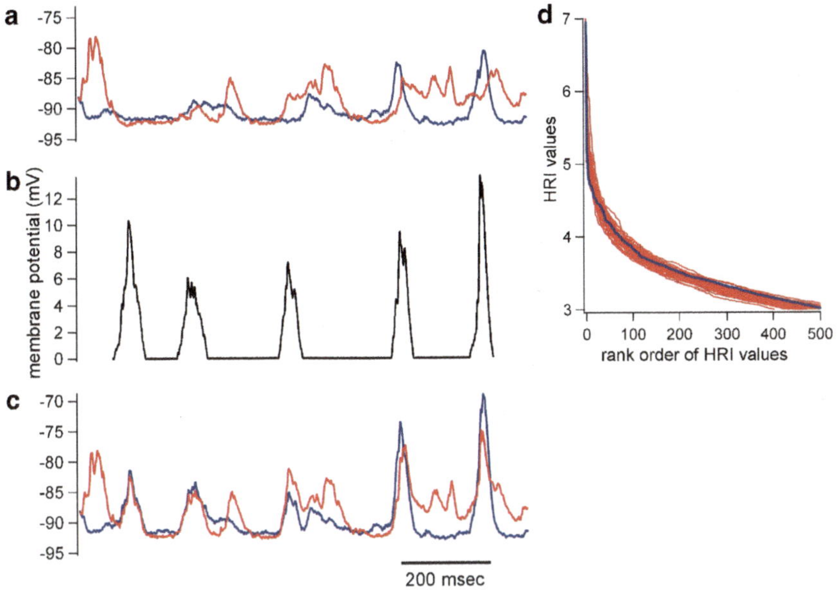

Fig. 5 Implanting an artificial repeating motif into a shuffled recording. (**a**) A 400-ms shuffled surrogate from an original cat in vivo current-clamp recording is composed. A 1-s segment from this shuffled surrogate recording is displayed (*blue*) with another 1-s segment from 9 s later superimposed (*red*). (**b**) The implant: a series of PSPs is constructed from the original recording, imposed on a 0-mV baseline. (**c**) The implant is summed into the 1-s segments, producing an implanted trace with recurring repeats. The implants are added approximate every 10 s into a 190-s recording, yielding 171 repeats. (**d**) Fifty 400-ms shuffle surrogates are constructed from the implanted recording, and the HRI values produced from those surrogates are compared with the values produced from the unshuffled implant recording. As shown, the LRI–HRI detection algorithm does not distinguish the implanted recording from the shuffled surrogates.

this produced a waveform whose peaks marked the onset of PSPs, and peaks passing a set threshold (typically, 0.9) were taken as the start times of PSPs. In some cases, an amplitude threshold was used in conjunction with the covariance function threshold. Thresholds were adjusted so that the fewest false positives and false negative results appeared, as can be judged in viewing Fig. 6. Importantly, the number of identified PSPs found in surrogate traces vs. original traces was unchanged by the creation of 400-ms shuffled surrogate traces.

The beginning of each PSP was then used as the points of alignment for comparing two different stretches of a recording, called here a putative motif-repeat (Fig. 7). A potential cost to this approach is that the identification of PSPs may not be precise, such that the realignment of traces based on those calculations could be impaired. In order to account for this, the initial alignment of the motif-repeat based on the two PSPs selected is altered according to the peak of the cross-covariance function produced by those PSPs. This is the same approach used to better align the traces during LRI analysis (Fig. 2), except that two individual PSPs are aligned, rather than the 1-s long traces.

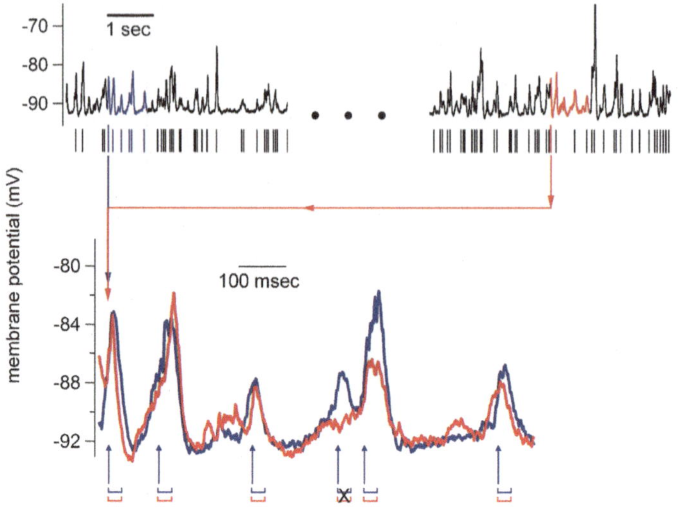

Fig. 6 Repeat detection with PHRI. The onset times of putative PSPs are estimated by calculating all cross-covariance values of an average risetime waveform against the entire recording. This yields correlation values for every point in the recording, and those points with a high cross-covariance value and minimum amplitude are marked as onset time of a PSP, as shown above by the tally marks below the recording. These onset times comprise the comparisons that will be performed – n onset times yields $n\,(n-1)\,/2$ comparisons. One such comparison is shown above: two putative PSPs are identified with the longest *blue tally* and longest *red tally*. These PSPs are then aligned such that they yield the highest T value (using 30-ms window, see (2)). This alignment is preserved with respect to the comparisons made between the subsequent PSPs in each respective trace extracted from the recording. The T values are calculated for the intervals dictated by the PSP onset times in the motif trace (*blue*), indicated with *blue arrows*. T values below a set threshold are discarded from the HRI calculation, thus the *black* "X." The minimum and maximum lengths of the motif-repeat traces that are included in the HRI calculation are 800 and 1200 ms, respectively. The minimum number of T values required for an HRI calculation is 3 (same as LRI–HRI criteria), and HRI is calculated as per (3). The HRI values for all lengths between 800 and 1200 ms are saved, and the motif-repeat length that yields the highest HRI value is saved. In the above example, the length of the motif and repeat is 853 ms, the PHRI = 5.3, and the delay between the motif and repeat is approximately 42 s.

The 190-s long in vivo cat recording used in Fig. 6 contained 1351 identified PSPs, yielding 911925 motif-repeat pairs to be examined for subsequent HRI analysis – more than 100× the number of pairs identified with LRI analysis (6750 pairs). In order to reduce this substantial increase in computation time, the HRI analysis in PHRI is reduced by computing T values only in the regions identified as having PSPs (Fig. 6). In contrast, the LRI-HRI technique measures T values for every 1-ms interval of the 1-s trace (yielding 900 T value calculations). These T values are then used just as before in the calculation of HRI (3). With this PHRI technique, the length of the motif-repeat is determined by length that yields the highest HRI value, and it is constrained by having a minimum of 800 ms and a maximum of 1200 ms. This constraint is enacted with respect to the shuffle surrogate technique described below: if motif-repeats are allowed that are the same length of the shuffle lengths

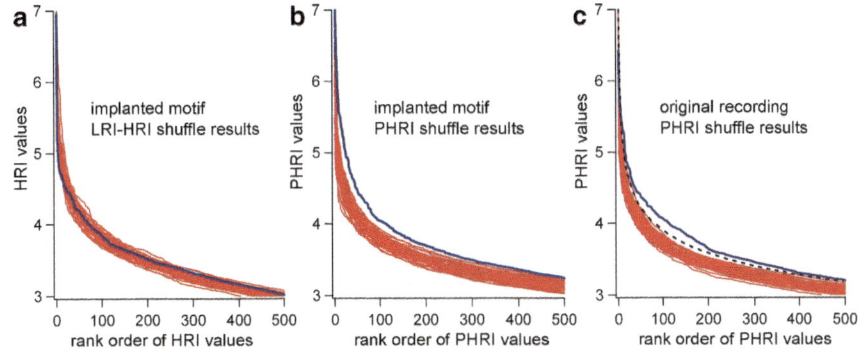

Fig. 7 The improved detector finds implanted motifs and distinguishes the original recording from its 400-ms shuffled surrogates. (**a**) The original LRI-HRI detector is unable to distinguish the implanted recording from its shuffled surrogates. (**b**) The PHRI detector, applied to the same data set as *A*, appears to distinguish the unshuffled (*blue*) from the shuffled surrogates (*red*). (**c**) The original 190-s cat in vivo current-clamp recording and fifty 400-ms shuffle surrogates are examined with the PHRI detector. The rank ordered values from the original are shown in *blue*, and shuffled surrogates in *red*. As these values were normally distributed for each rank order, it was possible to construct confidence intervals for the distribution, and the 99% confidence interval is shown (*dashed black line*). The original recording results (*blue line*) are clearly distinguished from the 99% confidence interval ($p < 0.01$).

(400 ms), then the shuffling is likely to keep many of the motif-repeats intact. (It would be analogous to using the LRI-HRI technique and shuffling with 1000-ms segments.) The mean length of the ten best repeats from the cat in vivo trace in Fig. 7, using PHRI, is 933 ± 32 ms. As in the LRI-HRI technique, a minimum of three T values that pass threshold is required.

The various parameters of the PHRI analysis were varied in order to enable it to distinguish the implanted trace (Fig. 5) from shuffled surrogates. When comparing PHRI values from 50 shuffled surrogates of the implanted trace to those of the unshuffled implanted trace there appears to be a significant difference in the distribution (Fig. 7b), or at least a much great difference in the difference compared with results obtained with the LRI-HRI method (Fig. 7a).

We show that this new detector is indeed sensitive enough to detect these artificially implanted repeats and distinguish the implanted trace from its respective shuffled surrogates (Fig. 6). Furthermore, when applying the PHRI detector to cat in vivo recordings (no implants), the number of repeats found is greater in the original vs. shuffled surrogates (Fig. 7). We have also found similar results with regards to in vivo recordings in mouse sensory neocortex.

Recording Conditions and Effects on Synaptic Repeat Detection

The experienced neuroscientist may notice that membrane potential of the in vivo recordings displayed in Figs. 5 and 6 is unusually hyperpolarized (see y-axes). During those recordings, a tonic DC hyperpolarization was applied to prevent action

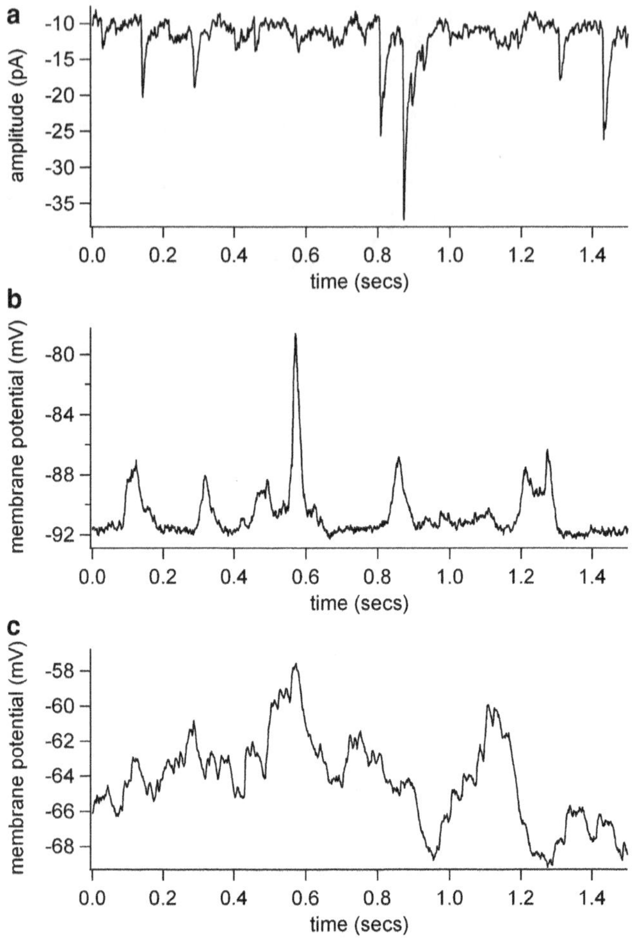

Fig. 8 Intracellular recordings in different conditions. (**a**) Whole cell voltage-clamp recording in vitro from a layer 5 pyramidal neuron, mouse V1 cortex. $V_{\text{clamp}} = -70\,\text{mV}$. (**b**) Sharp electrode current-clamp recording from cat visual cortex, supragranular layer, with a large tonic hyperpolarizing current. (**c**) Current-clamp recording from mouse cortex, in vivo and no tonic hyperpolarizing current. Note the similarities in recordings from A and B and how they both differ from C

potential generation, yielding a recording of only PSPs and possibly some voltage-activated currents. Interestingly, in current-clamp recordings that do not contain such a tonic hyperpolarization, it is always the case that almost any surrogate data set can produce as many repeats as the original [14]. That is, significant repeats cannot be detected in such recordings. Why would repeats be detected in voltage-clamp recordings, and in recordings that contain strong hyperpolarizations? One possibility is that the synaptic events that supposedly comprise the repeats are "hidden" in current-clamp recordings at more natural and depolarized membrane potentials. At a membrane potential of $-55\,\text{mV}$, for example, the EPSPs and IPSPs impinging on the neuron drive the membrane potential in opposite directions, possibly distorting the risetimes and waveforms comprising those individual events. In addition, many voltage activated currents are gated at depolarized values, and such currents intrinsic to the neuronal membrane itself may mask the alteration of membrane potential from synaptic currents alone. Hyperpolarizing a neuron to about -90 mV, however, removes many fast voltage-gated currents and also creates a condition where both IPSPs and EPSPs are depolarizing, allowing PSPs to rise out of a relatively steady baseline. Likewise, a voltage-clamp recording at -70 mV reduces almost all voltage-gated currents (voltage is clamped), and EPSCs and IPSCs are also deflecting in the same direction (Fig. 8).

Given the initial design of the experiment (Fig. 1), it seems critical that recording conditions allow identification of PSPs or PSCs in the recording, as these are the reflections of network activity that the technique attempts to analyze. We believe that the different recording conditions in recent investigations may account for conflicting results. Future experiments that record neurons in conditions that may favor a recording that reflects synaptic activity (hyperpolarized recordings, and/or voltage-clamp recordings) can be compared with recordings that do not. Ideally, future investigations will investigate the hypothesis directly by recording *the same neuron* under these two conditions sequentially and comparing these two parts of the recording and measuring repeats in both conditions. Furthermore, these recordings can be made in conjunction with recordings of several neurons via calcium imaging [1, 10, 16]. If the above hypothesis is correct, then the correlation of the intracellular recording with network dynamics should be greater during more hyperpolarized recordings.

References

1. Cossart R, Aronov D, Yuste R (2003) Attractor dynamics of network UP states in the neocortex. Nature 423: 283–288.
2. Shu Y, Hasenstaub A, McCormick DA (2003) Turning on and off recurrent balanced cortical activity. Nature 423: 288–293.
3. Sanchez-Vives MV, McCormick DA (2000) Cellular and network mechanisms of rhythmic recurrent activity in neocortex. Nat Neurosci 3: 1027–1034.
4. MacLean JN, Watson BO, Aaron GB, Yuste R (2005) Internal dynamics determine the cortical response to thalamic stimulation. Neuron 48: 811–823.
5. Yuste R, MacLean JN, Smith J, Lansner A (2005) The cortex as a central pattern generator. Nat Rev Neurosci 6: 477–483.
6. Abeles M (1991) Corticonics: Neural Circuits of the Cerebral Cortex. Cambridge, England: Cambridge University Press.
7. Abeles M, Bergman H, Margalit E, Vaadia E (1993) Spatiotemporal firing patterns in the frontal cortex of behaving monkeys. J Neurophysiol 70: 1629–1638.
8. Nadasdy Z, Hirase H, Czurko A, Csicsvari J, Buzsaki G (1999) Replay and time compression of recurring spike sequences in the hippocampus. J Neurosci 19: 9497–9507.
9. Shmiel T, Drori R, Shmiel O, Ben-Shaul Y, Nadasdy Z, Shemesh M, Teicher M, Abeles M (2006) Temporally precise cortical firing patterns are associated with distinct action segments. J Neurophysiol 96: 2645–2652.

10. Ikegaya Y, Aaron G, Cossart R, Aronov D, Lampl I, Ferster D, Yuste R (2004) Synfire chains and cortical songs: temporal modules of cortical activity. Science 304: 559–564.
11. Mao BQ, Hamzei-Sichani F, Aronov D, Froemke RC, Yuste R (2001) Dynamics of spontaneous activity in neocortical slices. Neuron 32: 883–898.
12. Oram MW, Wiener MC, Lestienne R, Richmond BJ (1999) Stochastic nature of precisely timed spike patterns in visual system neuronal responses. J Neurophysiol 81: 3021–3033.
13. Baker SN, Lemon RN (2000) Precise spatiotemporal repeating patterns in monkey primary and supplementary motor areas occur at chance levels. J Neurophysiol 84: 1770–1780.
14. Mokeichev A, Okun M, Barak O, Katz Y, Ben-Shahar O, Lampl I (2007) Stochastic emergence of repeating cortical motifs in spontaneous membrane potential fluctuations in vivo. Neuron 53: 413–425.
15. Ikegaya Y, Matsumoto W, Chiou HY, Yuste R, Aaron G (2008) Statistical significance of precisely repeated intracellular synaptic patterns. PLoS ONE 3: e3983.
16. Trevelyan AJ, Sussillo D, Watson BO, Yuste R (2006) Modular propagation of epileptiform activity: evidence for an inhibitory veto in neocortex. J Neurosci 26: 12447–12455.

Reverberatory Activity in Neuronal Networks*

Pak-Ming Lau and Guo-Qiang Bi

Abstract Reverberatory activity in neuronal cell assemblies has been proposed to carry "online" memory traces in the brain. However, the dynamics and cellular mechanism of such reverberation have been difficult to study because of the enormous complexity of intact circuits. To overcome this difficulty, small networks of interconnected neurons have been grown in culture dishes to provide a model system for studies using patch-clamp recording and fluorescent imaging approaches. In such networks, brief stimulation could elicit rhythmic reverberation that consists of repeating motifs of specific patterns of population activation in the network. Experimental and modeling analysis suggested that the reverberation is driven by recurrent excitation, is sustained by the oft-overlooked asynchronous synaptic transmission modulated by intracellular calcium, and is terminated by a slow component of short-term synaptic depression. More recent data suggest that Hebbian synaptic plasticity could underlie activity-induced emergence of reverberation. Thus, these in vitro networks may serve as prototypic Hebbian cell assemblies for the study of potential mechanisms of information representation and storage in brain circuits.

Background

The idea that neuronal activity could "reverberate" within closed-loop rings or "self-reexciting" chains of neurons was first proposed in the early 1930s to explain electrophysiological observations such as the reflex after-discharge [44]. In the late 1940s, Donald Hebb suggested that reverberation might exist in more elaborate

G.-Q. Bi (✉)
Department of Neurobiology, University of Pittsburgh School of Medicine, Pittsburgh, PA 15261, USA
e-mail: gqbi@pitt.edu

*This paper (Title: Reverberatory activity in neuronal networks *in vitro*; authors: LAU PakMing, BI GuoQiang), was first published in *Chinese Science Bulletin* (Chinese Sci Bull, 2009, 54(11): 1828–1835, doi: 10.1007/s11434-009-0135-1). It is reprinted here with permission from Science in China Press.

K. Josić et al. (eds.), *Coherent Behavior in Neuronal Networks*, Springer Series in Computational Neuroscience 3, DOI 10.1007/978-1-4419-0389-1_4,

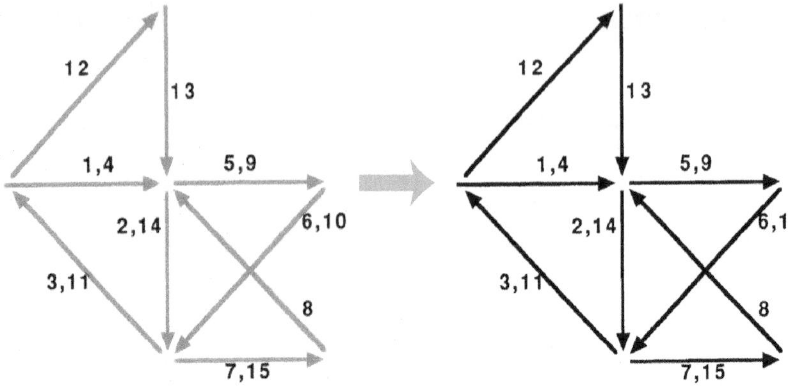

Fig. 1 Hebb's "cell assembly" and its activity-dependent growth. *Arrows* represent neural pathways firing sequentially according to the numbers on each. Reverberatory activity might persist in such a circuit. With repeated activation, a weakly-connected circuit (*gray arrows*) is strengthened (*black arrows*) so that the stability of the circuit is increased (Adapted from [34].)

circuits in the brain that he called the "cell assemblies" (Fig. 1). Such reverberation can persist for a period of time after the cessations of the input stimulus and serve as a short-term memory trace, which the brain can use to act on information prior to immediate sensory input [34].

In addition to suggesting that cell assemblies capable of holding reverberatory activity may function as elementary units underlying the thought process, Hebb further proposed that appropriate rules of synaptic modification can lead to the formation and stabilization of such assemblies, thereby converting short-term memory traces into long-term engrams [34], as stated in his famous "neurophysiological postulate":

"Let us assume then that the persistence or repetition of a reverberatory activity (or 'trace') tends to induce lasting cellular changes that adds to its stability. The assumption can be precisely stated as follows: *When an axon of cell* A *is near enough to excite a cell* B *and repeatedly or persistently takes part in firing it, some growth process or metabolic change takes place in one or both cells such that A's efficiency, as one of the cells firing* B, *is increased.*"

The cellular mechanism that Hebb postulated here, now known as "Hebb's learning rule" or "the Hebbian synapse," has gained popular supports from experimental studies of synaptic plasticity that has been a hot topic of extensive study over the past few decades [15, 16, 56, 65]. In particular, studies of long-term potentiation (LTP) and long-term depression (LTD) have revealed exquisite molecular mechanisms [4, 15, 23, 41, 48, 49, 66]. More recently, the discovery of a form of synaptic modification that depends on the precise timing of pre- and postsynaptic action potentials, i.e., spike-timing-dependent plasticity (STDP), has further provided a quantitative and more extensive notion for Hebb's rule [1, 10, 11, 14, 20, 21, 47, 51, 54, 79].

The major driving forces in the study of reverberatory activity have been theoretical analyses and simulations [24, 61, 75]. In fact, even in the early 1940s, McCulloch and Pitts have demonstrated the computational implications of similar reverberatory activity in their idealized few-cell networks [53]. In the past decades, reverberation dynamics and potential cellular mechanisms have been explored extensively in various attractor models [2,3,31,35,37,39,62,67,74,80]. In most models, recurrent excitation as an internal positive feedback mechanism keeps the network, or cell assembly, in distinct "up-states" of persistent firing activity without external drive, thus providing short-term memory traces of input stimuli.

Experimentally, it has long been known from in vivo experiments that during behavioral tasks such as working memory, neurons in the prefrontal cortex and other brain areas can fire persistently after the cessation of cue stimuli [24,27,28,38,75]. It has also been suggested that coordinated temporal fluctuations of neural activity observed in sensory modalities and in the hippocampus may reflect the organization and activation of cell assemblies during internal cognitive processing [33]. However, because of the complexity of native circuits, it has been difficult to demonstrate experimentally how reverberatory activity is implemented at the network level, and it has remained unclear whether network reverberation indeed underlies the observed persistent activity in vivo [75].

The application of various electrophysiological and optical techniques with in vitro preparations has allowed for new routes in the study of neuronal networks. Because of the simplicity and accessibility of the reduced system, patterns of population activity are more easily studied. In brain slices, stereotypic spatiotemporal patterns of spontaneous network activation have been observed [18,26,36,63]. Similar population bursting activity has also been characterized in cultured neurons and cultured slices [8,9,46,60,71,73]. Although in most cases such network activation does not last beyond tens of milliseconds, and thus differs from the reverberatory activity assumed to be involved in tasks such as working memory, it does reflect the existence of intrinsic network structures and their capability of self-organization reminiscent of the cell assembly.

In small networks of cultured hippocampal neurons, we recently observed a new form of coherence network activity [40]. Such network activity can last for seconds after brief stimulation, similar to the proposed reverberatory activity in the Hebbian cell assembly. The simplicity of in vitro systems provides a unique opportunity to address fundamental issues regarding the cellular mechanisms underlying reverberatory activity in neuronal networks [24,61,75]. For example, what is the minimum requirement for cellular substrates to support persistent reverberation? Can recurrent synaptic excitation alone sufficiently sustain persistent activity, or is bistability of single neurons needed? What components of synaptic currents are responsible for sustaining stable reverberation? How is synaptic plasticity involved in the emergence and evolution of reverberation? In this chapter, we will describe experimental and theoretical findings regarding network reverberation inside culture dishes, with an emphasis on the cellular mechanisms underlying its dynamics and plasticity.

Reverberatory Activity in Cultured Neuronal Networks

Neuronal cultures in most studies are prepared directly from animal brain tissues, and are thus called "primary cultures" in contrast to cell cultures derived from immortal cell lines [6]. Primary neuronal cultures have the advantage of preserving many biophysical and cell biological properties similar to their in vivo counterparts. However, it must be noted that even in a primary culture, cells are grown in conditions quite different from their native environment; and environmental factors (e.g., tropic factors, guidance cues, sensory inputs, etc.) play important roles in neuronal development. Furthermore, it is unlikely that cultured neurons can keep their native circuitry structure, which is important for their network functions. These must be kept in mind when interpreting phenomena observed in such in vitro preparations. Nevertheless, with cautions properly practiced, culture systems can help us gain valuable insights into the fundamental principles and mechanisms underlying the behavior of neurons and neuronal networks.

Rat hippocampal culture has been a good choice for in vitro studies partly because of the relative homogeneity of neuronal cell types [6]. To prepare such cultures, hippocampi are first dissected from rat embryos, followed by enzymatic and mechanical dissociation, before plating in petri dishes containing culture medium [6]. For many studies, neurons need to be grown on glass coverslips. These coverslips must be first coated with a layer of molecular substrate in order for the cells to adhere to the surface. In our experiments, the coverslips are precoated with patterns of poly-L-lysine spots of ~1-mm diameter using custom-made stamps. Confined by these patterns, neurons grow to form small networks that are relatively isolated from one another, each with an "island" of a monolayer of glial cells (Fig. 2).

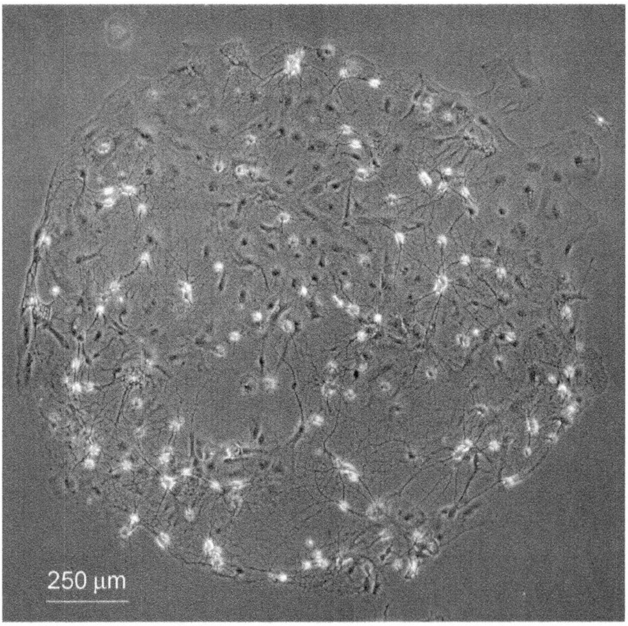

Fig. 2 A network of cultured hippocampal neurons grown on glass surface coated with poly-L-lysine dots. Under proper conditions, a monolayer of glial cells from the same source usually forms underneath the neurons (Adapted from [40].)

About 1 week after being plated in culture (7 days in vitro or DIV), hippocampal neurons begin to form synapses and neuronal activity begins to emerge and evolve. Such activities can be recorded by electrophysiological recordings and various imaging methods. In small (e.g., 10 neurons) and young (e.g., 10 DIV) networks when spontaneous activity is rare, brief external stimuli can often elicit sizable network responses. Typically, such a response is comprised of multiple polysynaptic voltage or current components and lasts for ~50 ms [12], suggesting the activation of a cascade of neurons via multiple pathways by a stimulus. In larger (e.g., 100 neurons) and more mature (e.g., 15 DIV) networks, in additions to short polysynaptic responses, the same brief stimulus could often elicit recurring or persistent responses lasting for seconds (Fig. 3). Because typical synaptic delay is only several milliseconds, cascade activation lasting for seconds will require thousands of neurons in such a pathway. Therefore, the long-lasting response of a network with ~100 neurons must involve repeated activation of the same neurons, similar to the reverberatory activity Hebb hypothesized for the cell assembly.

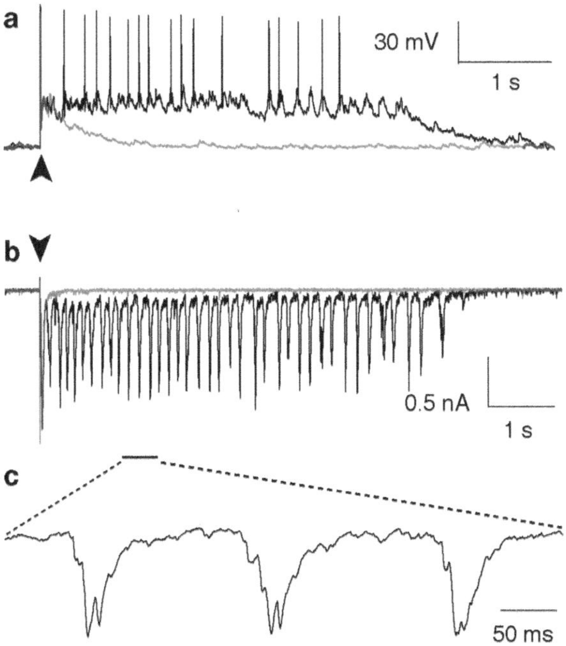

Fig. 3 Reverberatory activity in cultured neuronal networks. (**a**) Current-clamp recordings of transient polysynaptic potential (*gray*) and persistent reverberatory activity (*black*) evoked by brief (1 ms) stimulation of the recorded neuron in a network. (**b**) Voltage-clamp recordings of transient polysynaptic current (*gray*) and reverberatory activity (*black*) in response to 1-ms stimuli of the recorded neuron in the same network. (**c**) Expanded view of three reverberatory events (polysynaptic current clusters) within the above episode of network reverberation (Adapted from [40].).

Reverberatory activity in cultured networks exhibits several interesting dynamic features:

(1) In response to external stimuli, reverberatory activity is elicited in an all-or-none fashion. The distribution of the duration of network activity is highly bimodal: either reverberatory activity lasting for seconds or short polysynaptic responses lasting for tens of milliseconds are generated.
(2) For most neurons in a reverberatory network, their membrane potentials are generally maintained in a depolarized state for the duration of reverberation that usually lasts for several seconds, during which active neurons fire action potentials at a moderate rate of $\sim$10 Hz. This is to some extent similar to the in vivo persistent activity as well as the "up-state" observed in other preparations.
(3) Throughout an episode of reverberation, network activation happens in a rhythmic and coherent fashion: short periods ($\sim$50 ms each) of reverberatory "events" are interleaved by "silent" periods of similar or slightly longer durations. Neuronal firing occurs mostly during the periods of these reverberation events, as reflected by clusters of polysynaptic currents in voltage-clamp recording traces.
(4) Different reverberatory events in a network may have conserved spatiotemporal patterns of neuronal activation, as indicated by the similarity among polysynaptic current clusters (reflecting sequential neuronal firing presynaptic to the recorded cell).

Biophysical Mechanisms Underlying Persistent In Vitro Reverberation

The simplicity and accessibility to electrophysiological and pharmacological manipulations of in vitro systems has allowed for careful dissection of the cellular and biophysical mechanisms underlying reverberatory activity. In this chapter, we discuss the involvement of recurrent excitation and inhibition, and focus on the importance of oft-overlooked asynchronous synaptic transmission.

Intrinsic Bistability vs. Recurrent Excitation

In principle, a network of neurons can be maintained in an active state in the absence of continuous external drive by appropriate positive feedback. At least two distinct biophysical mechanisms could potentially achieve this: cellular bistability and network recurrent excitation. The first can involve a few "driver" neurons that are intrinsically capable of firing persistently once initiated. The second is similar to Hebb's scenario of the cell assembly, and is likely to underlie the observed reverberatory activity in cultured neuronal networks.

Intrinsic membrane bistability (or multistability) has been found in some neuronal types including layer V pyramidal neurons of the entorhinal cortex [25] and cerebellar Purkinje cells [43]. Upon proper initiation, these neurons could fire action potentials persistently without continued stimulation. Several ion channels, including the Ca^{2+}-sensitive cationic channels and noninactivating voltage-gated Na^{+} channels, could potentially underlie membrane bistability. It is conceivable that if a network contains some bistable neurons, then the persistent firing activity of these cells could drive the whole network into persistent activation, thereby carrying short-term memory traces.

In hippocampal cultures, however, none of the neurons examined were found to exhibit intrinsic bistability that was capable of driving persistent firing [40]. Furthermore, reverberation in cultured hippocampal neurons was highly sensitive to CNQX, an antagonist of AMPA-type glutamate receptors, suggesting the importance of excitatory synaptic connections. In fact, even partial inhibition of AMPA receptor-mediated synaptic current could dramatically reduce or abolish reverberatory activity. This is consistent with a mechanism of recurrent excitation and does not support the idea of persistent spiking of a few bistable driver neurons. Nonetheless, properties of intrinsic membrane excitability (not necessarily bistability) of individual neurons could play important roles in the robustness and dynamic characteristics of network reverberation.

It is estimated that ~10–20% cultured hippocampal neurons are GABAergic interneurons which are inhibitory at the stage of network experiments [6,40]. Blockade of GABA-A receptors facilitates the occurrence and prolongs the duration of evoked network reverberation [40]. In some larger networks, blockade of inhibition could lead to initial high-frequency firing (Lau and Bi, unpublished observations), similar to epileptiform activity observed in other systems [52]. This is consistent with a scenario that reverberation is driven by recurrent excitation, whereas GABA-A receptor-mediated inhibition plays a balancing role by suppressing reverberation. Additionally, it is also possible that GABAergic interneurons in the largely excitatory network could play a role in controlling spike timing and regulating the rhymicity of reverberation [50,76,77].

Asynchronous Synaptic Transmission

An intriguing feature in reverberatory activity observed in cultured hippocampal neurons is the existence of "silent" periods interleaving reverberatory events: rounds of network firing (see feature #3 in 1.2). The question is: if there is no neuronal firing during these periods, usually ranging from 50 to 100 ms, then some "signals" or "traces" must be left by each round of network firing to initiate the next. The major component of excitatory synaptic current mediated by AMPA type glutamate receptors typically lasts for ~10 to 20 ms, apparently not enough to carry out the task. Another type of glutamate receptor, the NMDA receptor, has a needed slow kinetics and has been suggested to play a crucial role in maintaining persistent network activity in the prefrontal cortex [74]. However, NMDA current constitutes only a very small portion of excitatory synaptic current in cultured hippocampal neurons. Furthermore, some networks are still capable of reverberation with complete blockade of NMDA receptors [40]. Therefore, there must be another player to bridge the gaps between adjacent reverberatory events.

A closer look at the voltage-clamp traces of evoked reverberation reveals a slow phase of the polysynaptic currents following the more synchronized component of the PSC cluster (Fig. 4a). Moreover, in networks where reverberation was not triggered by a single pulse stimulation of an input neuron, paired-pulse stimuli (PPS)

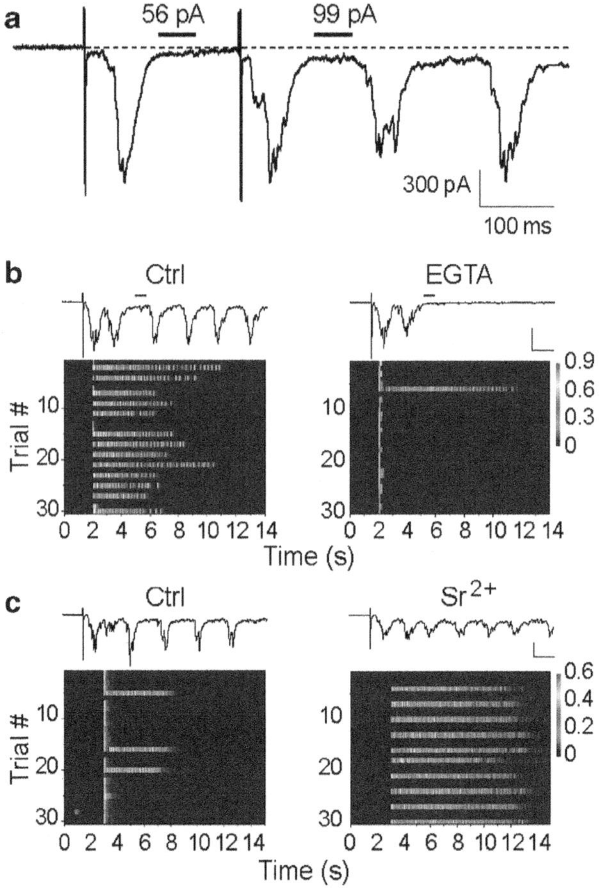

Fig. 4 Asynchronous phase of polysynaptic currents bridge reverberatory events. (**a**) Paired-pulse stimulation causes elevated asynchronous synaptic currents and elicits reverberation in a network that does not respond to a single pulse stimulus. (**b**) Buffering intracellular Ca^{2+} with EGTA-AM suppresses reverberatory activity in a network that normally exhibits reverberation. Note that asynchronous synaptic transmission is substantially suppressed in the presence of EGTA. (**c**) Elevating asynchronous release with extracellular Sr^{2+} substantially enhances both the occurrence and the duration of reverberation (Adapted from [40].)

could enhance this slow current component and often elicit reverberation (Fig. 4a). This slow phase of polysynaptic current is most likely a result of the so-called "asynchronous" synaptic transmitter release that was first observed by Katz and his colleagues in their original studies of the neuromuscular junctions and was later found in many preparations of central neurons [5, 7, 19, 22, 29, 45, 55, 58]. Through these classical studies, it has been well established that asynchronous release is caused by increased probability of synaptic vesicle fusion in response to residual calcium elevation after action potentials that trigger the "synchronous" phase of synaptic transmission.

Two pieces of experimental evidence confirm the critical role of asynchronous release in network reverberation. First, EGTA-AM, a slow calcium chelator that buffers residual calcium elevation and thereby suppresses asynchronous release while allowing normal synchronous transmission [19, 32] could virtually abolish reverberation (Fig. 4b). Second, when part of the extracellular calcium was replaced by strontium, a divalent cation that partially supports synaptic transmission and has been shown to enhance asynchronous transmitter release at various nerve terminals [29, 55, 78], both the duration and the occurrence probability of reverberation were substantially enhanced (Fig. 4c).

Compared with the fast decay of "synchronous" phase of synaptic transmission, asynchronous release typically has a long time constant of a few hundred milliseconds that is determined to a large extent by slow calcium clearance [19, 29, 32]. Meanwhile, at low activity levels, the contribution of asynchronous transmitter release to synaptic currents often appears to be insignificant compared with "synchronous" release [58]. Therefore, asynchronous release is often regarded as a "noise" of synaptic transmission. However, the slow kinetics of this oft-overlooked component of synaptic transmission makes it ideally situated for bridging the gap between adjacent reverberatory events. In addition, asynchronous release can be much more prominent following repeated presynaptic activation because of the accumulation of residual calcium [19, 32]. Therefore, during reverberation asynchronous transmission is "facilitating" so that once a reverberation episode gets started, more asynchronous release would happen, thereby more likely to reinitiate the next rounds of reverberatory events. This facilitation provides another level of positive feedback to stabilize reverberation and can explain the all-or-none nature of its activation.

At the functional level, the facilitation may also enable a network to detect physiologically relevant repetitive input stimuli. Several attractor models predict that a slow component of excitatory synaptic transmission is needed for the stability of various forms of network activity such as desynchronized persistent firing and robust graded persistent activity [62, 67, 74]. In these models, NMDA receptor-mediated excitatory synaptic current has been proposed to play this role by virtue of its slow time constant. It is possible that asynchronous transmitter release could also play a similar role in these other forms of functionally relevant persistent activities, especially in systems such as hippocampal neurons where excitatory transmission is mainly mediated by AMPA currents.

Asynchronous release is a fundamental component of synaptic transmission [5, 7, 19, 22, 29, 45, 55, 58] and is likely to exist in various synapses in the brain. It will be interesting to see whether it plays similar roles in the observed persistent activity in vivo. Moreover, asynchronous release can be affected by various cellular processes such as calcium uptake into and release from intracellular stores [42, 59, 64] that involves intricate inter- and intracellular signaling systems [70]. It is not unlikely that these regulatory processes may be involved in modulating persistent activity relevant to system behaviors such as attention. Meanwhile, defects in the regulation could lead to system dysfunctions such as epilepsy. Understanding these basic mechanisms and their relevance in vivo will provide important insights into both normal brain functions and related diseases.

Short-Term Synaptic Dynamics

Another important feature of synaptic transmission is short-term synaptic dynamics including short-term facilitation and short-term depression [81] that must have significant influence on collective network dynamics [68, 69]. This can be tested experimentally by manipulating key factors such as extracellular Ca^{2+} concentration [81]. The exact consequence of such manipulation on network reverberation can be complicated. For example, moderate decrease in extracellular Ca^{2+} could result in enhanced reverberation, apparently because of a reduction in short-term depression. Further decrease in Ca^{2+}, however, has the opposite effect because the synaptic drive is too low to initiate reverberation (P Lau & G Bi, unpublished results).

In cultured hippocampal neurons, short-term depression is the dominant form of synaptic dynamics [30, 32]. It is generally believed that short-term depression is because of the limited sizes of synaptic vesicle pools (e.g., the readily releasable pool) and the finite rates of their recovery after usage [81]. With the consideration of synaptic depression on multiple timescales together with presynaptic calcium dynamics that determines asynchronous vesicle release, a biophysically tractable model was constructed to explore the synaptic mechanisms underlying network reverberation [72]. The model successfully reproduced reverberatory activity in simulated networks of 50–100 neurons (Fig. 5).

Simulation results from the model confirm the key intuition obtained from experimental observations: asynchronous transmitter release plays the key role in reverberation by bridging the gap between adjacent reverberatory events. It was also found that a fast component of short-term depression in synchronous synaptic transmission ends each round of network activation (reverberatory event). Then asynchronous release "driven" by residual calcium was able to reinitiate the next round of network activation following sufficient recovery from the fast depression. Therefore, the interplay between synchronous and asynchronous transmission is responsible for the oscillatory nature of network activation during reverberation. Finally, the reverberation is terminated by a slow component of short-term depression (reflected by the "s" state in Fig. 5a) that outlasts residual calcium or asynchronous release.

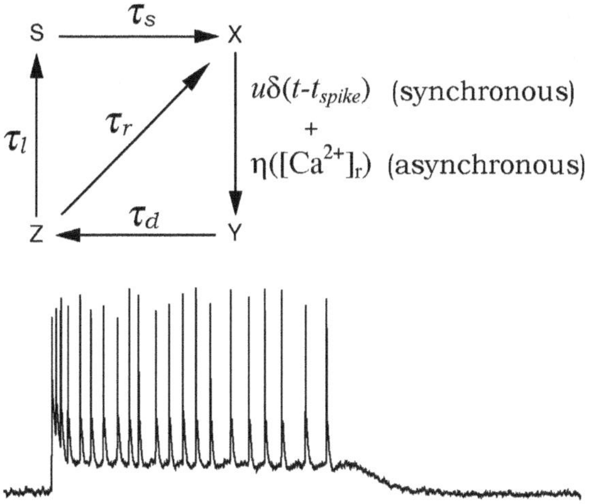

Fig. 5 Computational simulation of reverberatory activity in small neuronal networks. *Upper panel* illustrates synaptic mechanisms, including short-term synaptic depression and asynchronous release, which are the key factors influencing reverberatory activity. *Lower panel* shows simulated reverberation (compare with Fig. 3) (Courtesy of Dr. Vladislav Volman. For details, see [72].)

Summary and Outlook

The simplicity of the culture system has allowed us to answer several key questions regarding the mechanism of network reverberation: (1) Persistent reverberatory activity can exist in cultured neuronal networks that appear to form randomly without any predefined anatomical specializations; (2) Reverberation can be driven by recurrent excitation without special requirements for bistability of single neurons, although the latter could exist in other systems and play a role in ensuring the robustness of reverberation; (3) Asynchronous synaptic transmission plays a critical role in sustaining reverberation by bridging adjacent reverberatory events. Consequently, cellular mechanisms regulating intracellular calcium stores and thus asynchronous release could be novel functional nodes in the modulation of network reverberation; (4) multitimescale synaptic dynamics determines the dynamic features of reverberation: a fast component of short-term depression causes oscillatory activation of network reverberation, whereas a slow component of short-term depression leads to the termination of reverberation.

More questions remain to be answered. Relevant to Hebb's theory, a critical question is: how does reverberation emerge in a cultured network during its development? In particular, what are the roles for activity and activity-dependent Hebbian synaptic plasticity in the formation of reverberatory circuits? Furthermore, can a network maintain stable reverberation under the influence of Hebbian plasticity that is essentially a positive feedback process? From the point of view of information coding, the conserved spatiotemporal patterns of neuronal activation observed during reverberation is intriguing. The question is: how many different patterns can coexist in a network? Can a network be trained to store specific patterns?

Fortunately, in simplified cell culture system, these questions can be directly addressed with currently available technology, especially the combination of electrophysiology and imaging approaches. For example, our recent experiments suggest that Hebbian synaptic modification, similar to STDP [1, 11, 13, 17, 51], could indeed underlie activity-induced emergence of network reverberation (P. Lau & G. Bi, unpublished results). Therefore, to some extent, we can regard these cultured networks as prototypic Hebbian cell assemblies. Meanwhile, it was also found that reverberatory circuits could still develop in the absence of activity, as long as homeostatic plasticity scales up the overall strength of excitatory synapses in the network (R. Gerkin, P. Lau & G. Bi, unpublished results). However, reverberation developed in the absence of activity is often not stable and tends to evolve into uncontrolled spontaneous activity, whereas normally developed reverberation generally remains stable, suggesting the action of another layer of homeostatic mechanism that regulates the expression of synaptic plasticity.

Does reverberatory activity exist in native neuronal circuits in vivo as observed in culture? A well-known issue is that cultured neurons often form strong synaptic connections so that the activation of individual neurons is sufficient to trigger polysynaptic responses, whereas in native circuits, individual connections are generally subthreshold. Intriguingly, it has been observed recently that in human cerebral cortex, individual spikes could initiate complex polysynaptic responses, suggesting that single action potentials might be more important for the human brain than previously thought [57]. Still, dynamic properties of small networks do not simply scale up. Therefore, it is ultimately important to obtain direct evidence of reverberatory network dynamics in vivo during relevant behavior, probably using revolutionary new methodologies. Nevertheless, the existence of reverberation in generic, randomly connected neurons suggests that such activity could result from basic mechanisms and principles guiding the self-organization of neuronal circuits. Therefore, in the spirit of Hebb's legacy, we could speculate that collective network dynamics similar to reverberation might act as a basic unit of cognitive and perceptual processes such as online memory and decision making; deficiency in such dynamics might underlie pathological conditions such as epilepsy, chronic pain, and schizophrenia. Whereas the jury is still out on the physiological significance of network reverberation, it is likely that the basic principles and mechanisms learned from in vitro systems could provide new insights into circuit operation in vivo and could guide its study in the future.

References

1. Abbott LF and Nelson SB (2000) Synaptic plasticity: taming the beast. Nat Neurosci 3 Suppl: 1178–1183.
2. Amit DJ, Brunel N and Tsodyks MV (1994) Correlations of cortical Hebbian reverberations: theory versus experiment. J Neurosci 14: 6435–6445.
3. Amit DJ, Gutfreund H and Sompolinsky H (1985) Spin-glass models of neural networks. Phys Rev A 32: 1007–1018.
4. Artola A and Singer W (1993) Long-term depression of excitatory synaptic transmission and its relationship to long-term potentiation. Trends Neurosci 16: 480–487.
5. Atluri PP and Regehr WG (1998) Delayed release of neurotransmitter from cerebellar granule cells. J Neurosci 18: 8214–8227.

6. Banker G and Goslin K, eds. (1998) Culturing Nerve Cells, Second edn (Cambridge, MA: The MIT Press).
7. Barrett EF and Stevens CF (1972) The kinetics of transmitter release at the frog neuromuscular junction. J Physiol 227: 691–708.
8. Beggs JM and Plenz D (2003) Neuronal avalanches in neocortical circuits. J Neurosci 23: 11167–11177.
9. Beggs JM and Plenz D (2004) Neuronal avalanches are diverse and precise activity patterns that are stable for many hours in cortical slice cultures. J Neurosci 24: 5216–5229.
10. Bell CC, Han VZ, Sugawara Y and Grant K (1997) Synaptic plasticity in a cerebellum-like structure depends on temporal order. Nature 387: 278–281.
11. Bi GQ and Poo MM (1998) Synaptic modifications in cultured hippocampal neurons: dependence on spike timing, synaptic strength, and postsynaptic cell type. J Neurosci 18: 10464–10472.
12. Bi GQ and Poo MM (1999) Distributed synaptic modification in neural networks induced by patterned stimulation. Nature 401: 792–796.
13. Bi GQ and Poo MM (2001a) Synaptic modifications by correlated activity: Hebb's postulate revisited. Annu Rev Neurosci 24: 139–166.
14. Bi GQ and Poo MM (2001b) Synaptic modification by correlated activity: Hebb's postulate revisited. Annu Rev Neurosci 24: 139–166.
15. Bliss TV and Collingridge GL (1993) A synaptic model of memory: long-term potentiation in the hippocampus. Nature 361: 31–39.
16. Brown TH, Kairiss EW and Keenan CL (1990) Hebbian synapses: biophysical mechanisms and algorithms. Annu Rev Neurosci 13: 475–511.
17. Caporale N and Dan Y (2008) Spike timing-dependent plasticity: a Hebbian learning rule. Annu Rev Neurosci 31: 25–46.
18. Cossart R, Aronov D and Yuste R (2003) Attractor dynamics of network UP states in the neocortex. Nature 423: 283–288.
19. Cummings DD, Wilcox KS and Dichter MA (1996) Calcium-dependent paired-pulse facilitation of miniature EPSC frequency accompanies depression of EPSCs at hippocampal synapses in culture. J Neurosci 16: 5312–5323.
20. Dan Y and Poo MM (2006) Spike timing-dependent plasticity: from synapse to perception. Physiol Rev 86: 1033–1048.
21. Debanne D, Gahwiler BH and Thompson SM (1998) Long-term synaptic plasticity between pairs of individual CA3 pyramidal cells in rat hippocampal slice cultures. J Physiol (Lond) 507: 237–247.
22. Del Castillo J and Katz B (1954) Statistical factors involved in neuromuscular facilitation and depression. J Physiol 124: 574–585.
23. Derkach VA, Oh MC, Guire ES and Soderling TR (2007) Regulatory mechanisms of AMPA receptors in synaptic plasticity. Nat Rev Neurosci 8: 101–113.
24. Durstewitz D, Seamans JK and Sejnowski TJ (2000) Neurocomputational models of working memory. Nat Neurosci 3 Suppl: 1184–1191.
25. Egorov AV, Hamam BN, Fransen E, Hasselmo ME and Alonso AA (2002) Graded persistent activity in entorhinal cortex neurons. Nature 420: 173–178.
26. Eytan D, Brenner N and Marom S (2003) Selective adaptation in networks of cortical neurons. J Neurosci 23: 9349–9356.
27. Funahashi S, Bruce CJ and Goldman-Rakic PS (1989) Mnemonic coding of visual space in the monkey's dorsolateral prefrontal cortex. J Neurophysiol 61: 331–349.
28. Fuster JM and Alexander GE (1971) Neuron activity related to short-term memory. Science 173: 652–654.
29. Goda Y and Stevens CF (1994) Two components of transmitter release at a central synapse. Proc Natl Acad Sci U S A 91: 12942–12946.
30. Goda Y and Stevens CF (1996) Long-term depression properties in a simple system. Neuron 16: 103–111.

31. Gutkin BS, Laing CR, Colby CL, Chow CC and Ermentrout GB (2001) Turning on and off with excitation: the role of spike-timing asynchrony and synchrony in sustained neural activity. J comput neurosci 11: 121–134.
32. Hagler DJ, Jr. and Goda Y (2001) Properties of synchronous and asynchronous release during pulse train depression in cultured hippocampal neurons. J Neurophysiol 85: 2324–2334.
33. Harris KD (2005) Neural signatures of cell assembly organization. Nat Rev Neurosci 6: 399–407.
34. Hebb DO (1949) The Organization of Behavior (New York: Wiley).
35. Hopfield JJ (1982) Neural networks and physical systems with emergent collective computational abilities. Proc Natl Acad Sci U S A 79: 2554–2558.
36. Ikegaya Y, Aaron G, Cossart R, Aronov D, Lampl I, Ferster D and Yuste R (2004) Synfire chains and cortical songs: temporal modules of cortical activity. Science 304: 559–564.
37. Kleinfeld D and Sompolinsky H (1988) Associative neural network model for the generation of temporal patterns. Theory and application to central pattern generators. Biophys J 54: 1039–1051.
38. Kubota K and Niki H (1971) Prefrontal cortical unit activity and delayed alternation performance in monkeys. J Neurophysiol 34: 337–347.
39. Laing CR and Chow CC (2001) Stationary bumps in networks of spiking neurons. Neural Comput 13: 1473–1494.
40. Lau PM and Bi GQ (2005) Synaptic mechanisms of persistent reverberatory activity in neuronal networks. Proc Natl Acad Sci U S A 102: 10333–10338.
41. Lisman J, Lichtman JW and Sanes JR (2003) LTP: perils and progress. Nat Rev Neurosci 4: 926–929.
42. Llano I, Gonzalez J, Caputo C, Lai FA, Blayney LM, Tan YP and Marty A (2000) Presynaptic calcium stores underlie large-amplitude miniature IPSCs and spontaneous calcium transients. Nat Neurosci 3: 1256–1265.
43. Loewenstein Y, Mahon S, Chadderton P, Kitamura K, Sompolinsky H, Yarom Y and Hausser M (2005) Bistability of cerebellar Purkinje cells modulated by sensory stimulation. Nat Neurosci 8: 202–211.
44. Lorente de Nó R (1933) Vestibulo-ocular reflex arc. Arch Neurol Psychiatry 30: 245–291.
45. Lu T and Trussell LO (2000) Inhibitory transmission mediated by asynchronous transmitter release. Neuron 26: 683–694.
46. Maeda E, Robinson HP and Kawana A (1995) The mechanisms of generation and propagation of synchronized bursting in developing networks of cortical neurons. J Neurosci 15: 6834–6845.
47. Magee JC and Johnston D (1997) A synaptically controlled, associative signal for Hebbian plasticity in hippocampal neurons. Science 275: 209–213.
48. Malenka RC (2003) The long-term potential of LTP. Nat Rev Neurosci 4: 923–926.
49. Malenka RC and Nicoll RA (1999) Long-term potentiation – a decade of progress. Science 285: 1870–1874.
50. Mann EO and Paulsen O (2007) Role of GABAergic inhibition in hippocampal network oscillations. Trends Neurosci 30: 343–349.
51. Markram H, Lubke J, Frotscher M and Sakmann B (1997) Regulation of synaptic efficacy by coincidence of postsynaptic APs and EPSPs. Science 275: 213–215.
52. McCormick DA and Contreras D (2001) On the cellular and network bases of epileptic seizures. Annu Rev Physiol 63: 815–846.
53. McCulloch WS and Pitts W (1943) A logical calculus of the ideas immanent in nervous activity. Bull Math Biophys 5: 115–133.
54. Mehta MR, Barnes CA and McNaughton BL (1997) Experience-dependent, asymmetric expansion of hippocampal place fields. Proc Natl Acad Sci U S A 94: 8918–8921.
55. Miledi R (1966) Strontium as a substitute for calcium in the process of transmitter release at the neuromuscular junction. Nature 212: 1233–1234.
56. Milner B, Squire LR and Kandel ER (1998) Cognitive neuroscience and the study of memory. Neuron 20: 445–468.

57. Molnar G, Olah S, Komlosi G, Fule M, Szabadics J, Varga C, Barzo P and Tamas G (2008) Complex events initiated by individual spikes in the human cerebral cortex. PLoS biology 6: e222.
58. Nelson S (2000) Timing isn't everything. Neuron 26: 545–546.
59. Peng Y (1996) Ryanodine-sensitive component of calcium transients evoked by nerve firing at presynaptic nerve terminals. J Neurosci 16: 6703–6712.
60. Segev R, Shapira Y, Benveniste M and Ben-Jacob E (2001) Observations and modeling of synchronized bursting in two-dimensional neural networks. Phys Rev E Stat Nonlin Soft Matter Phys 64: 011920.
61. Seung HS (2000) Half a century of Hebb. Nat Neurosci 3 Suppl: 1166.
62. Seung HS, Lee DD, Reis BY and Tank DW (2000) Stability of the memory of eye position in a recurrent network of conductance-based model neurons. Neuron 26: 259–271.
63. Shu Y, Hasenstaub A and McCormick DA (2003) Turning on and off recurrent balanced cortical activity. Nature 423: 288–293.
64. Simkus CR and Stricker C (2002) The contribution of intracellular calcium stores to mEPSCs recorded in layer II neurones of rat barrel cortex. J Physiol 545: 521–535.
65. Stent GS (1973) A physiological mechanism for Hebb's postulate of learning. Proc Natl Acad Sci U S A 70: 997–1001.
66. Stevens CF (1996) Strengths and weaknesses in memory. Nature 381: 471–472.
67. Tegner J, Compte A and Wang XJ (2002) The dynamical stability of reverberatory neural circuits. Biol Cybern 87: 471–481.
68. Tsodyks M, Pawelzik K and Markram H (1998) Neural networks with dynamic synapses. Neural comput 10: 821–835.
69. Tsodyks M, Uziel A and Markram H (2000) Synchrony generation in recurrent networks with frequency-dependent synapses. J Neurosci 20: RC50.
70. Verkhratsky A (2002) The endoplasmic reticulum and neuronal calcium signalling. Cell Calcium 32: 393–404.
71. Volman V, Baruchi I and Ben-Jacob E (2005) Manifestation of function-follow-form in cultured neuronal networks. Phys Biol 2: 98–110.
72. Volman V, Gerkin RC, Lau PM, Ben-Jacob E and Bi GQ (2007) Calcium and synaptic dynamics underlying reverberatory activity in neuronal networks. Phys Biol 4: 91–103.
73. Wagenaar DA, Pine J and Potter SM (2006) An extremely rich repertoire of bursting patterns during the development of cortical cultures. BMC Neurosci 7: 11.
74. Wang XJ (1999) Synaptic basis of cortical persistent activity: the importance of NMDA receptors to working memory. J Neurosci 19: 9587–9603.
75. Wang XJ (2001) Synaptic reverberation underlying mnemonic persistent activity. Trends Neurosci 24: 455–463.
76. Whittington MA and Traub RD (2003) Interneuron diversity series: inhibitory interneurons and network oscillations in vitro. Trends Neurosci 26: 676–682.
77. Whittington MA, Traub RD, Kopell N, Ermentrout B and Buhl EH (2000) Inhibition-based rhythms: experimental and mathematical observations on network dynamics. Int J Psychophysiol 38: 315–336.
78. Xu-Friedman MA and Regehr WG (2000) Probing fundamental aspects of synaptic transmission with strontium. J Neurosci 20: 4414–4422.
79. Zhang LI, Tao HW, Holt CE, Harris WA and Poo MM (1998) A critical window for cooperation and competition among developing retinotectal synapses. Nature 395: 37–44.
80. Zipser D, Kehoe B, Littlewort G and Fuster J (1993) A spiking network model of short-term active memory. J Neurosci 13: 3406–3420.
81. Zucker RS and Regehr WG (2002) Short-term synaptic plasticity. Annu Rev Physiol 64: 355–405.

Gap Junctions and Emergent Rhythms

S. Coombes and M. Zachariou

Abstract Gap-junction coupling is ubiquitous in the brain, particularly between the dendritic trees of inhibitory interneurons. Such direct nonsynaptic interaction allows for direct electrical communication between cells. Unlike spike-time driven synaptic neural network models, which are event based, any model with gap junctions must necessarily involve a single neuron model that can represent the shape of an action potential. Indeed, not only do neurons communicating via gaps feel super-threshold spikes, but they also experience, and respond to, sub-threshold voltage signals. In this chapter, we show that the so-called *absolute* integrate-and-fire model is ideally suited to such studies. At the single neuron level voltage traces for the model may be obtained in closed form, and are shown to mimic those of fast-spiking inhibitory neurons. Interestingly, in the presence of a slow spike adaptation current, the model is shown to support periodic bursting oscillations. For both tonic and bursting modes, the phase response curve can be calculated in closed form. At the network level we focus on global gap junction coupling and show how to analyze the asynchronous firing state in large networks. Importantly, we are able to determine the emergence of nontrivial network rhythms due to strong coupling instabilities. To illustrate the use of our theoretical techniques (particularly the phase-density formalism used to determine stability) we focus on a spike adaptation induced transition from asynchronous tonic activity to synchronous bursting in a gap-junction coupled network.

Introduction

Gap-junction coupling is known to occur between many cell types, including for example pancreatic-β cells [13], heart cells [15], astrocytes [6], and neurons [22]. In this latter context, these junctions are primarily found between inhibitory cells

S. Coombes (✉)
School of Mathematical Sciences, University of Nottingham, Nottingham NG7 2RD, UK
e-mail: stephen.coombes@nottingham.ac.uk

K. Josić et al. (eds.), *Coherent Behavior in Neuronal Networks*, Springer Series in Computational Neuroscience 3, DOI 10.1007/978-1-4419-0389-1_5,

[26]. Interestingly, interneurons are known to play a key role in the generation of hippocampal and cortical rhythms, such as those at gamma frequency (30–100 Hz) [9, 21]. Gap junctions allow for the direct electrical communication between cells, and without the need for receptors to recognize chemical messengers are much faster than chemical synapses at relaying signals. The synaptic delay for a chemical synapse is typically in the range 1–100 ms, while the synaptic delay for an electrical synapse may be only about 0.2 ms. There is now little doubt that gap junctions play a substantial role in the generation of neural rhythms [5, 28], both functional [1, 5, 25, 28] and pathological [17, 51]. One natural question therefore is how does the presence of gap-junction coupling affect synchronous neuronal firing [4, 24, 40]. Independent experimental studies have proposed that gaps synchronize neuronal firing even in the absence of chemical synapses [16, 37]. However, other studies have demonstrated that synchrony can result from the interplay of electrical and chemical signaling and that gaps alone are not sufficient for obtaining synchronous activity [7, 47]. Contradictory results have been reported in the case of inspiratory motorneurons, where gaps desynchronize neural activity whereas synaptic inhibition alone promotes synchrony [8]. From a theoretical perspective the theory of weakly coupled oscillators has often been used to understand how gap junction coupling promotes synchrony or antisynchrony depending on the nature of the neural oscillator and the shape of the action potential [18, 31, 32, 35, 36, 41, 42, 46]. By its very nature, however, this sort of approach cannot tackle gap-induced variations in single neuron firing rate and is thus not ideally suited to answering questions about how the strength of gap junctions contributes to coherent neuronal behavior. Thus, we are led to the search for a tractable network model that can be analyzed in the strong coupling limit. In this chapter, we show how one can make progress in the strong coupling regime for a certain class of spiking neuron model that mimics the behavior of fast-spiking interneurons. Importantly, we are able to quantify a transition from asynchronous tonic spiking to synchronized bursting oscillations in a large globally gap-junction coupled network.

The layout of this chapter is as follows. In section "The Absolute Integrate-and-Fire Model," we introduce our single neuron model of choice, namely a nonlinear integrate-and-fire model, with a piece-wise linear nonlinearity. We show that this model can mimic the behavior of a fast-spiking interneuron whilst being analytically tractable. In illustration, we calculate periodic orbits and the phase response curve in closed form. A simple model of spike adaptation is used to augment this basic model so that it can also fire in a burst mode. Next in section "Gap-Junction Coupling," we pursue the analysis of large globally gap-junction coupled networks. The focus here is on asynchronous states that generate a constant mean field signal. These are calculated in closed form and provide the starting point for a subsequent stability analysis. This makes use of ideas originally developed by van Vreeswijk [48] for the study of synaptic interactions. Importantly, we are able to generate the instability borders in parameter space beyond which an asynchronous state is unstable to periodic temporal perturbations. Direct numerical simulations confirm the correctness of our calculations and show that the dominant solution to emerge beyond an instability is one where the mean-field signal shows a classical bursting signature. Moreover, neurons in this state are synchronized at the level of their firing rate, but not at the level of individual spikes. Finally in section "Discussion," we discuss natural extensions of the work in this chapter.

The *Absolute* Integrate-and-Fire Model

The presence of gap-junctional coupling in a neuronal network necessarily means that neurons directly "feel" the shape of action potentials from other neurons to which they are connected. From a modeling perspective one must therefore be careful to work with single neuron models that have an accurate representation of an action potential shape. On the other hand it is also desirable to work with a model that can be analyzed. A recent paper [12] advocates the use of piece-wise linear planar models. As an alternative we consider here the use of a nonlinear integrate-and-fire (IF) model. Indeed the quadratic IF model is a common choice for computational studies (and unlike the leaky IF model does generate an action potential shape). However for arbitrary time-dependent forcing formal closed solutions are not known. A somewhat overlooked tractable nonlinear IF model is that of Karbowski and Kopell [30], with a voltage dynamics given by

$$\dot{v} = f(v) + I, \tag{1}$$

subject to $v \to v_{\rm r}$ if $v = v_{\rm th}$. Here the function $f(v)$ has a shape like $|v - v_{\rm s}|$ and hence the name *absolute* integrate-and-fire (aif) model, for some *switch* value $v_{\rm s}$. The firing times T^n, $n \in \mathbb{Z}$, are defined according to

$$T^n = \inf\{t \mid v(t) \geq v_{\rm th} \; ; \; t \geq T^{n-1}\}. \tag{2}$$

Because of the choice of a piece-wise linear form of the nonlinearity, the aif model can be explicitly analyzed. To see that it is capable of generating behavior consistent with that of a fast-spiking interneuron we compare it with a more detailed biophysical model. A generic model for a neocortical fast-spiking interneuron is that of Wang and Buzsáki [52] (originally developed to describe CA1 hippocampal interneurons). The kinetics and maximal conductances, which are Hodgkin and Huxley style, are chosen so that the model displays two salient features of hippocampal and neocortical fast-spiking interneurons. The first being that the action potential is followed by a brief after-hyperpolarization, and the second that the model fires repetitive spikes at high frequencies. A plot of the response of this model to constant current injection is shown in Fig. 1. In the same figure we also show response of the aif model with the choice

$$f(v) = \begin{cases} (v - v_{\rm s}) & v > v_{\rm s} \\ -\alpha(v - v_{\rm s}) & v \leq v_{\rm s} \end{cases}, \qquad \alpha > 0. \tag{3}$$

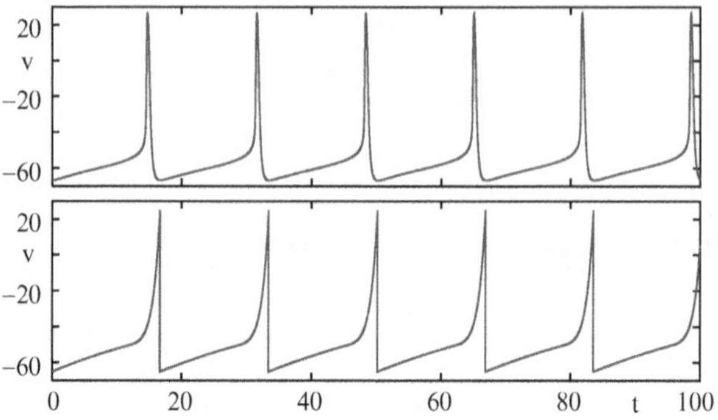

Fig. 1 *Top*: Periodic orbit in the Wang–Buzsáki model with constant current injection $I = 1$. *Bottom*: Periodic orbit in the aif model with $v_r = -65$, $v_s = -50$, $v_{th} = 25$, $\alpha = 0.03$, and $I = 1$

It is clear that an appropriately parametrized aif model can indeed capture the essential spike shape and frequency response of the more detailed biophysical model. Note that for accurate numerical computation of the spike times where $v \geq v_s$ (and solutions diverge as e^t) it is useful to consider the transformed variable $x = \ln(1 + v - v_s)$ and solve the dynamical system $\dot{x} = 1 + (I - 1)e^{-x}$ and then match to solutions with $v < v_s$.

Spike Adaptation

As well as supporting a tonic mode of spiking some interneurons have been reported to exhibit bursting [14, 38, 53]. With this in mind we show that by incorporating a form of spike adaptation [49] the aif model can exhibit both tonic and bursting behavior. For simplicity, we shall henceforth work with the explicit choice $f(v) = |v|$. In more detail we write

$$\dot{v} = |v| + I - a, \qquad \dot{a} = -a/\tau_a, \qquad \tau_a > 0, \tag{4}$$

subject to the usual IF reset mechanism as well as the adaptive step $a(T^m) \to a(T^m) + g_a/\tau_a$, for some $g_a > 0$. For sufficiently small g_a, the model fires tonically as shown in Fig. 2. Since the model is now a 2D (discontinuous) dynamical system it is also useful to view orbits in the (v, a) plane, where one can also plot the system nullclines, as shown in Fig. 3. For larger values of g_a, the model can also fire in a burst mode as shown in Fig. 4. The mechanism for this behavior is most easily understood in reference to the geometry of the phase-plane, as shown in Fig. 5. First consider that the dynamics after reset is such that the adaptive current is sufficiently strong so as to pull the trajectory toward the left-hand side of the voltage nullcline. If the separation of time-scales between the v and a variables is large (namely that

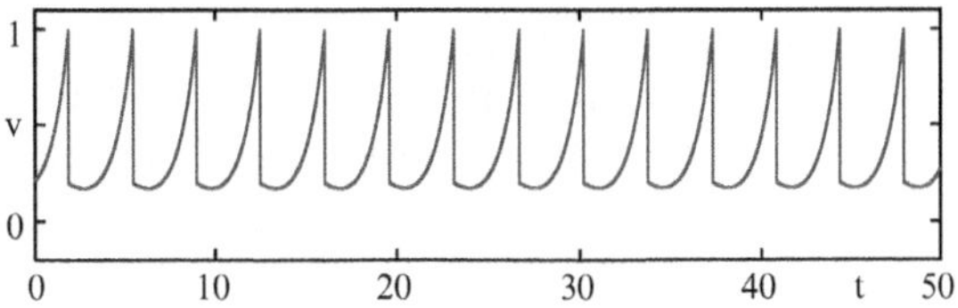

Fig. 2 Tonic firing in the aif model with spike adaptation. Here $\tau_a = 3$, $v_r = 0.2$, $v_{th} = 1$, $I = 0.1$, and $g_a = 0.75$

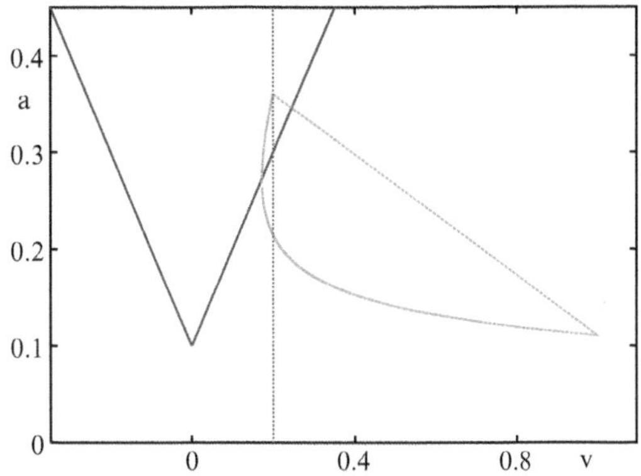

Fig. 3 A periodic orbit in the (v, a) plane corresponding to the tonic spiking trajectory shown in Fig. 2. Also shown is the voltage nullcline as well as the value of the reset

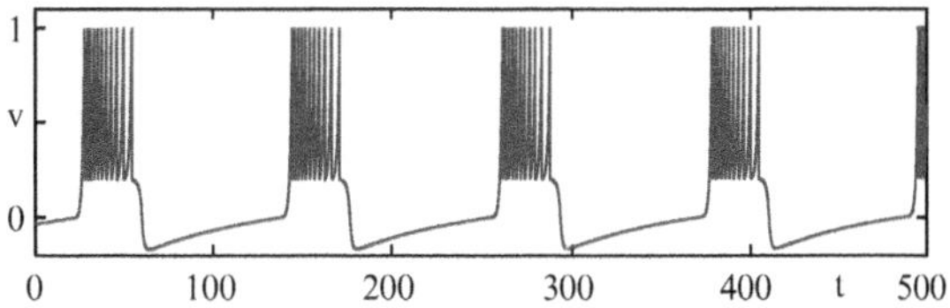

Fig. 4 Burst firing in the aif model with spike adaptation. Here $\tau_a = 75$, $v_r = 0.2$, $v_{th} = 1$, $I = 0.1$, and $g_a = 2$

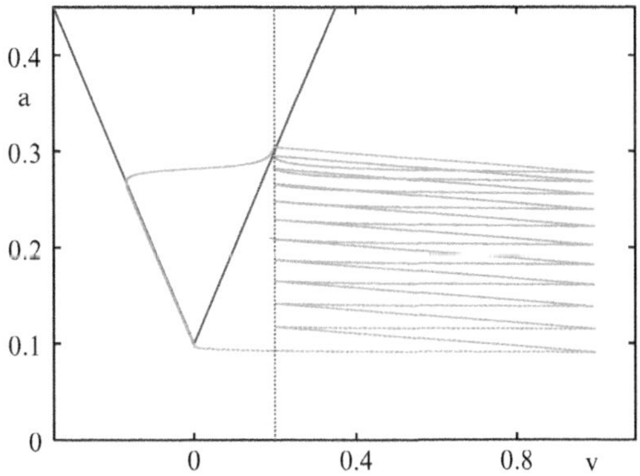

Fig. 5 A periodic orbit in the (v, a) plane corresponding to the bursting trajectory shown in Fig. 4

τ_a is large), then the trajectory will slowly track this nullcline ($a = I - v$) until it reaches $v = 0$, where there is a *switch* in the dynamics (from $f(v) = -v$ to $f(v) = +v$). After the switch the neuron is able to fire for as long as threshold can be reached – namely until a becomes so large as to preclude further firing. Thus, it is the negative feedback from the adaptive current that ultimately terminates a burst, and initiates a slow phase of subthreshold dynamics.

To solve the full nonlinear dynamical model, it is convenient to break the phase space into two regions separated by the line $v = 0$, so that in each region the dynamics (up to threshold and reset) is governed by a linear system. If we denote by v_+ and v_- the solution for $v > 0$ and $v < 0$, respectively, then variation of parameters gives us the closed form solution

$$v_\pm(t) = v_\pm(t_0)\mathrm{e}^{\pm(t-t_0)} + \int_{t_0}^{t} \mathrm{e}^{\mp(s-t)}[I - a(s)]\mathrm{d}s, \tag{5}$$

with initial data $v_\pm(t_0)$ and $t > t_0$. For example, considering the Δ-periodic tonic solution shown in Fig. 3, where $v > 0$ always, then we have that $a(t) = \overline{a}\mathrm{e}^{-t/\tau_a}$, with $\overline{a}$ determined self-consistently from $a(\Delta) + g_a/\tau_a = \overline{a}$, giving

$$\overline{a} = \frac{g_a}{\tau_a} \frac{1}{1 - \mathrm{e}^{-\Delta/\tau_a}}. \tag{6}$$

Hence, from (5), the voltage varies according to

$$v(t) = v_\mathrm{r}\mathrm{e}^t + I(\mathrm{e}^t - 1) - \frac{\overline{a}\tau_a}{1 + \tau_a}(\mathrm{e}^t - \mathrm{e}^{-t/\tau_a}). \tag{7}$$

The period is determined self-consistently by demanding that $v(\Delta) = v_{\mathrm{th}}$. A plot of the firing frequency $f = \Delta^{-1}$ as a function of g_a is shown in Fig. 6. From this we see that the frequency of tonic firing drops off with increasing adaptation, as expected. Note that one may also construct more complicated orbits (such as

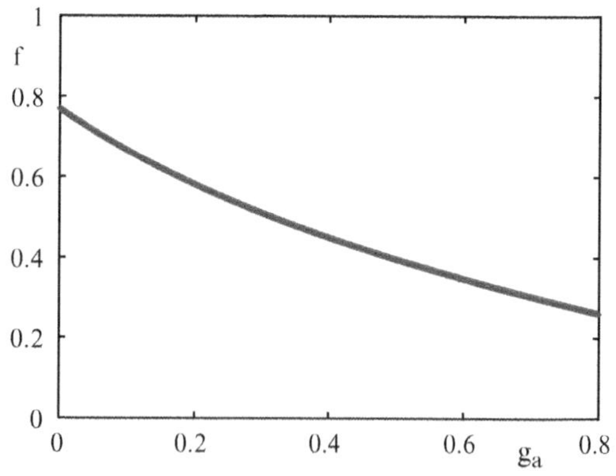

Fig. 6 Frequency of tonic firing as a function of the strength of adaptation g_a for the parameters of Fig. 2

tonic solutions which visit $v < 0$, period doubled tonic solutions, bursting states, etc.) using the ideas above. The main effort being in piecing together trajectories across $v = 0$.

Phase Response Curve

It is common practice to characterize a neuronal oscillator in terms of its phase response to a perturbation. This gives rise to the notion of a so-called phase response curve (PRC). For a detailed discussion of PRCs we refer the reader to [19, 20, 27]. Suffice to say that for a weak external perturbation, such that $(\dot{v}, \dot{a}) \to (\dot{v}, \dot{a}) + \epsilon(A_1(t), A_2(t))$, and ϵ small, then we can introduce a phase $\theta \in (0, 1)$ along a Δ-periodic orbit that evolves according to

$$\dot{\theta} = \frac{1}{\Delta} + \epsilon Q^T(A_1(t), A_2(t)). \tag{8}$$

The (vector) PRC, is given as $Q\Delta$, where Q obeys the so-called adjoint equation

$$\frac{\mathrm{d}Q}{\mathrm{d}t} = -DF^T(t)Q, \tag{9}$$

and $DF(t)$ is the Jacobian of the dynamical systems evaluated along the time-dependent orbit. To enforce the condition that $\dot{\theta} = 1/\Delta$ for $\epsilon = 0$ we must choose initial data for Q that guarantees $Q^T(\dot{v}, \dot{a}) = \Delta^{-1}$. For a continuous trajectory this normalization condition need only be enforced at a single point in time. However, for the aif model with adaptation there is a single discontinuity in the orbit (at reset) and so Q is not continuous. We therefore need to establish the conditions that ensure $Q(\Delta^+) = Q(0)$. Introducing components of Q as $Q = (q_1, q_2)$ this is equivalent to demanding continuity of $\mathrm{d}q_1/\mathrm{d}q_2$ at reset.

For the orbit given by (7) with $v > 0$ the Jacobian is simply the constant matrix

$$DF = \begin{bmatrix} 1 & -1 \\ 0 & -1/\tau_a \end{bmatrix}, \tag{10}$$

and the adjoint equation (9) may be solved in closed form as

$$q_1(t) = q_1(0)\mathrm{e}^{-t}, \qquad q_2(t) = q_2(0)\mathrm{e}^{t/\tau_a} + q_1(0)\frac{\tau_a}{1+\tau_a}[\mathrm{e}^{t/\tau_a} - \mathrm{e}^{-t}]. \tag{11}$$

The condition for continuity of $\mathrm{d}q_1/\mathrm{d}q_2$ at reset yields the relationship

$$\frac{q_2(0)}{q_1(0)} = \frac{q_2(\Delta)}{q_1(\Delta)} = -\frac{\tau_a}{1+\tau_a}, \tag{12}$$

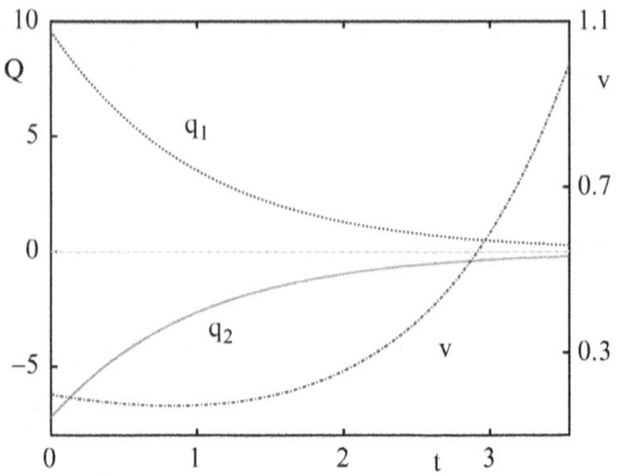

Fig. 7 Adjoint Q for the tonic spiking orbit shown in Fig. 3

whilst the normalization condition gives

$$q_1(0)[v_r + I - \overline{a}] - q_2(0)\frac{\overline{a}}{\tau_a} = \frac{1}{\Delta}. \tag{13}$$

The simultaneous solutions of (12) and (13) then gives the adjoint in the closed form

$$Q(t) = \frac{\kappa}{\Delta}e^{-t}\begin{bmatrix} 1 \\ -\tau_a/(1+\tau_a) \end{bmatrix}, \qquad t \in [0, \Delta), \tag{14}$$

and $\kappa = [v_r + I - \overline{a}\tau_a/(1+\tau_a)]^{-1}$. A plot of the adjoint for the tonic orbit (7) is shown in Fig. 7. Note that the orbit and PRC for other periodic solutions (crossing through $v = 0$) can be obtained in a similar fashion.

Gap-Junction Coupling

To model the direct gap-junction coupling between two cells, one labeled *post* and the other *pre*, we introduce an extra current to the right-hand side of equation (1) in the form

$$g_{\text{gap}}(v_{\text{pre}} - v_{\text{post}}), \tag{15}$$

where g_{gap} is the conductance of the gap junction. Indexing neurons in a network with the label $i=1,\ldots,N$ and defining a gap-junction conductance strength between neurons i and j as g_{ij} means that neuron i experiences a drive of the form $N^{-1}\sum_{j=1}^{N} g_{ij}(v_j - v_i)$. For a phase locked state then $(v_i(t), a_i(t)) = (v(t-\phi_i\Delta), a(t-\phi_i\Delta))$, $(v(t), a(t)) = (v(t+\Delta), a(t+\Delta))$, (for some constant phases $\phi_i \in [0,1)$) and we have N equations distinguished by the driving terms $N^{-1}\sum_{j=1}^{N} g_{ij}(v(t+(\phi_i-\phi_j)T) - v(t))$. For globally coupled networks with $g_{ij} = g$, maximally symmetric solutions describing synchronous, asynchronous, and cluster states are expected to be generic [2]. Here we shall focus on asynchronous states defined by $\phi_i = i/N$. Such solutions are often called splay or merry-go-round states, since all oscillators in the network pass through some fixed phase at regularly spaced time intervals of Δ/N.

Existence of the Asynchronous State

Here we will focus on a globally coupled network in the large N limit. In this case, we have the useful result that network averages may be replaced by time averages. In this case, the coupling term for an asynchronous state becomes

$$\lim_{N\to\infty} \frac{1}{N}\sum_{j=1}^{N} v(t + j\Delta/N) = \frac{1}{\Delta}\int_0^{\Delta} v(t)\mathrm{d}t, \tag{16}$$

which is independent of both i and t. Hence, for an asynchronous state every neuron in the network is described by the same dynamical system, namely

$$\dot{v} = |v| - gv + I - a + gv_0, \qquad \dot{a} = -a/\tau_a, \tag{17}$$

where

$$v_0 = \frac{1}{\Delta}\int_0^{\Delta} v(t)\mathrm{d}t. \tag{18}$$

Once again we may use variation of parameters to obtain a closed form solution for the trajectory:

$$v_\pm(t) = v_\pm(t_0)\mathrm{e}^{\pm(t-t_0)/\tau_\pm} + \int_{t_0}^{t} \mathrm{e}^{\mp(s-t)/\tau_\pm}[I_g - a(s)]\mathrm{d}s, \tag{19}$$

where $\tau_\pm = 1/(1 \mp g)$ and $I_g = I + gv_0$. A self-consistent solution for the pair (Δ, v_0) is now obtained from the simultaneous solution of the two equations $v(\Delta) = v_{\mathrm{th}}$ and $v_0 = \Delta^{-1}\int_0^{\Delta} v(t)\mathrm{d}t$. For example an orbit with $v > 0$ is easily constructed and generates the two equations

$$v_{\mathrm{th}} = v_{\mathrm{r}}\mathrm{e}^{\Delta/\tau_+} + I_g\tau_+(\mathrm{e}^{\Delta/\tau_+} - 1) - \overline{a}\tau(\mathrm{e}^{\Delta/\tau_+} - \mathrm{e}^{-\Delta/\tau_a}), \tag{20}$$

$$v_0 = -I_g\tau_+ + \frac{1}{\Delta}\left\{\tau_+[\mathrm{e}^{\Delta/\tau_+} - 1][v_{\mathrm{r}} + I_g\tau_+ - \overline{a}\tau] + \overline{a}\tau\tau_a[1 - \mathrm{e}^{-\Delta/\tau_a}]\right\}, \tag{21}$$

where $1/\tau = 1/\tau_+ + 1/\tau_a$. A plot of (Δ, v_0) as a function of the gap strength g is shown in Fig. 8.

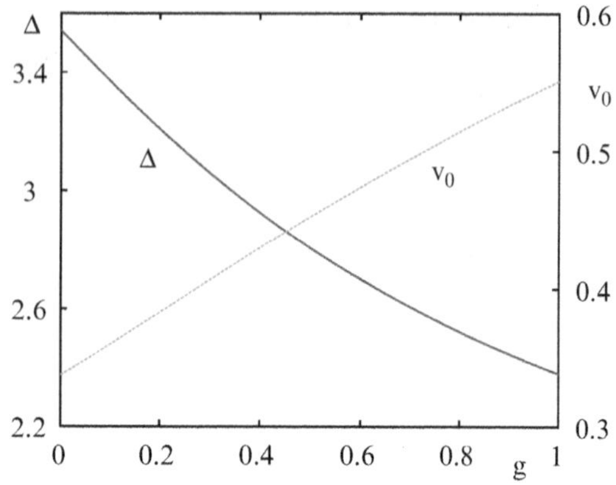

Fig. 8 Period $\varDelta$ and constant mean field signal v_0 as a function of gap strength g. Other parameters as in Fig. 3 left

Stability of the Asynchronous State

Here we use a phase reduction technique, first developed by van Vreeswijk [48] for synaptic coupling, to study the stability of the asynchronous state. To do this we write the coupling term $N^{-1}\sum_{j=1}^{N} v_j(t)$ in a more convenient form for studying perturbations of the mean field, namely we write

$$\lim_{N\to\infty}\frac{1}{N}\sum_{j=1}^{N} v_j(t) = \lim_{N\to\infty}\frac{1}{N}\sum_{j=1}^{N}\sum_{m\in\mathbb{Z}} u(t-T_j^m), \tag{22}$$

where $T_j^m = m\varDelta + j\varDelta/N$. Here $u(t) = 0$ for $t < 0$ and is chosen such that $v(t) = \sum_{m\in\mathbb{Z}} u(t - m\varDelta)$, ensuring that $v(t) = v(t+\varDelta)$. For arbitrary values of the *firing-times* T_j^m the coupling term (22) is time-dependent, and we may write it in the form

$$E(t) = \int_0^\infty f(t-s)u(s)\mathrm{d}s, \qquad f(t) = \lim_{N\to\infty}\frac{1}{N}\sum_{j,m}\delta(t-T_j^m), \tag{23}$$

where we recognize $f(t)$ as a firing rate. We now consider perturbations of the mean field such that $E(t)$ (the average membrane voltage) is split into a stationary part (arising from the asynchronous state) and an infinitesimal perturbation. Namely we write $E(t) = v_0 + \epsilon(t)$, with small $\epsilon(t)$. Since this perturbation to the *asynchronous* oscillator defined by (17) is small we may use phase reduction techniques to study the stability of the asynchronous state.

In terms of a phase $\theta \in (0,1)$ along the asynchronous state we can write the evolution of this phase variable in response to a perturbation in the mean field as

$$\frac{\mathrm{d}\theta}{\mathrm{d}t} = \frac{1}{\varDelta} + g\Gamma(\theta\varDelta)\epsilon(t), \tag{24}$$

where Γ is the g-dependent voltage component of the adjoint for the asynchronous state. This can again be calculated in closed form using the techniques developed in section "Phase Response Curve," and takes the explicit form

$$\Gamma(t) = \frac{\kappa(g)}{\Delta} \mathrm{e}^{-t/\tau_+}, \tag{25}$$

where $\kappa(g) = [v_r/\tau_+ + I_g - \overline{a}\tau_a/(1+\tau_a)]^{-1}$. In fact we need to treat N phase variables θ_i, each described by an equation of the form (24), which are coupled by the dependence of $\epsilon(t)$ on these variables. To make this more explicit we write

$$\epsilon(t) = \int_0^\infty \delta f(t-s) u(s) \mathrm{d}s, \tag{26}$$

and use a phase density description to calculate the dependence of the perturbed firing rate δf on the phases. We define a phase density function as the fraction of neurons in the interval $[\theta, \theta + \mathrm{d}\theta]$ namely $\rho(\theta,t) = N^{-1}\sum_j \delta(\theta_j(t) - \theta)$. Introducing the flux $J(\theta,t) = \rho(\theta,t)\dot{\theta}$, we have the continuity equation

$$\frac{\partial \rho}{\partial t} = -\frac{\partial J}{\partial \theta}, \tag{27}$$

with boundary condition $J(1,t) = J(0,t)$. The firing rate is the flux through $\theta = 1$, so that $f(t) = J(1,t)$. In the asynchronous state the phase density function is independent of time. Considering perturbations around this state, $(\rho, J) = (1, \Delta^{-1})$, means writing $\rho(\theta,t) = 1 + \delta\rho(\theta,t)$, with a corresponding perturbation of the flux that takes the form $\delta J(\theta,t) = \delta\rho(\theta,t)/\Delta + g\Gamma(\theta\Delta)\epsilon(t)$. Differentiation of $\delta J(\theta,t)$ gives the partial differential equation

$$\partial_t \delta J(\theta,t) = -\frac{1}{\Delta}\partial_\theta \delta J(\theta,t) + g\Gamma(\theta\Delta)\epsilon'(t), \tag{28}$$

where

$$\epsilon(t) = \int_0^\infty u(s)\delta J(1, t-s)\mathrm{d}s. \tag{29}$$

Assuming a solution of the form $\delta J(\theta,t) = \mathrm{e}^{\lambda t}\delta J(\theta)$, gives

$$\epsilon(t) = \delta J(1)\mathrm{e}^{\lambda t}\widetilde{u}(\lambda), \tag{30}$$

where $\widetilde{u}(\lambda) = \int_0^\infty u(t)\mathrm{e}^{-\lambda t}\mathrm{d}t$. In this case $\epsilon'(t) = \lambda\epsilon(t)$. Equation (28) then reduces to the ordinary differential equation

$$\frac{\mathrm{d}}{\mathrm{d}\theta}\delta J(\theta)\mathrm{e}^{\lambda\Delta\theta} = g\lambda\Delta\Gamma(\theta\Delta)\delta J(1)\widetilde{u}(\lambda)\mathrm{e}^{\lambda\Delta\theta}. \tag{31}$$

Integrating (31) from $\theta = 0$ to $\theta = 1$ and using the fact that $\delta J(1) = \delta J(0)$ yields an implicit equation for λ in the form $\mathcal{E}(\lambda) = 0$, where

$$\mathcal{E}(\lambda) = \mathrm{e}^{\lambda \Delta} - 1 - g\lambda \Delta \widetilde{u}(\lambda) \int_0^1 \Gamma(\theta \Delta)\mathrm{e}^{\lambda \theta \Delta} \mathrm{d}\theta. \tag{32}$$

We see that $\mathcal{E}(0) = 0$ so that $\lambda = 0$ is always an eigenvalue. Writing $\lambda = \nu + i\omega$ then the pair (ν, ω) may be found by the simultaneous solution of $\mathcal{E}_{\mathrm{R}}(\nu, \omega) = 0$ and $\mathcal{E}_{\mathrm{I}}(\nu, \omega) = 0$, where $\mathcal{E}_{\mathrm{R}}(\nu, \omega) = \mathrm{Re}\, \mathcal{E}(\nu + i\omega)$ and $\mathcal{E}_{\mathrm{I}}(\nu, \omega) = \mathrm{Im}\, \mathcal{E}(\nu + i\omega)$.

For the adjoint calculated given by (25) a simple calculation gives

$$\int_0^1 \Gamma(\theta \Delta)\mathrm{e}^{\lambda \theta \Delta} \mathrm{d}\theta = \frac{\kappa(g)}{\Delta} \frac{1}{\Delta} \frac{\mathrm{e}^{\Delta(\lambda - 1/\tau_+)} - 1}{(\lambda - 1/\tau_+)}. \tag{33}$$

For the calculation of $\widetilde{u}(\lambda)$ we use the result that $\int_0^\infty u(t)\mathrm{e}^{-\lambda t}\mathrm{d}t = \int_0^\Delta v(t+s)\mathrm{e}^{-\lambda t}\mathrm{d}t$, for some arbitrary time-translation $s \in (0, \Delta)$, with $v(t)$ the splay solution, defined for $t \in (0, \Delta)$. In contrast to the calculations in [12] for continuous periodic orbits, those of the aif model are discontinuous and so one must carefully treat this extra degree of freedom. Since we do not a priori know the phase of the signal $v(t)$ with respect to the time origin of the oscillator model we simply average over all possible phases and write

$$\widetilde{u}(\lambda) = \frac{1}{\Delta} \int_0^\Delta \left\{ \int_0^\Delta v(t + s)\mathrm{e}^{-\lambda t}\mathrm{d}t \right\} \mathrm{d}s. \tag{34}$$

For the splay solution of section "Existence of the Asynchronous State," a short calculation gives

$$\begin{aligned} \frac{\widetilde{u}(\lambda)}{\mathrm{e}^{\lambda\Delta} - 1} &= \frac{v_{\mathrm{r}} + I_g \tau_+ - \overline{a}\tau\, \tau_+}{\lambda - 1/\tau_+} \frac{\tau_+}{\Delta} (\mathrm{e}^{-\Delta(\lambda - 1/\tau_+)} - \mathrm{e}^{-\lambda\Delta}) - I_g \tau_+ \frac{\mathrm{e}^{-\lambda\Delta}}{\lambda} \\ &\quad - \frac{\overline{a}\tau}{\lambda + 1/\tau_a} \frac{\tau_a}{\Delta} (\mathrm{e}^{-\Delta(\lambda + 1/\tau_a)} - \mathrm{e}^{-\lambda\Delta}), \qquad \mathrm{Re}\,\lambda < 1/\tau_+. \end{aligned} \tag{35}$$

For $\lambda \in \mathbb{R}$ the condition for an eigenvalue to cross through zero from below is equivalent to the occurrence of a double zero of $\mathcal{E}(\lambda)$ at $\lambda = 0$. However, it is easy to show that $\mathcal{E}'(0) \neq 0$ so that no instabilities can arise in this fashion. Examples of the spectrum obtained from the zeros of $\mathcal{E}(\lambda)/(\mathrm{e}^{\lambda\Delta} - 1)$ are shown in Fig. 9 (the remaining zeros of $\mathcal{E}(\lambda)$ being at $\lambda\Delta = 2\pi i n$, $n \in \mathbb{Z}$).

Here we see that for fixed g and increasing g_a, a pair of complex conjugate eigenvalues crosses through the imaginary axis at a nonzero value of ω. This signals the onset of a dynamic instability, which is more easily quantified with the aid of Fig. 10 which tracks the first pair (ν, ω) to pass through $\nu = 0$ as a function of g_a. Until now we have assumed that the splay state exists for all parameters of choice. However, because the underlying model is described by a discontinuous flow then

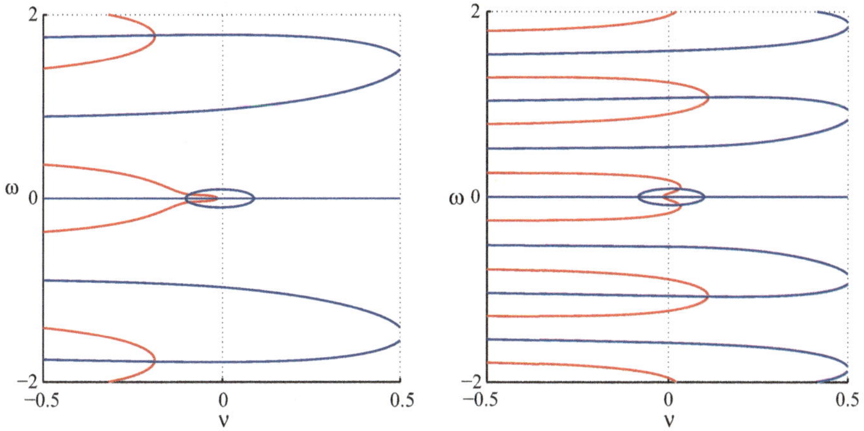

Fig. 9 Spectrum for the asynchronous state. Eigenvalues are at the positions where the red and blue curves intersect. Parameters as in Fig. 4 with $g = 0.5$. *Left*: $g_a = 1.5$, with $(\Delta, \nu_0) = (4.0575, 0.46685)$. *Right*: $g_a = 2.5$, with $(\Delta, \nu_0) = (6.6757, 0.39433)$. Note the unstable mode with $\omega \sim \pm 1$ in the right-hand figure.

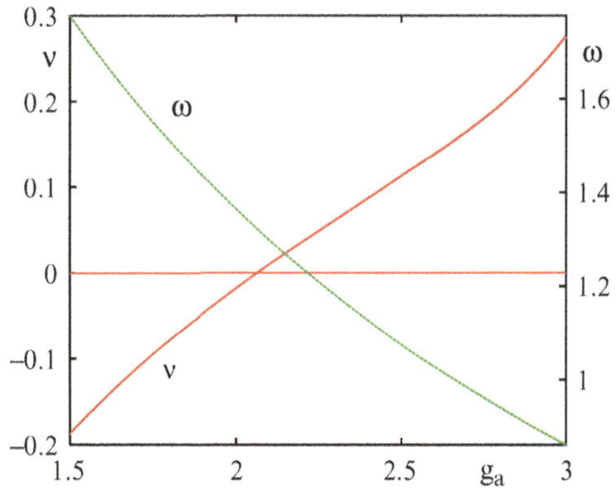

Fig. 10 A plot of (ν, ω), where $\mathcal{E}(\nu + i\omega) = 0$, as a function of g_a, with other parameters as in Fig. 9. Note the bifurcation at $g_a \sim 2.1$, where ν crosses zero from below with a nonzero value of ω.

there is also the possibility that a nonsmooth bifurcation can occur. For example a splay state with $v \geq 0$ may tangentially intersect the surface $v = 0$, where there is a switch in the dynamics for v. In this case, a new orbit will emerge that can either be tonic or bursting. The conditions defining this nonsmooth bifurcation are $v(t^*) = 0$ and $\dot{v}(t^*) = 0$ for some $t^* \in (0, \Delta)$. For the splay state considered here, we find that a dynamic instability, defined by $\mathcal{E}(i\omega) = 0$, is always met before the onset of a nonsmooth bifurcation.

By tracking the bifurcation point $\nu = 0$ in parameter space it is possible to map out those regions where the asynchronous state is stable. We do this in Fig. 11 which basically shows that if an asynchronous state is stable for fixed (g, τ_a) then it can always be destabilized by increasing g_a beyond some critical value.

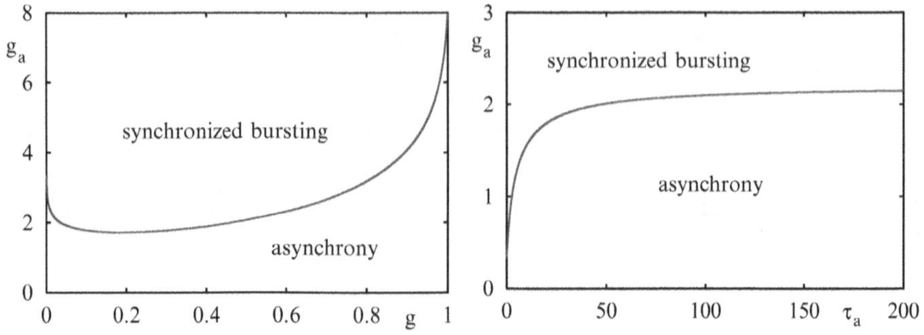

Fig. 11 Curves showing solutions of $\mathcal{E}(i\omega) = 0$ obtained by tracking the bifurcation point in Fig. 10. Parameters as in Fig. 9. *Left*: $\tau_a = 75$. *Right*: $g = 0.5$. Beyond an instability point of the asynchronous solution one typically sees the emergence of synchronized bursting states, as shown in Fig. 12

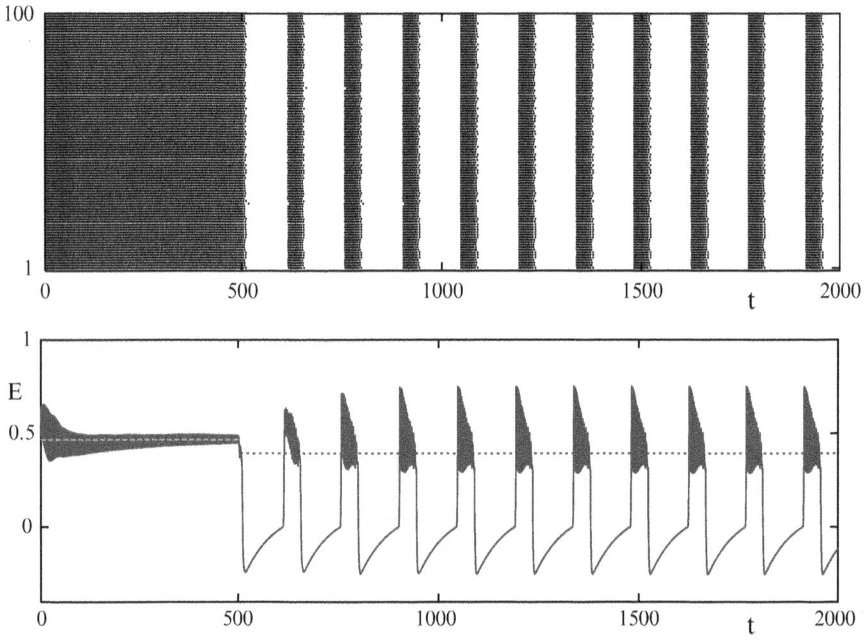

Fig. 12 A plot showing an instability of the asynchronous state in a network with $N = 100$ neurons, starting from random initial conditions. Here g_a is switched from the value in Fig. 9 left (asynchronous state stable) to that in Fig. 9 right (asynchronous state unstable) at $t = 500$. Note the emergence of a synchronized bursting state. The *lower plot* shows the time variation of the mean-field signal $E(t) = N^{-1}\sum_{i=1}^{N} v_i(t)$, as well as the value of v_0 – the mean field signal for the asynchronous state (*dashed and dotted lines*). Parameters as in Fig. 9

To determine the types of solutions that emerge beyond the instability borders we have performed direct numerical simulations. Not only do these confirm the correctness of our bifurcation theory, they show that a dominant emergent solution is a bursting mode in which neurons are synchronized at the level of their firing rates, but not at the level of individual spikes (within a burst). An example of a network state that switches from asynchronous tonic spiking to synchronized bursting with a switch in g_a across the bifurcation point is shown in Fig. 12. Here we plot both a raster diagram showing spike times as well as the mean field signal $E(t) = N^{-1} \sum_{i=1}^{N} v_i(t)$ for a network of 100 neurons. Interestingly the plot of the mean field signal suggests that bursting terminates roughly at the point where it reaches the value of v_0 for the unstable asynchronous orbit.

Discussion

In this chapter, we have shown how the absolute integrate-and-fire model is ideally suited for the theoretical study of gap-junction coupled networks. One such network where theory may help shed further light on function is that of the inferior olivary nucleus, which has extensive electrotonic coupling between dendrites. Chorev et al. [11] have shown that in vivo intracellular recordings from olivary neurons (of anesthetized rats) exhibit subthreshold oscillations of membrane voltage, organized in epochs, lasting from half a second to several seconds. If recorded, spikes were locked to the depolarized phase of these subthreshold oscillations. Thus it is of interest to probe the way in which neurons supporting both subthreshold oscillations and spikes use gap-junction coupling to coordinate spatiotemporal patterns for holding and then transferring rhythmic information to cerebellar circuits [50]. The techniques we have developed here are ideally suited to this task.

At the level of the single neuron we have shown how to construct both the periodic orbit and the phase response curve. This is particularly useful for the development of a weakly coupled oscillator theory for network studies, for both gap and synaptic coupling, as in the work of Kazanci and Ermentrout [31]. However, we have chosen here to instead pursue a strongly coupled network analysis. The tractability of the chosen model has allowed the explicit calculation of the asynchronous state, including the determination of its linear stability, in large globally gap-junction coupled networks. In the presence of a simple form of spike adaptation we have quantified a bifurcation from asynchrony to synchronized bursting. Interestingly, burst synchronization has been observed in both cell cultures and brain areas such as the basal ganglia. For a review of experiments and theory relating to burst synchronization we refer the reader to the article by Rubin [44]. One natural progression of the work in this chapter would be to analyze the properties of bursting in more detail, and in particular the synchronization properties of bursts relating to both gap and synaptic parameters. Techniques for doing this are relatively underdeveloped as compared to those for studying synchronized tonic spiking. However, it is well to point out the work of Izhikevich [29], de Vries and Sherman [13], and

Matveev et al. [39] in this area, as well as more recent numerical studies [43, 45]. The development of such a theory is especially relevant to so-called *neural signatures*, which consist of cell-specific spike timings in the bursting activity of neurons. These very precise intra-burst firing patterns may be quantified using computational techniques discussed in [33]. We refer the reader to [34] for a recent discussion of neural signatures in the context of the pyloric central pattern generator of the crustacean stomatogastric ganglion (where gaps are known to play a role in rhythm generation).

From a biological perspective it is important to emphasize that gaps are not the static structures that we have suggested here by treating gap strength as a single parameter. Indeed the connexin channels that underlie such junctions are dynamic and are in fact modulated by the voltage across the membrane. Baigent et al. [3] have developed a model of the dependency between the cell potentials and the *state* of the gap junctions. In this context the state of an individual channel corresponds to the conformation of the two connexons forming the pore. Of the four possible states (both open, both closed, or one open, and one closed), the scenario where both are closed is ignored. Because each cell–cell junction is composed of many channels, the state of the junction is determined by the distribution of channels amongst the three different states. Thus it would be interesting to combine the model we have presented here with this channel model and explore the consequences for coherent network behavior. Another form of gap-junction modulation can be traced to cannabinoids. Gap-junction coupling can be found among irregular spiking GABAergic interneurons that express cannabinoid receptors [23]. Interestingly, the potentiation of such coupling by cannabinoids has recently been reported [10]. All of the above are topics of current investigation and will be reported upon elsewhere.

Acknowledgments We would like to thank Bard Ermentrout for bringing our attention to the work of Karbowski and Kopell.

References

1. A. V. Alvarez, C. C. Chow, E. J. V. Bockstaele, and J. T. Williams. Frequency-dependent synchrony in locus ceruleus: Role of electrotonic coupling. *Proceedings of the National Academy of Sciences*, 99(6):4032–4036, 2002.
2. P. Ashwin and J. W. Swift. The dynamics of n weakly coupled identical oscillators. *Journal of Nonlinear Science*, 2:69–108, 1992.
3. S. Baigent, J. Stark, and A. Warner. Modelling the effect of gap junction nonlinearities in systems of coupled cells. *Journal of Theoretical Biology*, 186:223–239, Jan 1997.
4. M. Beierlein, J. R. Gibson, and B. W. Connors. A network of electrically coupled interneurons drives synchronized inhibition in neocortex. *Nature Neuroscience*, 3(9):904–910, 2000.
5. M. V. L. Bennet and R. S. Zukin. Electrical coupling and neuronal synchronization in the mammalian brain. *Neuron*, 41:495–511, 2004.
6. M. Bennett, J. Contreras, F. Bukauskas, and J. Sáez. New roles for astrocytes: gap junction hemichannels have something to communicate. *Trends in Neurosciences*, 26:610–617, 2003.

7. M. Blatow, A. Rozov, I. Katona, S. G. Hormuzdi, A. H. Meyer, M. A. Whittington, A. Caputi, and H. Monyer. A novel network of multipolar bursting interneurons generates theta frequency oscillations in neocortex. *Neuron*, 38(5):805–817, 2003.
8. C. Bou-Flores and A. J. Berger. Gap junctions and inhibitory synapses modulate inspiratory motoneuron synchronization. *Journal of Neurophysiology*, 85(4):1543–1551, 2001.
9. G. Buzsaki. *Rhythms of the Brain*. Oxford University Press, Inc., USA, 2006.
10. R. Cachope, K. Mackie, A. Triller, J. O'Brien, and A. E. Pereda. Potentiation of electrical and chemical synaptic transmission mediated by endocannabinoids. *Neuron*, 56(6):1034–1047, 2007.
11. E. Chorev, Y. Yarom, and I. Lampl. Rhythmic episodes of subthreshold membrane potential oscillations in the rat inferior olive nuclei *in vivo*. *The Journal of Neuroscience*, 27:5043–5052, 2007.
12. S. Coombes. Neuronal networks with gap junctions: a study of piece-wise linear planar neuron models. *SIAM Journal on Applied Dynamical Systems*, 7:1101–1129, 2008.
13. G. de Vries and A. Sherman. *Bursting: The Genesis of Rhythm in the Nervous System*, chapter Beyond synchronization: modulatory and emergent effects of coupling in square-wave bursting. World Scientific, 2005.
14. J. Deuchars and A. M. Thomson. Single axon fast inhibitory postsynaptic potentials elicited by a sparsely spiny interneuron in rat neocortex. *Neuroscience*, 65(4):935–942, 1995.
15. S. Dhein and J. S. Borer, editors. *Cardiovascular Gap Junctions (Advances in Cardiology)*. S Karger AG, 2006.
16. A. Draguhn, R. D. Traub, D. Schmitz, and J. G. Jefferys. Electrical coupling underlies high-frequency oscillations in the hippocampus in vitro. *Nature*, 394(6689):189–192, 1998.
17. F. E. Dudek. Gap junctions and fast oscillations: a role in seizures and epileptogenesis? *Epilepsy Currents*, 2:133–136, 2002.
18. B. Ermentrout. Gap junctions destroy persistent states in excitatory networks. *Physical Review E*, 74:031918(1–8), 2006.
19. G. B. Ermentrout and N. Kopell. Oscillator death in systems of coupled neural oscillators. *SIAM Journal on Applied Mathematics*, 50:125–146, 1990.
20. G. B. Ermentrout and N. Kopell. Multiple pulse interactions and averaging in systems of coupled neural oscillators. *Journal of Mathematical Biology*, 29:195–217, 1991.
21. T. F. Freund. Interneuron diversity series: rhythm and mood in perisomatic inhibition. *Trends in Neurosciences*, 26(9):489–495, 2003.
22. E. J. Furshpan and D. D. Potter. Mechanism of nerve-impulse transmission at a crayfish synapse. *Nature*, 180:342–343, 1957.
23. M. Galarreta, F. Erdelyi, G. Szabo, and S. Hestrin. Electrical coupling among irregular-spiking GABAergic interneurons expressing cannabinoid receptors. *Journal of Neuroscience*, 24:9770–9778, 2004.
24. M. Galarreta and S. Hestrin. A network of fast-spiking cells in the neocortex connected by electrical synapses. *Nature*, 402(6757):72–75, 1999.
25. J. R. Gibson, M. Beierlein, and B. W. Connors. Two networks of electrically coupled inhibitory neurons in neocortex. *Nature*, 402(6757):75–79, 1999.
26. J. R. Gibson, M. Beierlein, and B. W. Connors. Functional properties of electrical synapses between inhibitory interneurons of neocortical layer 4. *Journal of Neurophysiology*, 93:467–480, 2005.
27. F. C. Hoppensteadt and E. M. Izhikevich. *Weakly Connected Neural Networks*. Number 126 in Applied Mathematical Sciences. Springer-Verlag, New York, 1997.
28. S. G. Hormuzdi, M. A. Filippov, G. Mitropoulou, H. Monyer, and R. Bruzzone. Electrical synapses: a dynamic signaling system that shapes the activity of neuronal networks. *Biochimica et Biophysica Acta*, 1662:113–137, 2004.
29. E. M. Izhikevich. Synchronization of elliptic bursters. *SIAM Review*, 43:315–344, 2001.
30. J. Karbowski and N. Kopell. Multispikes and synchronization in a large-scale neural network with delays. *Neural Computation*, 12:1573–1606, 2000.
31. F. G. Kazanci and B. Ermentrout. Pattern formation in an array of oscillators with electrical and chemical coupling. *SIAM Journal on Applied Mathematics*, 67:512–529, 2007.

32. N. Kopell and B. Ermentrout. Chemical and electrical synapses perform complementary roles in the synchronization of interneuronal networks. *Proceedings of the National Academy of Sciences USA*, 101:15482–15487, 2004.
33. L. F. Lago-Fernández. Spike alignment in bursting neurons. *Neurocomputing*, 70:1788–1791, 2007.
34. R. Latorre, F. B. Rodríguez, and P. Varona. Neural signatures: multiple coding in spiking-bursting cells. *Biological Cybernetics*, 95:169–183, 2006.
35. T. J. Lewis and J. Rinzel. Dynamics of spiking neurons connected by both inhibitory and electrical coupling. *Journal of Computational Neuroscience*, 14:283–309, 2003.
36. J. G. Mancilla, T. J. Lewis, D. J. Pinto, J. Rinzel, and B. W. Connors. Synchronization of electrically coupled pairs of inhibitory interneurons in neocortex. *The Journal of Neuroscience*, 27:2058–2073, 2007.
37. P. Mann-Metzer and Y. Yarom. Electrotonic coupling interacts with intrinsic properties to generate synchronized activity in cerebellar networks of inhibitory interneurons. *Journal of Neuroscience*, 19(9):3298–3306, 1999.
38. H. Markram, M. Toledo-Rodriguez, Y. Wang, A. Gupta, G. Silberberg, and C. Wu. Interneurons of the neocortical inhibitory system. *Nature Reviews Neuroscience*, 5:793–807, 2004.
39. V. Matveev, A. Bose, and F. Nadim. Capturing the bursting dynamics of a two-cell inhibitory network using a one-dimensional map. *Journal of Computational Neuroscience*, 23:169–187, 2007.
40. H. B. Michelson and R. K. Wong. Synchronization of inhibitory neurones in the guinea-pig hippocampus in vitro. *Journal of Physiology*, 477(Pt 1):35–45, 1994.
41. B. Pfeuty, D. Golomb, G. Mato, and D. Hansel. Inhibition potentiates the synchronizing action of electrical synapses. *Frontiers in Computational Neuroscience*, 1(Article 8), 2007.
42. B. Pfeuty, G. Mato, D. Golomb, and D. Hansel. Electrical synapses and synchrony: the role of intrinsic currents. *The Journal of Neuroscience*, 23:6280–6294, 2003.
43. S. Postnova, K. Voigt, and H. A. Braun. Neural synchronization at tonic-to-bursting transitions. *Journal of Biological Physics*, 33:129–143, 2007.
44. J. E. Rubin. Burst synchronization. *Scholarpedia*, 2(10):1666, 2007.
45. Y. Shen, Z. Hou, and H. Xin. Transition to burst synchronization in coupled neuron networks. *Physical Review E*, 77:031920(1–5), 2008.
46. A. Sherman and J. Rinzel. Rhythmogenic effects of weak electrotonic coupling in neuronal models. *Proceedings of the National Academy of Sciences USA*, 89:2471–2474, 1992.
47. G. Tamas, E. H. Buhl, A. Lorincz, and P. Somogyi. Proximally targeted GABAergic synapses and gap junctions synchronize cortical interneurons. *Nature Neuroscience*, 3(4):366–371, 2000.
48. C. van Vreeswijk. Analysis of the asynchronous state in networks of strongly coupled oscillators. *Physical Review Letters*, 84:5110–5113, 2000.
49. C. van Vreeswijk and D. Hansel. Patterns of synchrony in neural networks with spike adaptation. *Neural Computation*, 13:959–992, Jan 2001.
50. P. Varona, C. Aguirre, J. J. Torres, H. D. I. Abarbanel, and M. I. Rabinovich. Spatio-temporal patterns of network activity in the inferior olive. *Neurocomputing*, 44–46:685–690, 2002.
51. J. L. P. Velazquez and P. L. Carlen. Gap junctions, synchrony and seizures. *Trends in Neurosciences*, 23:68–74, 2000.
52. X. J. Wang and G. Buzsaki. Gamma oscillation by synaptic inhibition in a hippocampal interneuronal network. *Journal of Neuroscience*, 16:6402–6413, 1996.
53. J. J. Zhu, D. J. Uhlrich, and W. W. Lytton. Burst firing in identified rat geniculate interneurons. *Neuroscience*, 91:1445–1460, 1999.

The Feed-Forward Chain as a Filter-Amplifier Motif

Martin Golubitsky, LieJune Shiau, Claire Postlethwaite, and Yanyan Zhang

Abstract Hudspeth, Magnasco, and collaborators have suggested that the auditory system works by tuning a collection of hair cells near Hopf bifurcation, but each with a different frequency. An incoming sound signal to the cochlea then resonates most strongly with one of these hair cells, which then informs the auditory neuronal system of the frequency of the incoming signal. In this chapter, we discuss two mathematical issues. First, we describe how periodic forcing of systems near a point of Hopf bifurcation is generally more complicated than the description given in these auditory system models. Second, we discuss how the periodic forcing of coupling identical systems whose internal dynamics is each tuned near a point of Hopf bifurcation leads naturally to successive amplification of the incoming signal. We call this coupled system a *feed-forward chain* and suggest that it is a mathematical candidate for a motif.

Introduction

In this chapter, we discuss how the periodic forcing of the first node in a chain of coupled identical systems, whose internal dynamics is each tuned near a point of Hopf bifurcation, can lead naturally to successive amplification of the incoming signal. We call this coupled system a *feed-forward chain* and suggest that it is a mathematical candidate for a motif [1]. Periodic forcing of these chains was considered experimentally by McCullen et al. [26]. That study contained observations concerning the amplitude response of solutions down the chain and the effectiveness of the chain as a filter amplifier. This chapter sheds light on these observations.

Our observations motivate the need for a theory of periodic forcing of systems tuned near a point of Hopf bifurcation. Given such a system with Hopf frequency ω_{H}, we periodically force this system at frequency ω_f. The *response curve* is a

M. Golubitsky (✉)
Mathematical Biosciences Institute, Ohio State University, 1735 Neil Avenue, Columbus, OH 43210, USA
e-mail: mg@mbi.ohio_state.edu

K. Josić et al. (eds.), *Coherent Behavior in Neuronal Networks*, Springer Series in Computational Neuroscience 3, DOI 10.1007/978-1-4419-0389-1_6,

graph of the amplitude of the resulting solution as a function of ω_f. In this chapter, we show that the response curve will, in general, be asymmetric and may even have regions of multiple responses when $\omega_f \approx \omega_{\mathrm{H}}$.

This second set of results has implication for certain models of the auditory system, in particular, models of the basilar membrane and attached hair bundles. Several authors [5–7, 21, 23, 27, 28] model the hair bundles by systems of differential equations tuned near a point of Hopf bifurcation; however, in their models they assume precisely the nongeneric condition that leads to a symmetric response curve. Since asymmetric response curves are seen experimentally, these authors then attempt to explain that the asymmetry follows from coupling of the hair bundles. Although this coupling may be reasonable on physiological grounds, our results show that it is not needed if one were only attempting to understand the observed response curve asymmetry.

Sections "Synchrony-Breaking Hopf Bifurcations" and "Periodic Forcing of Feed-Forward Chains" discuss the feed-forward chain and sections "Periodic Forcing near Hopf Bifurcation" and "Cochlear Modeling" discuss periodic forcing of systems near Hopf bifurcation and the auditory system. The remainder of this introduction describes our results in more detail.

The theory of coupled systems of identical differential equations [15, 16, 31] and their bifurcations [8, 11, 12, 24] singles out one three-cell network for both its simplicity and the surprising dynamics it produces via a synchrony-breaking Hopf bifurcation. We have called that network the *feed-forward chain* and it is pictured in Fig. 1. Note that the arrow from cell 1 to itself represents *self-coupling*.

The general coupled cell theory [16] associates to the feed-forward chain a class of differential equations of the form

$$\begin{aligned} \dot{x}_1 &= f(x_1, x_1, \lambda) \\ \dot{x}_2 &= f(x_2, x_1, \lambda) \\ \dot{x}_3 &= f(x_3, x_2, \lambda) \end{aligned} \qquad (1)$$

where $x_j \in \mathbf{R}^k$ is the vector of state variables of node j, $\lambda \in \mathbf{R}$ is a bifurcation parameter, and $f : \mathbf{R}^k \times \mathbf{R}^k \times \mathbf{R} \to \mathbf{R}^k$. We assume that the differential equations f in each cell are identical, and because of this the synchrony subspace $S = \{x_1 = x_2 = x_3\}$ is a *flow-invariant* subspace; that is, a solution with initial conditions in S stays in S for all time. Synchronous equilibria can be expected to occur in such systems and without loss of generality we may assume that such an equilibrium is at the origin; that is, we assume $f(0, 0, \lambda) = 0$. Because of the self-coupling in cell 1, each cell receives exactly one input and the function f can be the same in each equation in (1).

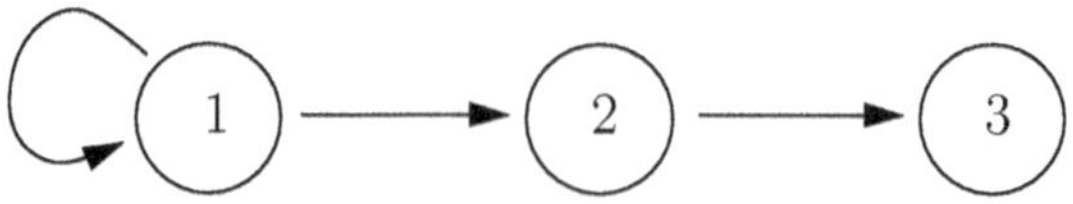

Fig. 1 The feed-forward chain

Recall that in generic Hopf bifurcation in a system with bifurcation parameter λ, the growth in amplitude of the bifurcating periodic solutions is of order $\lambda^{\frac{1}{2}}$. As reviewed in section "Synchrony-Breaking Hopf Bifurcations" synchrony-breaking Hopf bifurcation leads to a family of periodic solutions whose amplitude grows with the unexpectedly large growth rate of $\lambda^{\frac{1}{6}}$ [8, 12]. This growth rate suggests that when the feed-forward chain is tuned near a synchrony-breaking Hopf bifurcation, it can serve to amplify periodic signals whose frequency ω_f is near the frequency of Hopf bifurcation ω_{H} and dampen signals when ω_f is far from ω_{H}. This filter-amplifier motif-like behavior is described in section "Periodic Forcing near Hopf Bifurcation".

Experiments by McCullen et al. [26] with a feed-forward chain consisting of (approximately) identical coupled electronic circuits whose cells are decidedly not in normal form but with sinusoidal forcing confirm the band-pass filter role that a feed-forward chain can assume and the expected growth rates of the output. Additionally, simulations, when the system is in Hopf normal form and the forcing is spiking, also confirm the behavior predicted for the simplified setup. These results are discussed in sections "Simulations" and "Experiments" under "Periodic Forcing of Feed-Forward Chains", and motivate the need for a more general theory of periodic forcing of systems near Hopf bifurcation. We note that the lack of a general theory is more than just a question of mathematical rigor.

Analysis and simulation of periodic forcing of systems near Hopf bifurcation often assume that the forcing is small simple harmonic or sinusoidal forcing $\varepsilon e^{i\omega_f t}$ and that the system is in the simplest normal form for Hopf bifurcation (namely, the system is in third order truncated normal form and the cubic term is assumed to be real). A supercritical Hopf bifurcation vector field can always be transformed by a smooth change of coordinates to be in normal form to third order and the cubic term itself can be scaled to be $-1 + i\gamma$. Simulations of a system in normal form for Hopf bifurcation, but with $\gamma \neq 0$ show phenomena not present in the simplest case (see "Simulations" under "Periodic Forcing near Hopf Bifurcation"). In particular, the amplitude of the response as a function of ω_f can be asymmetric (if $\gamma \neq 0$) and have a region of multiple solutions (if $|\gamma|$ is large enough). Asymmetry and multiplicity have been noted by several authors. Bogoliubov and Mitropolsky [3] analyze the sinusoidally forced Duffing equation and find multiplicity as the frequency of the forcing is varied; Jordan and Smith [10] also analyze the forced Duffing equation and find multiplicity as the amplitude of the forcing is varied; and Montgomery et al. [27] see asymmetry in a forced system near Hopf bifurcation.

In section "Asymmetry and Multiplicity in Response Curve" we show that asymmetry in the response curve occurs as ω_f is varied whenever $\gamma \neq 0$ and that there are precisely two kinds of response curves. In Theorem 1 we use singularity theoretic methods to prove that multiple solutions occur in a neighborhood of the Hopf bifurcation precisely when $|\gamma| > \sqrt{3}$.

Additionally, when ω_f is sufficiently close to ω_H, Kern and Stoop [23] and Eguíluz et al. [7] together show that with a truncated normal form system and harmonic forcing the amplitude of the resulting periodic solution is of order $\varepsilon^{\frac{1}{3}}$. We make this result more precise in section "Scalings of Solution Amplitudes".

Consequently, in the feed-forward chain the amplitude of the periodic forcing can be expected to grow as $\varepsilon^{\frac{1}{3}}$ in the second cell and $\varepsilon^{\frac{1}{9}}$ in the third cell. This expectation is observed in the simulations in section "Simulations" under "Periodic Forcing of Feed-Forward Chains" even when the forcing is spiking. A general theory for the study of periodic solutions occurring in a periodically forced system near a point of Hopf bifurcation is being developed in [34].

The efficiency of band-pass filters is often measured by the Q-factor. We introduce this concept in section "Q-factor" and show, in forced normal form systems, that the Q-factor scales linearly with the Hopf frequency. We verify this point with simulations and note the perhaps surprising observation that spiking forcing seems to lead to higher Q factors than does sinusoidal forcing.

In recent years many proposed models for the auditory system have relied on the periodic forcing of systems near points of Hopf bifurcation, and a general theory for periodic forcing of such systems would have direct application in these models. In particular, Hudspeth and collaborators [6, 7, 18, 19] have considered models for the cochlea that consist of periodically forced components that are tuned near Hopf bifurcation. We discuss these models in section "Cochlear Modeling". In particular, we note that an asymmetry in the experimentally obtained response curves from cochlea is consistent with what would have been obtained in the models if the cubic term in the Hopf bifurcation was complex. Biophysically based cochlear models are sufficiently complicated that asymmetry could be caused by many factors. To our knowledge, multiple solutions in the cochlear response curve have not been observed; nevertheless, in section "Hopf Models of the Auditory System", we speculate briefly on the possible meaning of such multiplicity. In section "Two-Frequency Forcing", we briefly discuss some expectations for two-frequency forcing that are based on simulations.

Synchrony-Breaking Hopf Bifurcations

We begin with a discussion of Hopf bifurcations that can be expected in systems of the form (1). The coordinates in $f(u, v, \lambda)$ are arranged so that u is the vector of internal cell phase space coordinates and v is the vector of coordinates in the coupling cell. Thus, the $k \times k$ matrix $\alpha = f_u(0,0,0)$ is the *linearized internal dynamics* and the $k \times k$ matrix $\beta = f_v(0,0,0)$ is the *linearized coupling* matrix. The Jacobian matrix for (1) is

$$J = \begin{bmatrix} \alpha + \beta & 0 & 0 \\ \beta & \alpha & 0 \\ 0 & \beta & \alpha \end{bmatrix}. \tag{2}$$

Synchrony-breaking bifurcations correspond to bifurcations where the center subspace of J does not intersect the synchrony subspace S. Note that for $y \in \mathbf{R}^k$

$$J\begin{bmatrix} y \\ y \\ y \end{bmatrix} = \begin{bmatrix} (\alpha+\beta)y \\ (\alpha+\beta)y \\ (\alpha+\beta)y \end{bmatrix}.$$

Thus, the matrix of $J|S$ is just $\alpha + \beta$ and a synchrony-breaking bifurcation occurs if some eigenvalue of J has zero real part and no eigenvalue of $\alpha + \beta$ has zero real part. We focus on the case where the synchrony-breaking bifurcation occurs from a stable synchronous equilibrium; that is, we assume:

(H1) All eigenvalues of $\alpha + \beta$ have negative real part.

The lower diagonal block form of J shows that the remaining eigenvalues of J are precisely the eigenvalues of α repeated twice. The generic existence of double eigenvalues would be a surprise were it not for the the restrictions placed on J by the network architecture pictured in Fig. 1. *Synchrony-breaking Hopf bifurcation* occurs when

(H2) α has simple purely imaginary eigenvalues $\pm\omega_{\mathrm{H}}\mathrm{i}$, where $\omega_{\mathrm{H}} > 0$, and all other eigenvalues of α have negative real part.

The real part restriction on the remaining eigenvalues just ensures that bifurcation occurs from a stable equilibrium. In fact, in this chapter, we only consider the case where the internal dynamics in each cell is two-dimensional, that is, we assume $k=2$.

It was observed in [12] and proved in [8] that generically synchrony-breaking Hopf bifurcations lead to families of periodic solutions $x^{\lambda}(t) = (0, x_2^{\lambda}(t), x_3^{\lambda}(t))$, where the cell 2 amplitude $|x_2^{\lambda}|$ grows at the expected rate of $\lambda^{\frac{1}{2}}$ and the cell 3 amplitude $|x_3^{\lambda}|$ grows at the *unexpected* rate of $\lambda^{\frac{1}{6}}$. Thus, near bifurcation, the amplitude of the third cell oscillation is much bigger than the amplitude of the second cell oscillation. An example of a periodic solution obtained by simulation of such a coupled-cell system near a point of synchrony-breaking Hopf bifurcation is given in Fig. 2.

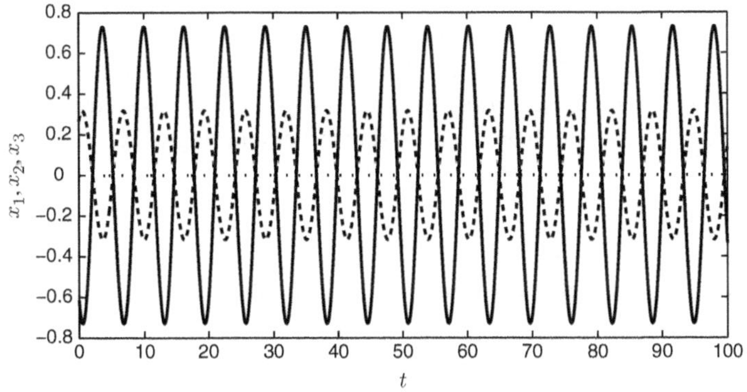

Fig. 2 Periodic solution near a synchrony-breaking Hopf bifurcation in the feed-forward chain. The first coordinate in each cell is plotted. Cell 1 is constant at 0 (*dotted curve*); cell 2 is the smaller signal (*dashed curve*); and cell 3 is the larger signal (*solid curve*) (see Figure 12 in [12])

The large growth in cell 3 can be understood as a result of resonance in a nonlinear system. To see this, observe that assumption (H1) implies that $x_1 = 0$ is a stable equilibrium for the first equation in (1). Thus, the asymptotic dynamics of the second cell is governed by the system of differential equations

$$\dot{x}_2 = f(x_2, 0, \lambda). \tag{3}$$

Assumption (H2) implies that the system (3) undergoes a standard Hopf bifurcation at $\lambda = 0$. In addition, we assume

(H3) (3) undergoes a generic supercritical Hopf bifurcation at $\lambda = 0$.

The consequence of assumption (H3) is that for $\lambda > 0$ the system (3) has a unique small amplitude stable periodic solution $x_2^\lambda(t)$ whose amplitude grows at the expected rate $\lambda^{\frac{1}{2}}$ and whose frequency is approximately ω_{H}.

It follows from (H3) that the asymptotic dynamics of the cell 3 system of differential equations reduces to the periodically forced system

$$\dot{x}_3 = f(x_3, x_2^\lambda(t), \lambda). \tag{4}$$

Since the system $\dot{x}_3 = f(x_3, 0, \lambda)$, which is identical to (3), is operating near a Hopf bifurcation with frequency ω_{H} and the periodic forcing itself has frequency near ω_{H}; it follows that (4) is being forced near resonance. Therefore, it is not surprising that the amplitude of cell 3 is greater than that of cell 2. It is not transparent, however, that cell 3 will undergo stable periodic oscillation and that the growth of the amplitude of that periodic solution will be $\lambda^{\frac{1}{6}}$. These facts are proved in [8, 12].

Remark 1. It is natural to ask what happens at synchrony-breaking Hopf bifurcation if extra cells are added to the feed-forward chain. The answer is simple: periodic solutions are found whose cell j amplitude grows at a rate that is the cube root of the growth in the amplitude of cell $j - 1$; that is, the amplitude of cell 4 grows at the rate $\lambda^{\frac{1}{18}}$, etc.

Remark 2. It was shown in [12] that the periodic solution in (1), that we have just described can itself undergo a secondary Hopf bifurcation to a quasiperiodic solution (see Fig. 3). This observation leads naturally to questions of frequency locking and Arnold tongues, which are discussed in Broer and Vegter [4].

Periodic Forcing of Feed-Forward Chains

An important characteristic of a network motif is that it performs some function [1]. Numerical simulations and experiments [26] with identical coupled circuits support the notion that the feed-forward chain can act as an efficient filter-amplifier, and

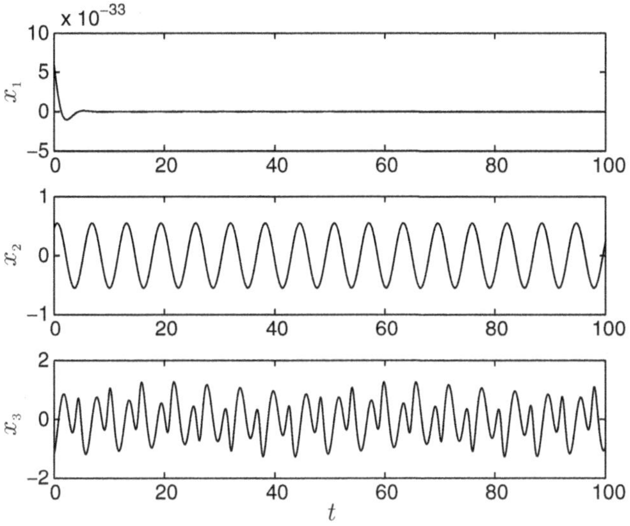

Fig. 3 ([12, Figure 13]) Quasiperiodic solution near a secondary bifurcation from a periodic solution obtained by synchrony-breaking Hopf bifurcation in (1)

Fig. 4 The feed-forward chain

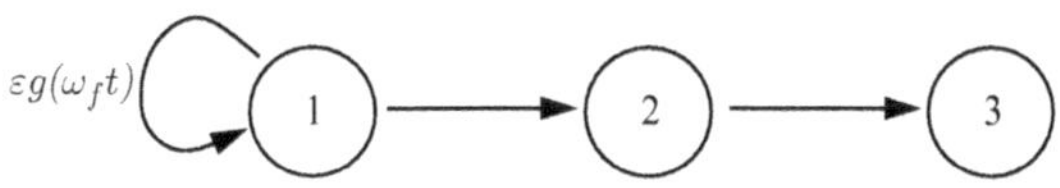

hence be a motif. However, the general theoretical results that support this assertion have been proved only under restrictive assumptions. It this section, we present numerical and experimental evidence in favor of the feed-forward chain being a motif.

We assume that the feed-forward chain in Fig. 1 is modified so that a small amplitude ε periodic forcing of frequency ω_f is added to the coupling in the first cell (see Fig. 4). We assume further that there is a bifurcation parameter λ for the internal cell dynamics that is tuned near a point of Hopf bifurcation. The question we address is: What are the amplitudes of the responses in cells 2 and 3 as a function of the forcing frequency ω_f? Due to resonance that response should be large when the forcing frequency is near the Hopf frequency and small otherwise.

Simulations

The general form of the differential equations for the periodically forced feed-forward chain is

$$\begin{aligned} \dot{x}_1 &= f(x_1, x_1 + \varepsilon g(\omega_f t), \lambda) \\ \dot{x}_2 &= f(x_2, x_1, \lambda) \\ \dot{x}_3 &= f(x_3, x_2, \lambda) \end{aligned} \tag{5}$$

where $x_j \in \mathbf{R}^k$ is the phase variable of cell j, $g : \mathbf{R} \to \mathbf{R}^k$ is a 2π periodic forcing function, and λ is a bifurcation parameter for a Hopf bifurcation.

To proceed with the simulations we need to specify f and g. Specifically, we assume that the cell dynamics satisfy:

(B1) The internal cell phase space is two-dimensional and identified with $\mathbf{C}$,
(B2) The internal cell dynamics is in truncated normal form for Hopf bifurcation,
(B3) The Hopf bifurcation is supercritical so that the origin is stable for $\lambda < 0$,
(B4) The cubic term in this normal form is real, and
(B5) The coupling is linear.

In addition, we normalize the cubic term to be -1 and simplify the coupling to be $-y$; that is, we assume

$$f(z, y, \lambda) = (\lambda + \omega_{\mathrm{H}} \mathrm{i} - |z|^2)z - y \tag{6}$$

where $z, y \in \mathbf{C}$. We assume that $\lambda < 0$ is small so that the internal dynamics is tuned near the point of Hopf bifurcation.

We will perform simulations with two types of forcing: simple harmonic and spiking (see Fig. 5). In simple harmonic forcing $g(t) = \mathrm{e}^{it}$. In spike forcing g is obtained numerically as a solution to the Fitzhugh–Nagumo equations

$$\begin{aligned} \dot{v} &= 6.4 - 120m^3h(v - 115) - 36n^4(v + 12) - 0.3(v - 10.5989), \\ \dot{n} &= \alpha_n(1 - n) - \beta_n n, \end{aligned} \tag{7}$$

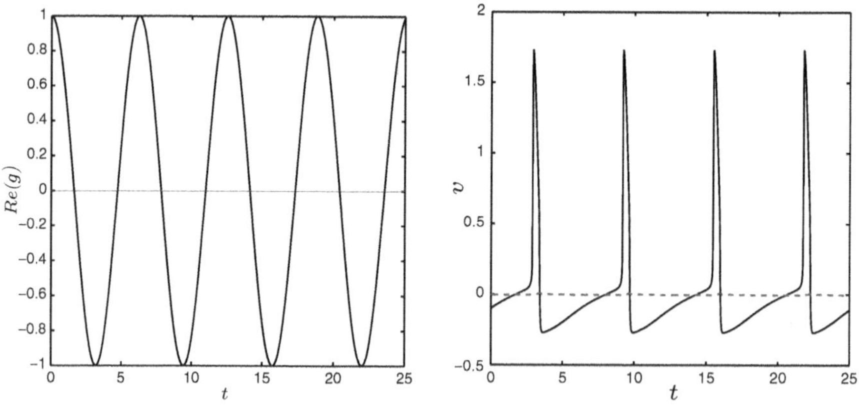

Fig. 5 First coordinate of time series of 2π-periodic forcings. (*Left*) Simple harmonic forcing $\cos t$. (*Right*) Spike forcing obtained from the Fitzhugh–Nagumo equations (7).

where

$$h = 0.8 - n \quad m = \alpha_m/(\alpha_m + \beta_m) \qquad \alpha_m = 0.1(25 - v)\mathrm{e}^{1-(25-v)/10}$$
$$\beta_m = 4\mathrm{e}^{-v/18} \quad \alpha_n = 0.01(10 - v)\mathrm{e}^{1-(10-v)/10} \quad \beta_n = 0.125\mathrm{e}^{-v/80}.$$

To obtain $g(t)$ we normalize $(v(t), n(t))$ so that it has mean zero and diameter 2. The first coordinate of the time series for the spiking forcing is shown in Fig. 5 (right). This time series is compared to simple harmonic forcing in Fig. 5 (left).

Recall that for sufficiently small ε, periodic forcing of amplitude ε, of a system of ODEs near a stable equilibrium, always produces an order ε periodic response. The frequency of the response equals that of the forcing. Hence, (H1) implies that the periodic output $x_1(t)$ from cell 1 will be of order ε with frequency ω_f.

The periodic output $x_1(t)$ is fed into cell 2. Although $\lambda < 0$ implies that the origin in the cell 2 equation is stable, the fact that λ is near a bifurcation point implies that the rate of attraction of that equilibrium will be small. Thus, only if ε is very small will the periodic output of cell 2 be of order ε.

Because of resonance, we expect that the amplitude of $x_2(t)$ will be large when ω_f is near ω_{H}. Indeed, Kern and Stoop [23] observe that when the differential equation f is (6) with $\varepsilon > |\lambda|^{\frac{3}{2}}$, then the growth of the periodic output will be of order $\varepsilon^{\frac{1}{3}}$. We revisit this point in section "Periodic Forcing near Hopf Bifurcation" when we discuss some of the theory behind the amplification. Moreover, we can expect the amplitude of $x_3(t)$ to be even larger in this range; that is, we can expect the amplitude of cell 3 to grow at the rate $\varepsilon^{\frac{1}{9}}$.

To illustrate these statements we perform the following simulation. Fix $\varepsilon > 0$ and $\lambda < 0$, and plot the amplitudes of the periodic states in cells 1, 2, and 3 as a function of the forcing frequency ω_f. The results are given in Fig. 6. Note that the input forcing is amplified when $\omega_f \approx \omega_{\mathrm{H}}$ and that the qualitative results do not depend particularly on the form of g. In particular, note that the response curves are symmetric in $\omega_f = \omega_{\mathrm{H}}$.

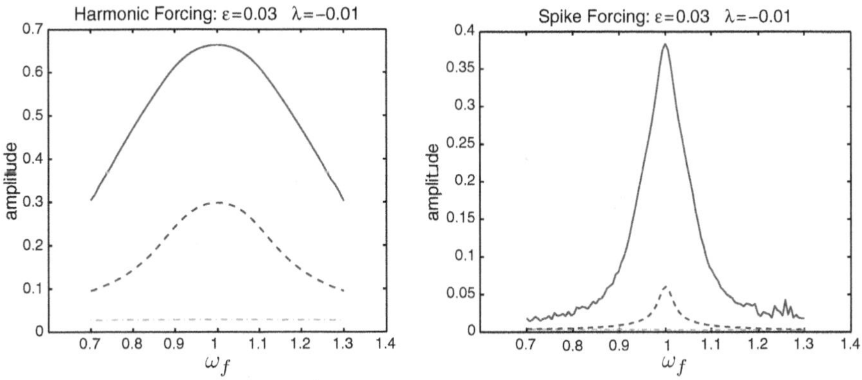

Fig. 6 Amplitudes of cells 1 (*dotted curve*), 2 (*dashed curve*) and 3 (*solid curve*) as a function of forcing frequency; $\lambda = -0.01$, $\varepsilon = 0.03$, $\omega_{\mathrm{H}} = 1$, $0.7 \leq \omega_f \leq 1.3$. (*Left*) simple harmonic forcing; (*right*) spike forcing

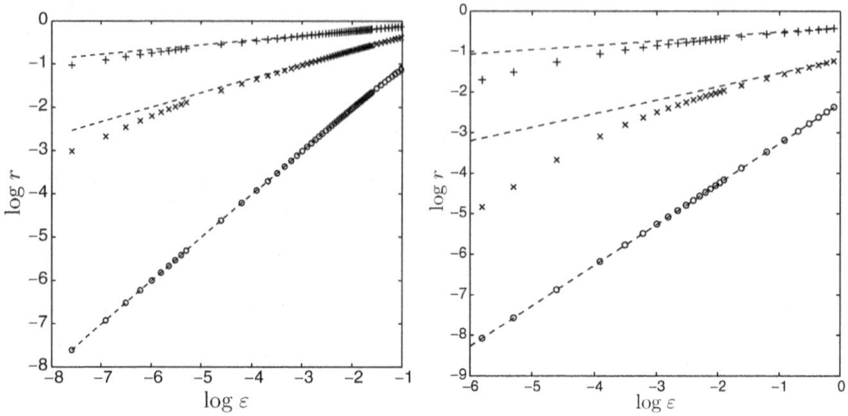

(a) Harmonic forcing **(b)** Spiking forcing

Fig. 7 Log–log plot of amplitudes of response in cells 1 (○), 2 (×), and 3 (+), as a function of ε, for harmonic forcing and spiking forcing. Also shown are lines of slope 1, 1/3, and 1/9 (from *bottom* to *top*). Parameters are $\omega_f = \omega_H = 1$, $\lambda = -0.01$, (**a**) $0.0005 < \varepsilon < 0.36$, (**b**) $0.0025 < \varepsilon < 0.9$

In Fig. 7, we show the amplitudes of the responses in the three cells as a function of ε, for both harmonic and spiking forcing. In both cases we see a similar pattern of growth rate of amplitude. The amplitude in the first cell grows linearly with ε. In the second cell, as ε increases, the growth rate tends toward "cube root," that is $r \sim \varepsilon^{1/3}$. Similarly in the third cell, for large enough ε, we see $r \sim \varepsilon^{1/9}$. As ε increases from zero, there is a transition region into these regimes. This appears to occur for different values of ε for the different types of forcing. However, since it is not clear how one should define the "amplitude" of the spiking forcing, and we have arbitrarily chosen to set the diameters of the two forcings equal, this is not unexpected. We investigate the transition region for harmonic forcing more fully in section "Periodic Forcing near Hopf Bifurcation".

Experiments

McCullen et al. [26] performed experiments on a feed-forward chain of coupled nonlinear modified van der Pol autonomous oscillators. Even though the McCullen experiments were performed with a system that was not in normal form, the results conform well with the simulations. The responses to a simple harmonic forcing with varying frequency are shown in Fig. 8. Note the similarity with the simulation results in Fig. 6. The plot on the right of Fig. 8 shows the expected cube root scaling in the amplitude of cell 3 as a function of the amplitude cell 2.

Recall that a band-pass filter allows signals in a certain range of frequencies to pass, whereas signals with frequencies outside this range are attenuated. As we

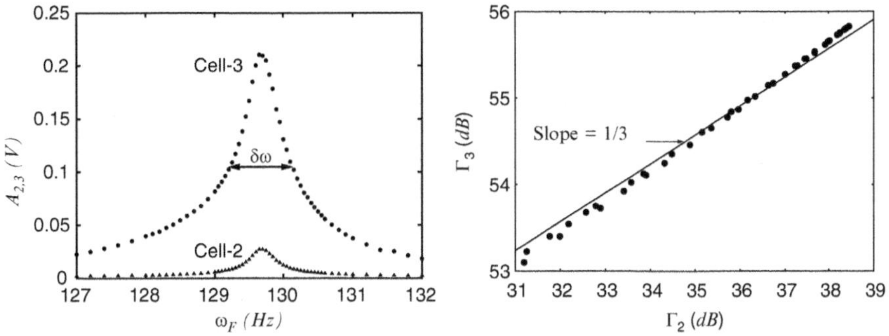

Fig. 8 (*Left*) [26, Fig. 2]: Cells 2 and 3 amplitudes as a function of forcing frequency in oscillator experiment. (*Right*) [26, Fig. 5]: Log–log plot of amplitudes of oscillations in cells 2 and 3 as a function of forcing frequency near Hopf bifurcation point

have seen, the feed-forward chain can act as a kind of band-pass filter by exciting small amplitude signals to an amplitude larger than some threshold only if the input frequency is near enough to the Hopf frequency. To determine the frequency of an incoming sound, the human auditory system should have the capability of acting like a band-pass filter. As noted, several authors have suggested that the structure of outer hair cells on the basilar membrane is tuned to be a linear array of coupled cells each tuned near a Hopf bifurcation point but at different frequencies (see "Cochlear Modeling").

Periodic Forcing near Hopf Bifurcation

In section "Periodic Forcing of Feed-Forward Chains," we discussed numerical simulations and experiments which suggest that the amplification results for forced feed-forward chains near normal form Hopf bifurcation with sinusoidal forcing appear to hold even when the forcing is not sinusoidal or the system is not in normal form. These observations motivate the need for a general theory of periodic forcing of systems near Hopf bifurcation. In this section, we make a transition from studying feed-forward chains to the simpler situation of periodic forcing of systems near Hopf bifurcation.

In particular we show that when the cubic term in Hopf bifurcation has a sufficiently large complex part, then multiple periodic solutions will occur as ω_f is varied near ω_{H}. The importance of the complex part of the cubic term in different aspects of forced Hopf bifurcation systems was noted previously by Wang and Young [33]. The existence of regions of multiplicity motivates the need for a general theory of periodic forcing of systems near Hopf bifurcation. A detailed study of these forced systems, based on equivariant singularity theory, is being developed in [34].

Simulations

As in section "Periodic Forcing of Feed-Forward Chains," we assume that the system we are forcing is in truncated normal form. More precisely, we assume that this system satisfies (B1–B3), but we do not assume that the cubic term is real. We also assume that the forcing is additive so that the equation is

$$\dot{z} = (\lambda + \mathrm{i}\omega_{\mathrm{H}})z + c|z|^2 z + \varepsilon \mathrm{e}^{\mathrm{i}\omega_f t}, \tag{8}$$

where $\lambda < 0$ and $\varepsilon > 0$ are small, $c = c_R + ic_I$ and $c_R < 0$. We can rescale z to set $c_R = -1$. The scaled equation has the form

$$\dot{z} = (\lambda + i\omega_{\mathrm{H}})z + (-1 + i\gamma)|z|^2 z + \varepsilon \mathrm{e}^{\mathrm{i}\omega_f t}, \tag{9}$$

where $\gamma = -c_I/c_R$.

We show the results of simulation of (9) when $\gamma = 0$ (the case that is most often analyzed in the literature) and when $\gamma = 10$. Both simulations show amplification of the forcing when $\omega_f \approx \omega_{\mathrm{H}} = 1$. However, when $\gamma \neq 0$, we find that there can be bistability of periodic solutions. Figure 9 (right) shows results of two sets of simulations, with different initial conditions. For a range of ω_f, there are two stable solutions with different amplitude $r = |z|$.

Asymmetry and Multiplicity in Response Curve

It is well known that the normal form for Hopf bifurcation has phase shift symmetry and hence that the normal form equations can be solved in rotating coordinates.

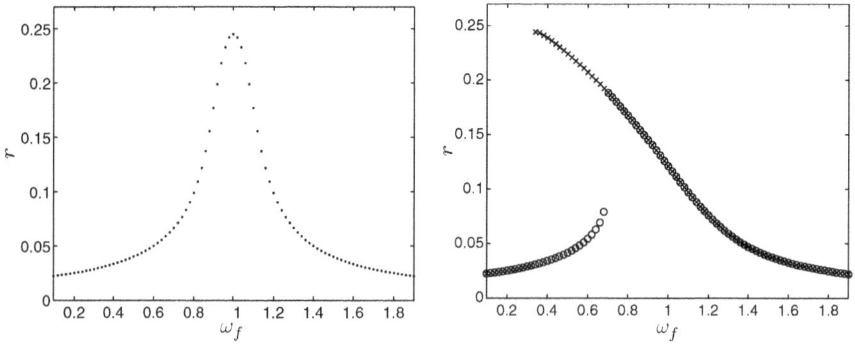

Fig. 9 Amplitudes of solutions as function of Hopf frequency of (9), with $\omega_{\mathrm{H}} = 1, \lambda = -0.0218$, $\varepsilon = 0.02$. (*Left*) $\gamma = 0$. (*Right*) $\gamma = 10$; The o's and $\times$'s indicate two separate sets of simulations with different initial conditions. For $0.35 < \omega_f < 0.7$, there are two stable solutions

Rotating coordinates can also be used in the forced system. Write (9) in rotating coordinates $z = ue^{i(\omega_f t - \theta)}$, where θ is an arbitrary phase shift, to obtain

$$\dot{u} = (\lambda + \mathrm{i}\omega)u + (-1 + \mathrm{i}\gamma)|u|^2 u - \varepsilon e^{\mathrm{i}\theta}$$

where $\omega = \omega_{\mathrm{H}} - \omega_f$. Note that stationary solutions in u, for any θ, correspond to periodic solutions $z(t)$ with frequency ω_f. We set $\dot{u} = 0$ and solve

$$g(u) \equiv (\lambda + \mathrm{i}\omega)u + (-1 + \mathrm{i}\gamma)|u|^2 u = \varepsilon e^{\mathrm{i}\theta} \tag{10}$$

for any u and θ. Note that finding a solution to (10) for some θ is equivalent to finding u such that

$$|g(u)|^2 = \varepsilon^2. \tag{11}$$

Note also that

$$|g(u)|^2 = (\lambda^2 + \omega^2)|u|^2 + 2(\omega\gamma - \lambda)|u|^4 + (1 + \gamma^2)|u|^6.$$

That is, $|g(u)|^2$ depends only on $|u|^2$.

Set $\delta = \varepsilon^2$ and $R = |u|^2$. We can write (11) as

$$G(R; \lambda, \omega, \gamma, \delta) \equiv (1 + \gamma^2)R^3 + 2(\omega\gamma - \lambda)R^2 + (\lambda^2 + \omega^2)R - \delta = 0. \tag{12}$$

Since $G(R; \lambda, \omega, \gamma, \delta)$ is invariant under the parameter symmetry $(\omega, \gamma) \to (-\omega, -\gamma)$, we can assume $\gamma \geq 0$. Additionally, if $\gamma = 0$, then $G(R; \lambda, \omega, \gamma, \delta)$ is invariant under the parameter symmetry $\omega \to -\omega$.

Fix $\lambda < 0$, $\delta > 0$ and $\gamma \geq 0$. We seek to answer the following question. Determine the bifurcation diagram consisting of solutions $R > 0$ to (12) as ω varies. Note that variation of ω corresponds to variation of either ω_f or ω_{H} in the original forced equation (8). In Fig. 10, we plot sample bifurcation diagrams of (12) for three values of γ. We see that as γ is increased, asymmetry occurs in the bifurcation diagram, ultimately leading to multiple solutions.

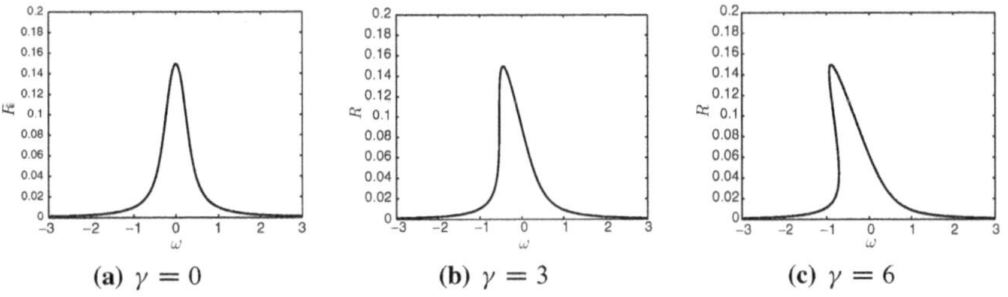

(a) $\gamma = 0$ **(b)** $\gamma = 3$ **(c)** $\gamma = 6$

Fig. 10 Bifurcation diagrams of solutions to (12) for varying γ. As γ is increased, the response curve becomes asymmetric, and as it is increased further, for some values of ω there are multiple solutions. Parameters used are $\delta = 0.01$, $\lambda = -0.109$

We use bifurcation theory, in particular hysteresis points, to prove that multiplicity occurs for arbitrarily small $\lambda < 0$ and $\delta > 0$. Hysteresis points correspond to points where the bifurcation diagram has a vertical cubic order tangent and such points are defined by

$$G = G_R = G_{RR} = 0 \quad \text{and} \quad G_\omega \neq 0 \neq G_{RRR}$$

See [13, Proposition 9.1, p. 94]. Multiplicity of solutions occurs if variation of γ leads to a universal unfolding of the hysteresis point. It is shown in [13, Proposition 4.3, p. 136] that γ is a universal unfolding parameter if and only if

$$\det \begin{pmatrix} G_\omega & G_{\omega R} \\ G_\gamma & G_{\gamma R} \end{pmatrix} \neq 0. \tag{13}$$

In this application of singularity theory we will need the following:

$$G = (1+\gamma^2)R^3 + 2(\omega\gamma - \lambda)R^2 + (\lambda^2 + \omega^2)R - \delta = 0, \tag{14}$$

$$G_R = 3(1+\gamma^2)R^2 + 4(\omega\gamma - \lambda)R + (\lambda^2 + \omega^2) = 0, \tag{15}$$

$$G_{RR} = 6(1+\gamma^2)R + 4(\omega\gamma - \lambda) = 0, \tag{16}$$

$$G_{RRR} = 6(1+\gamma^2) > 0, \tag{17}$$

$$G_\omega = 2R(\gamma R + \omega) \neq 0, \tag{18}$$

and

$$G_{\omega R} = 2(2\gamma R + \omega), \tag{19}$$

$$G_\gamma = 2R^2(\gamma R + \omega), \tag{20}$$

$$G_{\gamma R} = 2R(3\gamma R + 2\omega). \tag{21}$$

Note that the determinant in (13) is just G_ω^2, which is nonzero at any hysteresis point. Hence, variation of γ will always lead to a universal unfolding of a hysteresis point and to multiple solutions for fixed ω.

Theorem 1 *For every small $\lambda < 0$ and $\delta > 0$ there exists a unique hysteresis point of G at $R = R_c(\delta,\lambda)$, $\omega = \omega_c(\delta,\lambda)$, $\gamma = \gamma_c(\delta,\lambda)$. Moreover,*

$$\omega_c(\delta,0) = -\sqrt{3}(2\delta)^{\frac{1}{3}} \qquad \gamma_c(\delta,0) = \sqrt{3} \qquad R_c(\delta,0) = \left(\frac{\delta}{4}\right)^{\frac{1}{3}}. \tag{22}$$

Proof. We assert that (14)–(16) define R_c, ω_c, γ_c uniquely in terms of δ and λ. Specifically, we show that

$$R_c(\delta,\lambda) = \left(\frac{\delta}{1+\gamma_c^2}\right)^{\frac{1}{3}} > 0. \tag{23}$$

$$\gamma_c(\delta, \lambda) = \frac{\lambda + \sqrt{3}\omega_c}{\omega_c - \sqrt{3}\lambda}. \tag{24}$$

Moreover, let

$$p(\omega) = \omega^3 - \sqrt{3}\lambda\omega^2 + \lambda^2\omega - \sqrt{3}\lambda^3. \tag{25}$$

Then $\omega_c(\delta, \lambda)$ is the unique solution to the equation

$$p(\omega_c) = -6\sqrt{3}\delta. \tag{26}$$

We can compute these quantities explicitly when $\lambda = 0$. Specifically, (26) reduces $\omega_c^3 = -6\sqrt{3}\delta$. It is now straightforward to verify (22).

To verify (24) combine (15) and (16) to yield

$$R = -\frac{2(\omega\gamma - \lambda)}{3(1 + \gamma^2)} \quad \text{and} \quad R^2 = \frac{\lambda^2 + \omega^2}{3(1 + \gamma^2)}. \tag{27}$$

More precisely, the first equation is obtained by solving $G_{RR} = 0$ and the second by solving $RG_{RR} - G_R = 0$. Multiplying the first equation in (27) by R and substituting for R^2 in the second equation yields

$$2(\omega\gamma - \lambda)R = -(\lambda^2 + \omega^2). \tag{28}$$

Substituting (28) into (14) yields

$$R^3 = \frac{\delta}{1 + \gamma^2} \tag{29}$$

thus verifying (23).

We eliminate R from (27) in two ways, obtaining

$$-\frac{8(\omega\gamma - \lambda)^3}{27(1 + \gamma^2)^2} = \delta \quad \text{and} \quad \frac{4(\omega\gamma - \lambda)^2}{3(1 + \gamma^2)} = \lambda^2 + \omega^2 \tag{30}$$

To verify the first equation in (30), cube the first equation in (27) and use (29) to substitute for R^3. To verify the second equation, square the first equation in (27) and use the second equation in (27) to substitute for R^2.

Next we derive (24). Rewrite the second equation in (30) to obtain

$$\omega^2(\gamma^2 - 3) - 8\omega\gamma\lambda + \lambda^2(1 - 3\gamma^2) = 0, \tag{31}$$

which can be factored as

$$\left(\gamma(\omega - \sqrt{3}\lambda) - (\lambda + \sqrt{3}\omega)\right)\left(\gamma(\omega + \sqrt{3}\lambda) - (\lambda - \sqrt{3}\omega)\right) = 0. \tag{32}$$

Thus, potentially, there are two solutions for γ_c; namely,

$$\gamma = \frac{\lambda + \sqrt{3}\sigma\omega}{\omega - \sigma\sqrt{3}\lambda}, \tag{33}$$

where $\sigma = \pm 1$, depending on which bracket is chosen in (32).

Next use (32) to show that

$$\omega\gamma - \lambda = \sqrt{3}\frac{\lambda^2 + \omega^2}{\sigma\omega - \sqrt{3}\lambda}. \tag{34}$$

Squaring the second equation in (30) and substituting the first yields

$$(\lambda^2 + \omega^2)^2 = 6(\omega\gamma - \lambda)\frac{8(\omega\gamma - \lambda)^3}{27(1 + \gamma^2)^2} = -6\delta(\omega\gamma - \lambda). \tag{35}$$

Next use (34) to eliminate $\omega\gamma - \lambda$. A short calculation leads to

$$(\omega^2 + \lambda^2)(\sigma\omega - \sqrt{3}\lambda) = -6\sqrt{3}\delta. \tag{36}$$

Since $\delta > 0$, we must have $\sigma\omega - \sqrt{3}\lambda < 0$, or $\sigma\omega < \sqrt{3}\lambda < 0$.

We claim that for $\gamma \geq 0$, we must choose $\sigma = +1$. Since $\lambda < 0$ and $\sigma\omega < 0$, the numerator of (33) is negative. If $\sigma = -1$, then the denominator of (33) is $\sigma(\sigma\omega - \sqrt{3}\lambda) > 0$, since $\sigma\omega - \sqrt{3}\lambda < 0$. Hence $\gamma < 0$. We thus write $\sigma = +1$ and verify (24).

We claim that given δ and λ there is a unique solution ω to (36). Observe that the cubic polynomial in ω on the left side of (36) is (25). Since

$$p'(\omega) = (\sqrt{3}\omega - \lambda)^2, \tag{37}$$

$p(\omega)$ is monotonic; and there is a unique solution ω_c, as claimed.

Finally, we must show that G_ω is nonzero at R_c, ω_c, γ_c. We do this by showing that

$$\gamma_c(\delta, \lambda)R_c(\delta, \lambda) + \omega_c(\delta, \lambda) < 0.$$

By cubing the first equation in (30) and dividing by the square of the second equation, we can eliminate the $\omega\gamma - \lambda$ factor and show that

$$1 + \gamma_c^2 = \frac{(\lambda^2 + \omega_c^2)^3}{27\delta^2}.$$

Using this alongside (29) we write R_c as

$$R_c = \frac{\delta^{1/3}}{(1 + \gamma_c^2)^{1/3}} = \delta^{1/3}\frac{3\delta^{2/3}}{\lambda^2 + \omega_c^2} = \frac{3\delta}{\lambda^2 + \omega_c^2}.$$

Then use (34) to substitute for γ_c to find

$$R_c\gamma_c = \frac{3\delta}{\lambda^2+\omega_c^2}\frac{\lambda+\sqrt{3}\omega_c}{\omega_c-\sqrt{3}\lambda} = -\frac{\lambda+\sqrt{3}\omega_c}{2\sqrt{3}},$$

where we have used (36) to simplify the denominator. Therefore

$$R_c\gamma_c+\omega_c = \frac{\sqrt{3}\omega_c-\lambda}{2\sqrt{3}} = \frac{1}{2}\left(\omega_c-\frac{1}{\sqrt{3}}\lambda\right) < \frac{1}{2}\left(\omega_c-\sqrt{3}\lambda\right) < 0,$$

where the penultimate inequality follows because $\frac{1}{\sqrt{3}}\lambda > \sqrt{3}\lambda$. □

Scalings of Solution Amplitudes

Kern and Stoop [23] and Eguíluz et. al [7] consider the system (9) with $\gamma = 0$, and observe that there are regions of parameter space in which the input signal (forcing) is amplified – that is, the solution $z(t) = re^{i(\omega t+\theta)}$ has an amplitude r which scales like $\varepsilon^{1/3}$.

Specifically, Eguíluz et. al [7] specialize (9) exactly at the bifurcation point ($\lambda = 0$) and show that the solution has an amplitude $r \sim \varepsilon^{1/3}$ when $\omega_{\mathrm{H}} = \omega_f$. Away from resonance, ($\omega_{\mathrm{H}} \neq \omega_f$), they show that for ε small enough (small enough forcing), the response $r \sim \varepsilon/|\omega_{\mathrm{H}} - \omega_f|$. Kern and Stoop [23] consider forcing exactly at the Hopf frequency (i.e., $\omega_{\mathrm{H}} = \omega_f$), and show that the solution has an amplitude $r \sim \varepsilon^{1/3}$ when $\lambda = 0$, and when $\lambda < 0$ the amplitude $r \sim \varepsilon/|\lambda|$.

In the following, we make precise the meaning of "cube-root growth," and additionally, do not assume $\gamma = 0$. We show that, in some parameter regime, the response r can be bounded between two curves, specifically, that

$$\left(\frac{\varepsilon}{\sqrt{2}}\right)^{\frac{1}{3}} < r < \varepsilon^{\frac{1}{3}},$$

that is, r lies between two lines in log–log plots. In Fig. 11, we show the result of numerical simulations of (9) as ε is varied along with the two lines given above. For large enough ε, the response amplitude lies between these lines. Compare also with Fig. 7 – we could perform a similar process here of bounding the amplitudes of response to determine regions of different growth rates.

The width of this region is in some sense arbitrary – choosing a different lower boundary would merely result in different constants in the proof of the lemmas given below. We consider here only the scaling of the amplitude of the maximum response (as a function of ω), but note that our calculations can easily be extended beyond this regime.

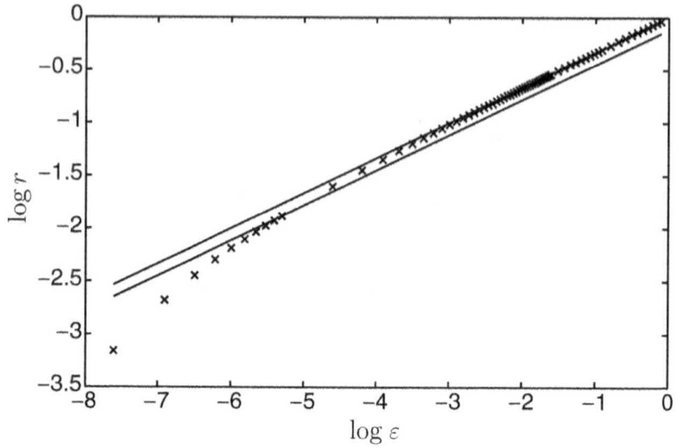

Fig. 11 For "cube root growth," the amplitude r is bounded by two straight lines in a plot of $\log r$ vs. $\log \varepsilon$

For consistency with the previous section, we work with $R = r^2$ and $\delta = \varepsilon^2$; it is clear that similar relations will hold between R and δ. Recall

$$G(R;\lambda,\omega,\gamma,\delta) = (1+\gamma^2)R^3 + 2(\omega\gamma-\lambda)R^2 + (\lambda^2+\omega^2)R - \delta,$$

where $\omega = \omega_{\mathrm{H}} - \omega_f$. The amplitude of solutions is given by $G(R;\lambda,\omega,\gamma,\delta) = 0$. Consider the amplitude as ω is varied, then the maximum response R occurs when $G_\omega = 0$, that is, at $\omega = -\gamma R$, which is nonzero for $\gamma \neq 0$. At $\gamma = 0$, the response curve is symmetric in ω, and so the maximum must occur at $\omega = 0$.

Write

$$G(R;\lambda,\omega,\gamma,\delta) = \Gamma(R;\lambda,\omega,\gamma) - \delta, \tag{38}$$

so the amplitude squared of the response, R, is related to the amplitude squared of the forcing, δ, by

$$\Gamma(R;\lambda,\omega,\gamma) = \delta.$$

Consider the function $\Gamma(R;\lambda,\omega,\gamma)$ evaluated at the value of ω for which the maximum response occurs, that is, compute

$$\Gamma(R;\lambda,-\gamma R,\gamma) = R^3 - 2\lambda R^2 + \lambda^2 R,$$

which turns out to be independent of γ, and so write $\mathcal{G}(R;\lambda) \equiv \Gamma(R;\lambda,-\gamma R,\gamma)$. Moreover, since $\lambda < 0$, $\mathcal{G}(R;\lambda)$ is monotonically increasing in R and hence invertible. Therefore, the response curve has a unique maxima for all γ.

Write $\mathcal{H}(\delta;\lambda) \equiv \mathcal{G}(R;\lambda)^{-1}$. Then for given δ, λ, the maximum R satisfies $R = \mathcal{H}(\delta;\lambda)$.

Observe that for $|R|$ small, $\mathcal{G}(R;\lambda) \approx \lambda^2 R$, and for $|R|$ large, $\mathcal{G}(R;\lambda) \approx R^3$. Therefore, for $|\delta|$ small we expect $R = \mathcal{H}(\delta;\lambda) \approx \delta/\lambda^2$, and for $|\delta|$ large, $R = \mathcal{H}(\delta;\lambda) \approx \delta^{1/3}$. We make these statements precise in the following lemmas.

Lemma 1. *If* $|\lambda| < 0.33\,\delta^{1/3}$, *then*

$$\left(\frac{\delta}{2}\right)^{1/3} < \mathcal{H}(\delta;\lambda) < \delta^{1/3}.$$

Remark 3. The constant 0.33 in the statement of Lemma 1 can be replaced by k_1, the unique positive root of $y^2 + 2^{2/3}y - 2^{-2/3} = 0$. The hypothesis in this lemma can then read $|\lambda| < k_1\delta^{1/3}$. In the proof we use k_1 rather than 0.33.

Proof. Since $\mathcal{G}(R;\lambda)$ is monotonic increasing, we need to show that

$$\mathcal{G}\left(\left(\frac{\delta}{2}\right)^{1/3};\lambda\right) < \delta < \mathcal{G}\left(\delta^{1/3};\lambda\right).$$

Since $\lambda < 0$ and $\delta > 0$ the second inequality follows from

$$\mathcal{G}\left(\delta^{1/3};\lambda\right) = \delta - 2\lambda\delta^{2/3} + \lambda^2\delta^{1/3} > \delta.$$

For the first inequality, we have

$$\mathcal{G}\left(\left(\frac{\delta}{2}\right)^{1/3};\lambda\right) = \frac{\delta}{2} - 2^{1/3}\lambda\delta^{2/3} + \lambda^2\frac{\delta^{1/3}}{2^{1/3}}.$$

We have assumed $-\lambda < k_1\delta^{1/3}$; so

$$\mathcal{G}\left(\left(\frac{\delta}{2}\right);\lambda\right) < \frac{\delta}{2} + 2^{1/3}k_1\delta + k_1^2\frac{\delta}{2^{1/3}} = \frac{1}{2^{1/3}}\left(\frac{1}{2^{2/3}} + 2^{2/3}k_1 + k_1^2\right)\delta = \delta,$$

since $k_1^2 + 2^{2/3}k_1 = 2^{-2/3}$. □

Lemma 2. *If* $|\lambda| > 1.06\,\delta^{1/3}$, *then*

$$\frac{\delta}{2\lambda^2} < \mathcal{H}(\delta;\lambda) < \frac{\delta}{\lambda^2}.$$

Remark 4. The constant 1.06 in the statement of Lemma 2 can be replaced by k_2, where $y = k_2^3$ is the unique positive root of $4y^2 - 4y - 1 = 0$. The hypothesis in this lemma can then read $|\lambda| > k_2\delta^{1/3}$. In the proof we use k_2 rather than 1.06.

Proof. Since $\mathcal{G}(R;\lambda)$ is monotonic increasing in R, we have to show that

$$\mathcal{G}\left(\frac{\delta}{2\lambda^2};\lambda\right) < \delta < \mathcal{G}\left(\frac{\delta}{\lambda^2};\lambda\right).$$

Since $\lambda < 0$ and $\delta > 0$ the second inequality follows from

$$\mathcal{G}\left(\frac{\delta}{\lambda^2};\lambda\right) = \frac{\delta^3}{\lambda^6} - 2\frac{\delta^2}{\lambda^3} + \delta > \delta.$$

For the first inequality, we have

$$\mathcal{G}\left(\frac{\delta}{2\lambda^2};\lambda\right) = \frac{\delta^3}{8\lambda^6} - \frac{\delta^2}{2\lambda^3} + \frac{\delta}{2}.$$

We assumed $\lambda^6 > k_2^6\delta^2$ and $-\lambda^3 > k_2^3\delta$. So

$$\mathcal{G}\left(\frac{\delta}{2\lambda^2};\lambda\right) < \frac{\delta^3}{8k_2^6\delta^2} + \frac{\delta^2}{2k_2^3\delta} + \frac{\delta}{2} = \frac{1}{8k_2^6}(4k_2^6 + 4k_2^3 + 1)\delta = \delta,$$

since $4k_2^3 + 1 = 4k_2^6$. □

Remark 5. Note that $k_1 \approx 0.33$ and $k_2 \approx 1.06$, so $k_1 < k_2$. It follows that the region of linear amplitude (ε) growth is very small, whereas the region of cube root growth is quite large. In Fig. 12, we illustrate this point by graphing the curves that separate the regions; namely $\varepsilon = (\lambda/k_1)^{3/2}$ (dashed curve for cube root growth) and $\varepsilon = (\lambda/k_2)^{3/2}$ (continuous curve for linear growth). Since $\frac{1}{k_2} < \frac{1}{k_1}$ the linear and cube root growth regions are disjoint.

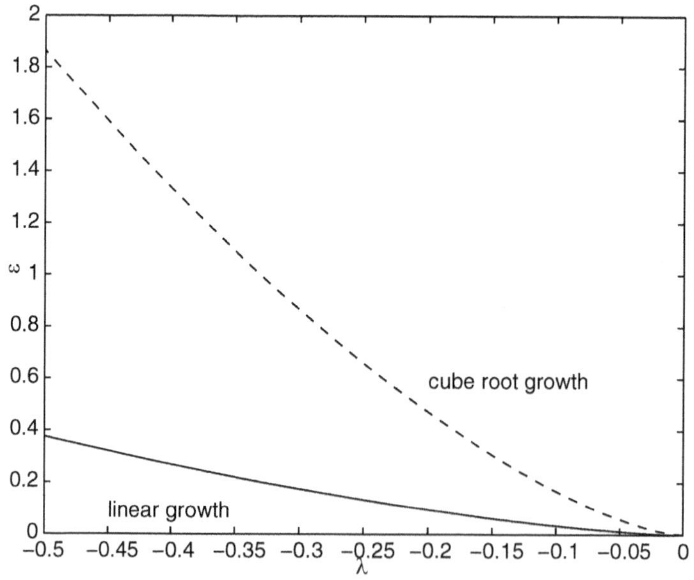

Fig. 12 Regions of linear and cube root growth in the λ-ε plane

Remark 6. The maximum amplitude r satisfies $\mathcal{G}\left(r^2;\lambda\right)=\varepsilon^2$, where $\mathcal{G}\left(R;\lambda\right)$ is an increasing function of R. Hence r increases as ε increases. Recall that the maximum r occurs at $\omega = -\gamma r^2$. Hence with fixed parameters $\lambda < 0$, $\gamma \neq 0$, the forcing frequency for which the maximum amplitude occurs, varies as the amplitude of the forcing (ε) increases.

Remark 7. It is simple to extend this type of reasoning into regions away from the maximum amplitude of response, to find linear growth rates for ω far from the maximum response. However, the algebra is rather messy and so we do not include the details here.

Q-Factor

Engineers use the *Q-factor* to measure the efficiency of a band-pass filter. The Q-factor is nondimensional and defined by

$$Q=\frac{\omega_{\max}}{\mathrm{d}\omega},$$

where $\omega_{\max}$ is the forcing frequency at which maximum response amplitude $r_{\max}$ occurs, and $\mathrm{d}\omega$ is the width of the amplitude response curve when it has half the maximum height. In Fig. 13, we give a schematic of a response curve and show how Q is calculated.

The larger the Q, the better the filter. Quantitatively, there is a curious observation. It can be shown that for our Hopf normal form, Q varies linearly with ω_{H}. The response curve in ω–r space is defined implicitly by (12) (recall $R=r^2$, $\delta=\varepsilon^2$). This curve depends only on ω (not on ω_f or ω_{H} independently). Therefore,

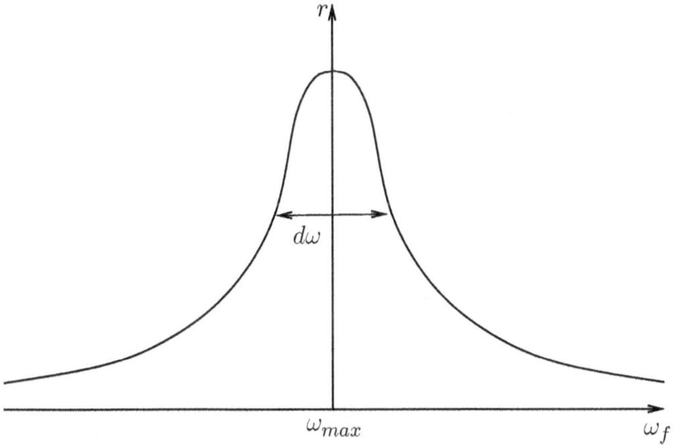

Fig. 13 Schematic of a response curve as ω_f is varied, for fixed ω_{H}. The Q-factor is $\omega_{\max}/\mathrm{d}\omega$

as ω_{H} varies, the curve will be translated, but its shape will not change. Hence the width of the curve, $d\omega$, is independent of ω_{H}.

As shown in section "Scalings of Solution Amplitudes," the maximum response $r_{\max}$ is independent of ω_{H}. The position of the maximum response is given by $\omega = -\gamma r_{\max}^2$, or $\omega_f = \omega_{\mathrm{H}} + \gamma r_{\max}^2$. Hence

$$Q = \frac{\omega_{\mathrm{H}} + \gamma r_{\max}^2}{d\omega}$$

and so depends linearly on ω_{H}.

In addition, we find from simulations that for any given ω_{H} the Q-factor of the system is better for spiking forcing than for sinusoidal forcing. Figure 14 shows results of simulations of the three cell feed-forward network from section "Simulations" under "Periodic Forcing of Feed-Forward Chains" using sinusoidal and spiking forcing. From these figures, we see two, perhaps surprising, results. First, that the Q-factor for spiking forcing is almost five times higher than that of sinusoidal forcing. Second, that for both forcings, the Q-factor for cell 3 is less than that of cell 2.

We explain the first observation by the following analogy. Consider the limit of very narrow spiking forcing, on a damped harmonic oscillator, for example, pushing a swing. Resonant amplification can only be achieved if the frequency of the forcing and the oscillations exactly match. If they are slightly off, then the forcing occurs at a time when the swing is not in the correct position and so only a small amplitude solution can occur.

We further note that although the output from a cell receiving spiking forcing is not sinusoidal, it is closer to sinusoidal than the input. That is, as the signal proceeds along the feed-forward chain, at each cell the output is closer to sinusoidal than the last. Combining this observation with the first explains why the Q-factor of cell 3 should be less than that for cell 2.

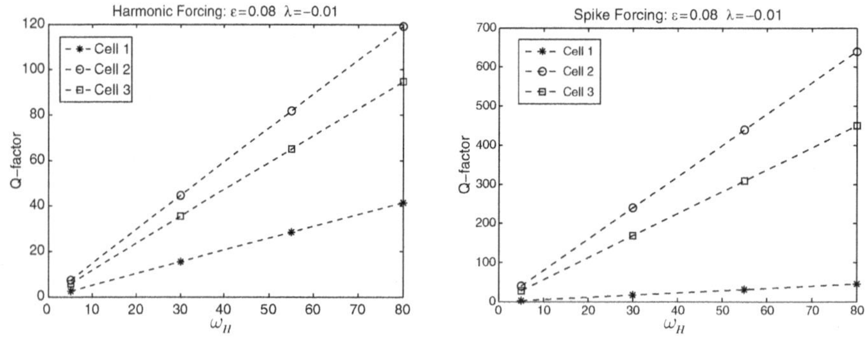

Fig. 14 The figures show the Q factor as ω_{H} is varied for sinusoidal forcing (*left*) and spiking forcing (*right*) for each cell in the feed-forward network from section "Simulations" under "Periodic Forcing of Feed-Forward Chains". In these simulations $\lambda = -0.01$ and $\varepsilon = 0.08$

Cochlear Modeling

The cochlea in the inner ear is a fluid filled tube divided lengthwise into three chambers. The basilar membrane (BM) divides two of these chambers and is central to auditory perception. Auditory receptor cells, or hair cells, sit on the BM. Hair bundles (cilia) protrude from these cells, and some of the cilia are embedded in the tectorial membrane in the middle chamber. For reviews of the mechanics of the auditory system, see [2, 17, 29].

When a sound wave enters the cochlea, a pressure wave in the fluid perturbs the BM near its base. This initiates a wave along the BM, with varying amplitude, that propagates toward the apex of the cochlea. The envelope of this wave has a maximum amplitude, the position of which depends on the frequency of the input. High frequencies lead to maximum vibrations at the stiffer base of the BM, and low frequencies lead to maximum vibrations at the floppier apex of the BM. As discussed in [22], each point along the BM oscillates at the input frequency. As the sound wave bends the BM, the hair cells convert the mechanical energy into neuronal signals. There is evidence [6,23,25] that the oscillations of the hair cells have a natural frequency which varies with the position of the hair cell along the BM.

Experiments have shown that the ear has a sharp frequency tuning mechanism along with a nonlinear amplification system – there is no audible sound soft enough to suggest that the cochlear response is linear. Many authors [5–7,21,23,27,28] have suggested that these two phenomena indicate that the auditory system may be tuned near a Hopf bifurcation. Detailed models of parts of the auditory system (Hudspeth and Lewis [18, 19], Choe, Magnasco, and Hudspeth [6]) have been shown to contain Hopf bifurcations for biologically realistic parameter values.

Hopf Models of the Auditory System

Most simplified models model a single hair cell as a forced Hopf oscillator, similar to (9), but with the imaginary part of the cubic term (γ) set equal to zero. As we have shown in section "Periodic Forcing near Hopf Bifurcation", this assumption leads to nongeneric behavior, in particular, that the response curve is symmetric in ω. In fact, a center manifold reduction of the model of Hudspeth and Lewis [18, 19] by Montgomery et al. [27] finds that $\gamma \neq 0$. Specifically, they find $\gamma = -1.07$.

Furthermore, the response curve in the auditory system has been shown experimentally (see [29] and references within) to be asymmetric. Two papers [23, 25] have considered the dynamics of an array of Hopf oscillators (rather than the single oscillators studied by most other authors). They achieve the aforementioned asymmetry through couplings between the oscillators via a traveling wave which supplies the forcing terms. This complicates the matter significantly, so that analytical results cannot be obtained.

However, we note that merely having a complex, rather than real, cubic term in the Hopf oscillator model would have a similar effect. The value of γ found by Montgomery et al. [27] is in the regime where we observe asymmetry, but not multiplicity of solutions. Multiple solutions in this model could correspond to perception of a sound of either low or high amplitude for the same input forcing. We have seen no mention of this phenomena in the literature.

Two-Frequency Forcing

It is clear that stimuli received by the auditory system are not single frequency, but contain multiple frequencies. If each hair cell is to be modeled as a Hopf oscillator, we are interested in the effect of multifrequency forcing on an array of Hopf oscillators. We give here some numerical results from an array of N uncoupled Hopf oscillators:

$$\dot{z}_j = (\lambda + \mathrm{i}\omega_{\mathrm{H}}(j))z + (-1 + \mathrm{i}\gamma)|z_j|^2 z_j + \varepsilon g(t), \quad j = 1, \ldots, N, \tag{39}$$

where $\omega_{\mathrm{H}}(j) = \omega_1 + j\Delta\omega$, for some ω_1, $\Delta\omega$, that is, the Hopf frequency increases linearly along the array of oscillators. Note that all oscillators receive the same forcing.

Consider forcing which contains two frequency components, for instance:

$$g(t) = \mathrm{e}^{\mathrm{i}t} + \mathrm{e}^{\sqrt{5}\mathrm{i}t}. \tag{40}$$

In Fig. 15, we plot the mean amplitude of the responses of each oscillator in the array. The response clearly has two peaks, one close to each frequency component of the input.

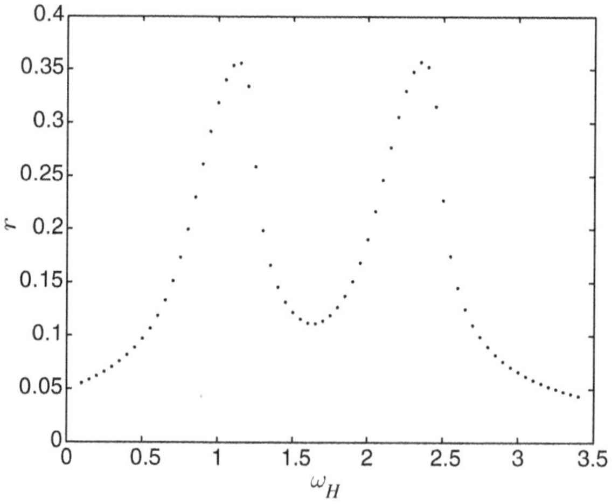

Fig. 15 The mean amplitude for each of an array of forced Hopf oscillators (39), with forcing given in (40). The phase plane portraits for the outputs of the oscillators with $\omega_{\mathrm{H}} = 1$ and $\omega_{\mathrm{H}} = 1.6$ are shown in Fig. 16. Remaining parameters are $\lambda = -0.01$, $\varepsilon = 0.05$, $\gamma = -1$

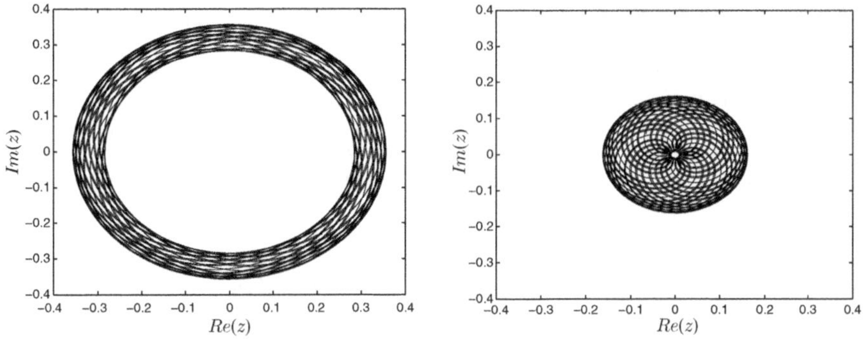

Fig. 16 The phase plane portraits for Hopf oscillators with $\omega_{\mathrm{H}} = 1$ (*left*) and $\omega_{\mathrm{H}} = 1.6$ (*right*), with forcing as given in (40). The left figure is almost periodic, but the right is clearly quasiperiodic. Remaining parameters are $\lambda = -0.01$, $\varepsilon = 0.05$, $\gamma = -1$

Note also that the forcing $g(t)$ is quasiperiodic. In those oscillators which have a Hopf frequency close to one component of the forcing, only that component is amplified. This results in an output which is close to periodic. In Fig. 16 we show the resulting phase plane solutions from two of the Hopf oscillators. The first has $\omega_{\mathrm{H}} = 1$, so the first component of the forcing is amplified, and the solution is close to periodic. The second has $\omega_{\mathrm{H}} = 1.6$, which is far from both 1 and $\sqrt{5}$. Hence neither component is amplified and the output is quasiperiodic.

Acknowledgments We thank Gemunu Gunaratne, Krešimir Josić, Edgar Knobloch, Mary Silber, Jean-Jacques Slotine, and Ian Stewart for helpful conversations. This research was supported in part by NSF Grant DMS-0604429 and ARP Grant 003652-0009-2006.

References

1. U. Alon. *An Introduction to Systems Biology: Design Principles of Biological Circuits.* CRC, Boca Raton, 2006.
2. M. Bear, B. Connors, and M. Paradiso. *Neuroscience: Exploring the Brain.* Lippincott Williams & Wilkins, Philadelphia PA, 2006.
3. N.N. Bogoliubov and Y.A. Mitropolsky. *Asymptotic Methods in the Theory of Non-linear Oscillations.* Hindustan Publ. Corp., Delhi, 1961.
4. H.W. Broer and G. Vector. Generic Hopf-Neimark-Sacker bifurcations in feed-forward systems. *Nonlinearity* **21** (2008) 1547–1578.
5. S. Camalet, T. Duke, F. Jülicher, and J. Prost. Auditory sensitivity provided by self-tuned oscillations of hair cells. *Proc. Natl. Acad. Sci.* **97** (2000) 3183–3188.
6. Y. Choe, M.O. Magnasco, and A.J. Hudspeth. A model for amplification of hair-bundle motion by cyclical binding of Ca^{2+} to mechanoelectrical-transduction channels. *Proc. Natl. Acad. Sci. USA* **95** (1998) 15321–15326.
7. V.M. Eguíluz, M. Ospeck, Y. Choe, A.J. Hudspeth, and M.O. Magnasco. Essential nonlinearities in hearing. *Phys. Rev. Lett.*, **84** (2000) 5232–5235.
8. T. Elmhirst and M. Golubitsky. Nilpotent Hopf bifurcations in coupled cell systems. *J. Appl. Dynam. Sys.* **5** (2006) 205–251.

9. C. D. Geisler and C. Sang. A cochlear model using feed-forward outer-hair-cell forces. *Hearing Res.* **86** (1995) 132–146.
10. D.W. Jordan and P. Smith. *Nonlinear Ordinary Differential Equations*, fourth ed. Oxford University Press, Oxford, 2007.
11. M. Golubitsky and R. Lauterbach. Bifurcations from Synchrony in Homogeneous Networks: Linear Theory. *SIAM J. Appl. Dynam. Sys.* **8** (1) (2009) 40–75.
12. M. Golubitsky, M. Nicol, and I. Stewart. Some curious phenomena in coupled cell networks. *J. Nonlinear Sci.* **14** (2) (2004) 207–236.
13. M. Golubitsky and D.G. Schaeffer. *Singularities and Groups in Bifurcation Theory: Vol. I. Appl. Math. Sci.* **51**, Springer-Verlag, New York, 1984.
14. M. Golubitsky and I. Stewart. *The Symmetry Perspective: From Equilibrium to Chaos in Phase Space and Physical Space*. Birkhäuser, Basel 2002.
15. M. Golubitsky and I. Stewart. Nonlinear dynamics of networks: the groupoid formalism. *Bull. Amer. Math. Soc.* **43** No. 3 (2006) 305–364.
16. M. Golubitsky, I. Stewart, and A. Török. Patterns of synchrony in coupled cell networks with multiple arrows. *SIAM J. Appl. Dynam. Sys.* **4** (1) (2005) 78–100.
17. A.J. Hudspeth. Mechanical amplification of stimuli by hair cells, *Curr. Opin. Neurobiol.* **7** (1997) 480–486.
18. A.J. Hudspeth and R.S. Lewis. Kinetic-analysis of voltage-dependent and ion-dependent conductances in saccular hair-cells of the bull frog, *rana catesbeiana*. *J. Physiol.* **400** (1988) 237–274.
19. A.J. Hudspeth and R.S. Lewis. A model for electrical resonance and frequency tuning in saccular hair cells of the bull frog, *rana catesbeiana*. *J. Physiol.* **400** (1988) 275–297.
20. T.S.A. Jaffer, H. Kunov, and W. Wong. A model cochlear partition involving longitudinal elasticity. *J. Acoust. Soc. Am.* **112** No. 2 (2002) 576–589.
21. F. Jülicher, D. Andor, and T. Duke. Physical basis of two-tone interference in hearing. *Proc. Natl. Acad. Sci.* **98** (2001) 9080–9085.
22. J. Keener and J. Sneyd. *Mathematical Physiology* Interdisciplinary. Applied Mathematics **8**, Springer-Verlag, New York, 1998.
23. A. Kern and R. Stoop. Essential role of couplings between hearing nonlinearities. *Phys. Rev. Lett.* **91** No. 12 (2003) 128101.
24. M.C.A. Leite and M. Golubitsky. Homogeneous three-cell networks. *Nonlinearity* **19** (2006) 2313–2363. DOI: 10.1088/0951-7715/19/10/04
25. M. Magnasco. A wave traveling over a Hopf instability shapes the Cochlea tuning curve. *Phys. Rev. E* **90** No. 5 (2003) 058101-1.
26. N.J. McCullen, T. Mullin, and M. Golubitsky. Sensitive signal detection using a feed-forward oscillator network. *Phys. Rev. Lett.* **98** (2007) 254101.
27. K.A. Montgomery, M. Silber, and S.A. Solla. Amplification in the auditory periphery: The effect of coupled tuning mechanisms. *Phys. Rev. E* **75** (2007) 051924.
28. M. Ospeck, V. M. Eguíluz, and M. O. Magnasco. Evidence of a Hopf bifurcation in frog hair cells. *Biophys. J.* **80** (2001) 2597–2607.
29. L. Robles and M. A. Ruggero. Mechanics of the mammalian cochlea. *Physiol. Rev.* **81** (3) (2001) 1305–1352.
30. L. Robles, M. A. Ruggero and N. C. Rich. Two-tone distortion in the basilar membrane of the cochlea. *Nature* **349** (1991) 413.
31. I. Stewart, M. Golubitsky, and M. Pivato. Symmetry groupoids and patterns of synchrony in coupled cell networks. *SIAM J. Appl. Dynam. Sys.* **2** No. 4 (2003) 609–646.
32. R. Stoop and A. Kern. Two-tone suppression and combination tone generation as computations performed by the Hopf cochlea. *Phys. Rev. Lett.* **93** (2004) 268103.
33. Q. Wang and L.S. Young. Strange attractors in periodically-kicked limit cycles and Hopf bifurcations. *Commun. Math. Phys.* **240** No. 3 (2003) 509–529.
34. Y. Zhang. PhD Thesis, Ohio State University.

Gain Modulation as a Mechanism for Switching Reference Frames, Tasks, and Targets

Emilio Salinas and Nicholas M. Bentley

Abstract In the mammalian brain, gain modulation is a ubiquitous mechanism for integrating information from various sources. When a parameter modulates the gain of a neuron, the cell's overall response amplitude changes, but the relative effectiveness with which different stimuli are able to excite the cell does not. Thus, modulating the gain of a neuron is akin to turning up or down its "loudness". A well-known example is that of visually sensitive neurons in parietal cortex, which are gain-modulated by proprioceptive signals such as eye and head position. Theoretical work has shown that, in a network, even relatively weak modulation by a parameter P has an effect that is functionally equivalent to turning on and off different subsets of neurons as a function of P. Equipped with this capacity to switch, a neural circuit can change its functional connectivity very quickly. Gain modulation thus allows an organism to respond in multiple ways to a given stimulus, so it serves as a basis for flexible, nonreflexive behavior. Here we discuss a variety of tasks, and their corresponding neural circuits, in which such flexibility is paramount and where gain modulation could play a key role.

The Problem of Behavioral Flexibility

The appropriate response to a stimulus depends heavily on the context in which the stimulus appears. For example, when James Brown yells "Help me!" in the middle of a song, it calls for a different response than when someone yells the same thing from inside a burning building. Normally, the two situations are not confused: concert goers don't rush onstage to try to save James Brown and firemen don't take to dancing outside of burning buildings. This is an extreme example, but to a certain degree, even the most basic sensory stimuli are subject to similar interpretations that depend on the ongoing behavioral context. For instance, the sight and smell

E. Salinas (✉)
Department of Neurobiology and Anatomy, Wake Forest University School of Medicine, Winston-Salem, NC 27157, USA
e-mail: esalinas@wfubmc.edu

K. Josić et al. (eds.), *Coherent Behavior in Neuronal Networks*, Springer Series in Computational Neuroscience 3, DOI 10.1007/978-1-4419-0389-1_7,

of food may elicit very different responses depending on how hungry an animal is. Such context sensitivity is tremendously advantageous, but what are the neuronal mechanisms behind it?

Gain modulation is a mechanism whereby neurons integrate information from various sources [47, 48]. With it, contextual and sensory signals may be combined in such a way that the same sensory information can lead to different behavioral outcomes. In essence, gain modulation is the neural correlate of a switch that can turn on or off different subnetworks in a circuit, and thus provides, at least in part, a potential solution to the problem of generating flexible, context-sensitive behavior illustrated in the previous paragraph. This chapter will address three main questions: What is gain modulation? What are some of its experimental manifestations? And, what computational operations does it enable? We will show that relatively weak modulatory influences like those observed experimentally may effectively change the functional connectivity of a network in a drastic way.

What is Gain Modulation?

When a parameter modulates the gain of a neuron, the cell's response amplitude changes, but its selectivity with respect to other parameters does not. Let's elaborate on this. To describe gain modulation, we must consider the response of a neuron as a function of two quantities; call them x and y. To take a concrete example, suppose that x is the location of a visual stimulus with respect to the fixation point and y is the position of the eye along the horizontal axis. The schematic in Fig. 1 shows the relationship between these variables. First, fix the value of y, say it is equal to 10° to the left, and measure the response of a visual neuron as a function of stimulus location, x. Assume that the response to a stimulus is quantified by the number of action potentials evoked by that stimulus within a fixed time window, which is a standard procedure. When the evoked firing rate r is plotted against x, the result will be a curve like those shown in Fig. 2, with a peak at the preferred stimulus location of the cell. Now repeat the measurements with a different value of y, say 10° to the right this time. How does the second curve (r vs. x) compare with the first one?

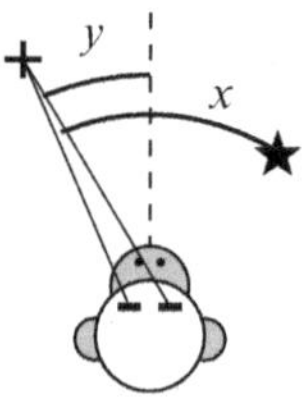

Fig. 1 Reference points for locating an object. The *cross* indicates the fixation point and the *star* indicates a visual stimulus. The angle of the eye in the orbit is y. The position of the object in eye-centered coordinates (i.e., relative to the fixation point) is x

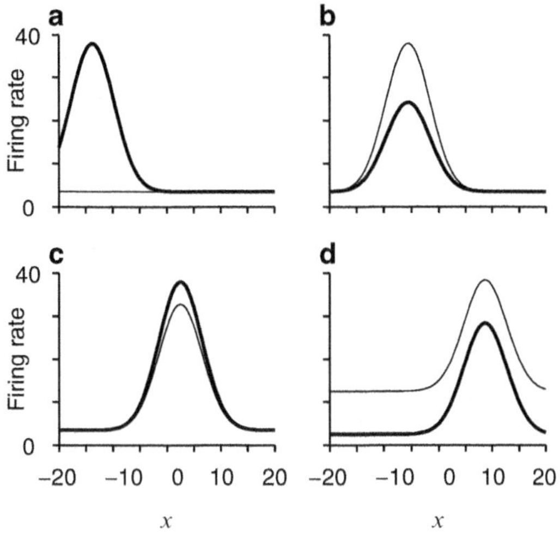

Fig. 2 Hypothetical responses of four neurons that are sensitive to two quantities, x and y. Each panel shows the firing rate as a function of x for two values of y, which are distinguished by *thin* and *thick lines*. (**a–c**), Three neurons that are gain-modulated by y. (**d**), A neuron that combines x- and y- dependencies linearly, and so is not gainmodulated

If the neuron is not sensitive to eye position at all, then the new curve will be the same as the old one. This would be the case, for instance, with the ganglion cells of the retina, which respond depending on where light falls on the retina. On the other hand, if the neuron is gain-modulated by eye position, then the second curve will be a scaled version of the first one. That is, the two curves will have the same shapes but different amplitudes. This is illustrated in Fig. 2b and c, where thin and thick lines are used to distinguish the two y conditions. This is precisely what happens in the posterior parietal cortex [1, 2, 12], where this phenomenon was first documented by Andersen and Mountcastle [3]. To describe the dependency on the modulatory quantity y, these authors coined the term "gain field," in analogy with the receptive field, which describes the dependency on the sensory stimulus x. Of course, to properly characterize the gain field of a cell, more values of y need to be tested [2], but for the moment we will keep discussing only two.

To quantify the strength of the modulation, consider the decrease in response amplitude observed as a neuron goes from its preferred to its nonpreferred y condition. In panels a–c of Fig. 2, the resulting percent suppression is, respectively, 100%, 40%, and 15%. In the parietal cortex, and in experimental data in general, 100% suppression is very rarely seen; on the contrary, the effects are often quite subtle (more on this below). More importantly, however, the firing rate r of the neuron can be described as a product of two factors, a function of x, which determines the shape of the curves, times a gain factor that determines their amplitude:

$$r_j(x, y) = g_j(y) f_j(x) \tag{1}$$

where the index j indicates that this applies to neuron j. The functions f and g may have different forms, so this is a very general expression. Much of the data reported by Andersen and colleagues, as well as data from other laboratories, can be fit quite well using (1) [2,12,30,31,57]. Note, however, that for the theoretical results discussed later on, the key property of the above expression is that it combines x and y in a nonlinear way. Thus, in general, the theoretical results are still valid if the relationship between x- and y-dependencies is not exactly multiplicative, as long as it remains nonlinear. A crucial consequence of this is that a response that combines x and y dependencies linearly, for instance,

$$r_j(x, y) = g_j(y) + f_j(x) \tag{2}$$

does *not* qualify as a gain-modulated response. Note how different this case is from the multiplicative one: when y changes, a neuron that responds according to (2) increases or decreases its firing rate by the same amount for all x values. This is illustrated in Fig. 2d, where the peaked curve moves up and down as y is varied but nothing else changes other than the baseline. Such a neuron is sensitive to both x and y, but (1) it cannot be described as gain-modulated by y, because it combines f and g linearly, and (2) it cannot perform the powerful computations that gain-modulated neurons can.

Experimental Evidence for Gain Modulation

To give an idea of the diversity of neural circuits in which gain modulation may play an important functional role, in this section we present a short survey of experimental preparations in which gain modulation has been observed. In all of these cases the reported effects are roughly in accordance with (1). Note that, although the x and y variables are different in all of these examples, all experiments are organized like the one discussed in the previous section: the firing rate of a neuron is plotted as a function of x for a fixed value of y; then y is varied and the response function or tuning curve (r vs. x) of the cell is plotted again; finally, the results with different y values are compared.

Modulation by Proprioceptive Information

To begin, we mention three representative examples of proprioceptive signals that are capable of modulating the visually evoked activity of cortical neurons. First of all, although initial work described responses tuned to stimulus location (x) with gain fields that depend on eye position, an additional dependence on head position was discovered later, such that the combination of head and eye angles – the gaze angle – seems to be the relevant quantity [12]. Thus, the appropriate modulatory parameter y is likely to be the gaze angle.

Another example of modulation by proprioceptive input is the effect of eye and head velocity (which now play the role of y) on the responses of neurons that are sensitive to heading direction (which now corresponds to x). Neurons in area MSTd respond to the patterns of optic flow that are generated by self-motion; for instance, when a person is walking, or when you are inside a moving train and look out the window. Under those conditions, the MSTd population encodes the direction of heading. Tuning curves with a variety of shapes are observed when the responses of these neurons are plotted as functions of heading direction, and crucially, for many units those curves are gain-modulated by eye and head velocity [50].

A third example of proprioceptive modulation occurs in an area called the parietal reach region, or PRR, where cells are tuned to the location of a reach target. The reach may be directed toward the location of a visible object or toward its remembered location; in either case, PRR cells are activated only when there is an intention to reach for a point in space. The response of a typical PRR cell as a function of target location (the x parameter in this case) is peaked. A key modulatory quantity for these cells is the initial location of the hand (this is y now): the amplitude of the peaked responses to target location depends on the position of the hand *before* the reach [7, 13]. We will come back to this example later on.

Attentional Modulation

Another modulatory signal that has been widely studied is attention, and its effects on sensory-driven activity are often very close to multiplicative. For example, visual neurons in area V4 are sensitive to the orientation of a bar shown in their receptive field. The corresponding tuning curves of firing rate vs. bar orientation change depending on where attention is directed by the subject. In particular, the tuning curves obtained with attention directed inside vs. outside the receptive field are very nearly scaled versions of each other, with an average suppression of about 20% in the non-preferred condition [31]. The magnitude of this effect is not very large, but this is with rather simple displays that include only two stimuli, a target and a distractor; results reported using more crowded displays and more demanding tasks are two-to-three times as large [14, 21].

Another cortical area where attentional effects have been carefully characterized is area MT, which processes visual motion. There, neurons are sensitive to the direction in which an object, or a cloud of objects, moves across their receptive fields. Thus, for each neuron the relevant curve in this case is that of firing rate vs. movement direction. What the experiments show is that the location where attention is directed has an almost exactly multiplicative effect on the direction tuning curves of MT cells. But interestingly, the amplitude of these curves also depends on the specific movement direction that is being attended [30, 57]. Thus, as in V4 [21], stronger suppression (of about 50%) can be observed when an appropriate combination of attended location and attended feature is chosen.

Nonlinear Interactions between Multiple Stimuli

In addition to studies in which two distinct variables are manipulated, there is a vast literature demonstrating other nonlinear interactions that arise when multiple stimuli are displayed simultaneously. For instance, the response of a V1 neuron to a single oriented bar or grating shown inside its receptive field may be enhanced or suppressed by neighboring stimuli outside the receptive field [20, 25, 28, 51]. Thus, stimuli that are ignored by a neuron when presented in isolation may have a strong influence when displayed together with an effective stimulus. These are called "extra classical receptive field" effects, because they involve stimuli outside the normal activation window of a neuron.

Nonlinear summation effects, however, may also be observed when multiple stimuli are shown inside the receptive field of a cell. For example, suppose that a moving dot is presented inside the receptive field of an MT neuron and the evoked response is r_1. Then another moving dot is shown moving in the same direction but at a different location in the receptive field, and the response is now r_2. What happens if the two moving dots are displayed simultaneously? In that case, the evoked response is well described by

$$r = \alpha \left(r_1^n + r_2^n\right)^{1/n} \tag{3}$$

where α is a scale factor and n is an exponent that determines the strength of the nonlinearity [11]. This expression is interesting because it captures a variety of possible effects, from simple response averaging ($\alpha = 0.5$, $n = 1$) or summation ($\alpha = 1$, $n = 1$), to winner-take-all behavior ($\alpha = 1$, n large). Neurons in MT generally show a wide variety of n exponents [11], typically larger than 1, meaning that the presence of one stimulus modulates nonlinearly the response to the other. Neurons in V4 show similar nonlinear summation [21].

In general, many interactions between stimuli are well described by divisive normalization models [52, 53], of which (3) is an example. Such models implement another form of gain control where two or more inputs are combined nonlinearly.

Context- and Task-Dependent Modulation

Finally, there are numerous experiments in which more general contextual effects have been observed. This means that activity that is clearly driven by the physical attributes of a sensory stimulus is also affected by more abstract factors, such as the task that the subject needs to perform, or the subject's motivation. These phenomena are often encountered in higher-order areas, such as the prefrontal cortex. For example, Lauwereyns and colleagues designed a task in which a monkey observed a cloud of moving dots and then, depending on a cue at the beginning of each trial, had to respond according to either the color or the direction of movement of the dots [27]. Prefrontal neurons were clearly activated by the moving dots; however, the firing rates evoked by identical stimuli varied quite dramatically depending on which feature was imporant in a given trial, color or movement direction.

Another striking case of task-dependent modulation was revealed by Koida and Komatsu [26] using color as a variable. They trained monkeys to perform two color-based tasks. In one of them, the monkeys had to classify a single color patch into one of two categories, reddish or greenish. In the other task, the monkeys had to indicate which of two color patches matched a previously shown sample. Neuronal responses in inferior-temporal cortex (IT) were recorded while the monkeys performed these two tasks plus a passive fixation task, in which the same stimuli were presented but no motor reactions were required. The responses of many IT neurons were strongly gain-modulated by the task: their color tuning curves had similar shapes in all cases, but the overall amplitude of their responses depended on which task was performed. This means that their activity was well described by (1), with x being color and y being the current task. Note that, in these experiments, the exact same stimuli were used in all tasks – what varied was the way the sensory information was processed. Furthermore, the classification and discrimination tasks were of approximately equal difficulty, so it is unlikely that the observed modulation was due to factors such as motivation or reward rate. Nevertheless, these two variables are known to modulate many types of neurons quite substantially [36, 49], so the extent of their influence is hard to determine.

A last example of modulation due to task contingencies comes from neurophysiological work in the supplementary and presupplementary motor areas, which are involved in fine aspects of motor control such as the planning of movement sequences. Sohn and Lee made recordings in these areas while monkeys performed sequences of joystick movements [54]. Each individual movement was instructed by a visual cue, but importantly, the monkey was rewarded only after correct completion of a full sequence, which consisted of 5–10 movements. The activity of the recorded neurons varied as a function of movement direction, as expected in a motor area, but interestingly, it also varied quite strongly according to the number of remaining movements in a sequence. It is possible that the true relevant variable here is the amount of time or the number of movements until a *reward* is obtained. Regardless of this, remarkably, the interaction between movement direction (the x variable in this case) and the number of remaining movements in a sequence (the y variable) was predominantly multiplicative, in accordance with (1).

In summary, there is an extremely wide range of experimental preparations and neuronal circuits that reveal strongly nonlinear modulation of evoked activity. The next sections discuss (1) some computational operations that neuronal networks can implement relatively easily based on such nonlinearities, and (2) a variety of behaviors and tasks where such operations could play crucial functional roles.

Computations Based on Gain Modulation

The previous section reviewed a large variety of experimental conditions under which gain modulation – understood as a nonlinear interaction between x- and y-dependencies – is observed. What is interesting about the chosen examples, and is not immediately obvious, is that all of those effects may be related to similar computational operations. That is precisely what theoretical and modeling studies have shown, that gain modulation serves to implement a type of mathematical transformation that arises under many circumstances. It is relatively well established that gain modulation plays an important role in carrying out coordinate transformations [4, 38, 46, 48], so that is the first type of operation that we will discuss. However, the applicability of gain modulation goes way beyond the traditional notion of coordinate transformation [41].

Coordinate Transformations

The first clue that gain modulation is a powerful computing mechanism was provided by Zipser and Andersen [63], who trained an artificial neural network to transform the neural representation of stimulus location from one reference frame to another. Their network had three layers of model neurons. The activities in the input and output layers were known – they were given to the model – and the job of the model was to determine the responses of the units in the middle layer. The input layer had two types of neurons, ones that encoded the location of a visual stimulus in eye-centered coordinates, i.e., with respect to the fixation point, and others that encoded the gaze angle. Thus, stimulus location was x and gaze angle was y, exactly as discussed in the section "What is Gain Modulation." In contrast, the neurons in the output layer encoded the location of the stimulus in head-centered coordinates, which for this problem means that they responded as functions of $x + y$. It can be seen from Fig. 1 that $x + y$ in this case gives the location of the stimulus relative to the head.

Adding x plus y seems deceivingly simple, but the problem is rather hard because x, y, and $x + y$ are encoded by populations of neurons where each unit responds nonlinearly to one of those three quantities. The middle neurons in the model network had to respond to each pattern of evoked activity (in the first layer) and produce the correct output pattern (in the third layer). The way the authors were able to "train" the middle layer units was by using a synaptic modification rule, the backpropagation algorithm, which modified the network connections each time an input and a matching output pattern were presented, and thousands of input and output examples were necessary to learn the mapping with high accuracy. But the training procedure is not very important; what is noteworthy is that, after learning, the model units in the middle layer had developed gaze-dependent gain fields, much like those found in parietal cortex. This indicated that the visual responses modulated by gaze angle are an efficient means to compute the required coordinate transformation.

Further intuition about the relation between gain fields and coordinate transformations came from work by Salinas and Abbott [43], who asked the following question. Consider a population of parietal neurons that respond to a visual stimulus at retinal location x and are gain-modulated by gaze angle y. Suppose that the gain-modulated neurons drive, through synaptic connections, a downstream population of neurons involved in generating an arm movement toward a target. To actually reach the target with any possible combination of target location and gaze angle, these downstream neurons must encode target location in body-centered coordinates, that is, they must respond as functions of $x + y$ (approximately). Under what conditions will this happen? Salinas and Abbott developed a mathematical description of the problem, found those conditions, and showed that, when driven by gain-modulated neurons, the downstream units can explicitly encode the sum $x + y$ or any other linear combination of x and y. In other words, the gain-field representation is powerful because downstream neurons can extract *from the same modulated neurons* any quantity $c_1x + c_2y$ with arbitrary coefficients c_1 and c_2, and this can be done quite easily – "easily" meaning through correlation-based synaptic modification rules [43, 44]. Work by Pouget and Sejnowski [37] extended these results by showing that a population of neurons tuned to x and gain-modulated by y can be used to generate a very wide variety of functions of x and y downstream, not only linear combinations.

A crucial consequence of these theoretical results is that when gain-modulated neurons are found in area A, it is likely that neurons in a downstream area B will have response curves as functions of x that shift whenever y is varied. This is because a tuning curve that is a function of $c_1x + c_2y$ will shift along the x-axis when y changes. To illustrate this point, consider another example of a coordinate transformation; this case goes back to the responses of PRR cells mentioned in the section "Modulation by Proprioceptive Information."

The activity of a typical PRR cell depends on the location of a target that is to be reached, but is also gain-modulated by the initial position of the hand before the reach [7, 13]. Thus, now x is the target's location and y is the initial hand position (both quantities in eye-centered coordinates). Fig. 3a–c illustrates the corresponding experimental setup with three values of y: with the hand initially to the left, in the middle, or to the right. In each diagram, the small dots indicate the possible locations of the reach target, and the two gray disks represent the receptive fields of two neurons. The response of a hypothetical PRR neuron is shown in Fig. 3e, which includes three plots, one for each of the three initial hand positions. In accordance with the reported data [7, 13], the gain of this PRR cell changes as a function of initial hand position. The other two cells depicted in the figure correspond to idealized responses upstream and downstream from the PRR. Figure 3d shows the firing rate of an upstream cell that is insensitive to hand position y, so its receptive field (light disk in Fig. 3a–c) does not change. This cell is representative of a neural population that responds only as a function of x, and simply encodes target location in eye-centered coordinates. In contrast, Fig. 3f plots the tuning curve of a hypothetical neuron downstream from the PRR. The curve shifts when the initial hand position changes because it is a function of $x - y$ (as explained above, but with $c_1 = 1$ and $c_2 = -1$). For such neurons, the receptive field (dark disk in Fig. 3a–c) moves as a function of hand position. This cell is representative of a population that encodes target location in hand-centered coordinates: the firing responses are the same whenever the target maintains the same spatial relationship relative to the hand. Neurons like this one are found in parietal area 5, which is downstream from the PRR [13]. They can be constructed by combining the responses of multiple gain-modulated neurons like that in Fig. 3e, but with diverse target and hand-location preferences.

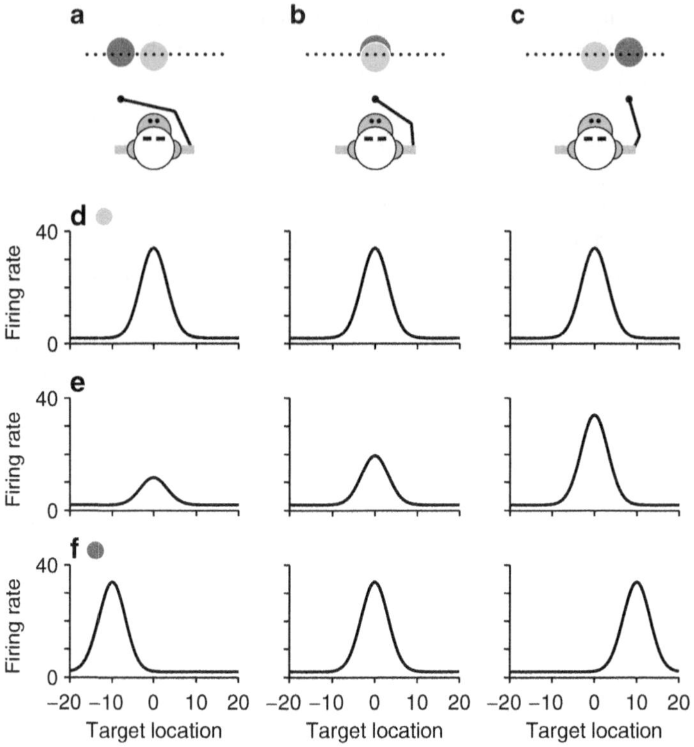

Fig. 3 Hypothetical responses of three neurons that play a role in reaching a target. (**a–c**), Three experimental conditions tested. *Dots* indicate the possible locations of a reach target. Conditions differ in the initial hand position (y). *Circles* indicate visual receptive fields of two hypothetical neurons, as marked by gray disks below. (**d–f**), Firing rate as a function of target location (x) for three idealized neurons. Each neuron is tested in the three conditions shown above. Gaze is assumed to be fixed straight ahead in all cases

Why are such shifting tuning curves useful? Because the transformation in the representation of target location from Fig. 3d to Fig. 3f partially solves the problem of how to acquire a desired target. In order to reach an object, the brain must combine information about target location with information about hand position to produce a vector. This vector, sometimes called the "motor error" vector, encodes how much the hand needs to move and in what direction. Cells like the one in Fig. 3d do not encode this vector at all, because they have no knowledge of hand position. Cells that are gain modulated by hand position, as in Fig. 3e, combine information about x and y, so they do encode the motor vector, but they do so implicitly. This means that actually reading out the motor vector from their evoked activity requires considerable computational processing. In contrast, neurons like the one in Fig. 3f encode the motor vector explicitly: reading out the motor vector from their evoked activity requires much simpler operations.

To see this, draw an arrow from the hand to the center of the shifting receptive field (the dark disk) in Fig. 3a–c. You will notice that the vector is the same regardless of hand position; it has the same length L and points in the same direction, straight ahead, in the three conditions shown. Thus, a strong activation of the idealized neuron of Fig. 3f, which is elicited when x is equal to the peak value of the tuning curve, can be easily translated into a specific motor command; such activation means "move the hand L centimeters straight ahead and you'll reach the target." Conversely, for this command to be valid all the time, the preferred target position of the cell must move whenever the hand moves.

In reality, area 5 neurons typically do not show full shifts [13]; that is, their curves do shift but by an amount that is smaller than the change in hand position. This is because their activity still depends to a certain extent on eye position. It may be that areas further downstream from area 5 encode target position in a fully hand-centered representation, but it is also possible that such a representation is not necessary for the accurate generation of motor commands. In fact, although full tuning curve shifts associated with a variety of coordinate transformations have been documented in many brain areas, partial shifts are more common [6, 19, 22, 24, 50, 55]. This may actually reflect efficiency in the underlying neural circuits: according to modeling work [16, 62], partial shifts should occur when more than two quantities x and y are simultaneously involved in a transformation, which is typically the case. Anyhow, the point is that a gain-modulated representation allows a circuit to construct shifting curves downstream, which encode information in a format that is typically more accessible to the motor apparatus.

A final note before closing this section. Many coordinate transformations make sense as a way to facilitate the interaction with objects in the world, as in the examples just discussed. But the shifting of receptive fields, and in general the transformation of visual information from an eye-centered representation to a reference frame that is independent of the eye may be an important operation for perception as well. For instance, to a certain degree, humans are able to recognize objects regardless of their size, perspective, and position in the visual field [15, 18]. This phenomenon is paralleled by neurons in area IT, which respond primarily depending on the type of image presented, and to some extent are insensitive to object location, scale or perspective [17, 56, 64]. For constructing such invariant responses, visual information must be transformed from its original retinocentered representation, and modeling work suggests that gain modulation could play an important role in such perceptual transformations too, because the underlying operations are similar. In particular, directing attention to different locations in space alters the gain of many visual neurons that are sensitive to local features of an image, such as orientation [14]. According to the theory [44, 45], this attentional modulation could be used to generate selective visual responses similar to those in IT, which are scale- and translation-invariant. Such responses would represent visual information in an attention-centered reference frame, and there is evidence from psychophysical experiments in humans [29] indicating that object recognition indeed operates in a coordinate frame centered on the currently attended location.

Arbitrary Sensory-Motor Remapping

In the original modeling studies on gain modulation, proprioceptive input combined with spatial sensory information led to changes in reference frame. It is also possible, however, to use gain modulation to integrate sensory information with other types of signals. Indeed, the mechanism also works for establishing arbitrary associations between sensory stimuli and motor actions on the basis of more abstract contextual information.

In many tasks and behaviors, a given stimulus is arbitrarily associated with two or more motor responses, depending on separate cues that we will refer to as "the context." Figure 4 schematizes one such task. Here, the shape of the fixation point indicates whether the correct response to a bar is a movement to the left or to the right (arrows in Fig. 4a and b). Importantly, the same set of bars (Fig. 4c) can be partitioned in many ways; for instance, according to orientation (horizontal vs. vertical), color (filled vs. not filled), or depending on the presence of a feature (gap vs. no gap). Therefore, there are many possible maps between the eight stimuli and the two motor responses, and the contextual cue that determines the correct map is rather arbitrary.

Neurophysiological recordings during such tasks typically reveal neuronal responses that are sensitive to both the ongoing sensory information and the current context or valid cue, with the interaction between them being nonlinear [5, 27, 58, 59]. Some of these effects were mentioned in the section "Context- and Task-Dependent Modulation." The key point is that, as the context changes in such tasks, there is (1) a nonlinear modulation of the sensory-triggered activity, and (2) a functional reconnection between sensory and motor networks that must be very fast. How is this reconnection accomplished by the nervous system?

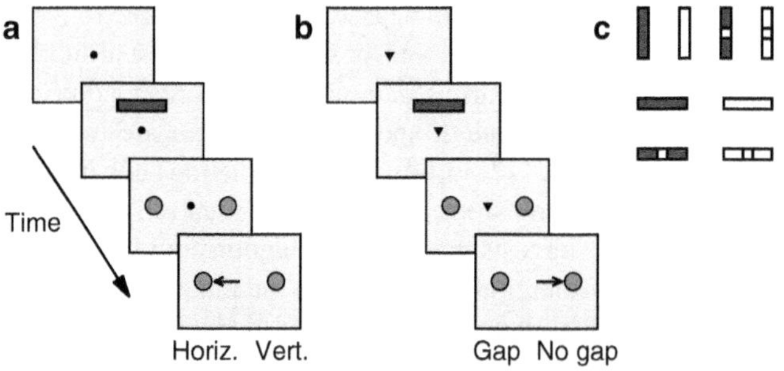

Fig. 4 A context-dependent classification task. The shape of the fixation point determines the correct response to a stimulus. (**a**), Events in a single trial in which a bar is classified according to orientation. (**b**), Events in a single trial in which a bar is classified according to whether it has a gap or not. (**c**), Stimulus set of eight bars

How the Contextual Switch Works

The answer is that such abstract transformations can be easily generated if the sensory information is gain-modulated by the context. According to modeling work [40,41], when this happens, multiple maps between sensory stimuli and motor actions are possible but only one map, depending on the context, is implemented at any given time. Let's sketch how this works.

In essence, the mechanism for solving the task in Fig. 4, or other tasks like it, is very similar to the mechanism for generating coordinate transformations (i.e., shifting receptive fields) described in section "Coordinate Transformations," except that x now describes the stimulus set and y the possible contexts. Instead of continuous quantities, such as stimulus location and eye position, these two variables may now be considered indices that point to individual elements of a set. For instance, $x = 1$ may indicate that the first bar in Fig. 4c was shown, $x = 2$ may indicate that the second bar was shown, etc. In this way, the expression $f_j(x)$ still makes sense: it stands for the firing rate of cell j in response to each stimulus in the set. Similarly, $y = 1$ now means that the current context is context 1, $y = 2$ means that the current context is context 2, and so on. Therefore, the response of a sensory neuron j that is gain-modulated by context can be written exactly as in (1),

$$r_j(x, y) = g_j(y) f_j(x) \tag{4}$$

except that now $r_j(x = 3, y = 1) = 10$ means "neuron j fires at a rate of 10 spikes/s when stimulus number 3 is shown and the context is context 1." Although x and y may represent quantities with extremely different physical properties, practically all of the important mathematical properties of (1) remain the same regardless of whether x and y vary smoothly or discretely. As a consequence, although the coordinate transformations in Figs. 1 and 3 may feel different from the remapping task of Fig. 4, mathematically they represent the very same problem, and the respective neural implementations may thus have a lot in common.

Now consider the activity of a downstream neuron that is driven by a population of gain-modulated responses through a set of synaptic weights,

$$R(x, y) = \sum_j w_j g_j(y) f_j(x) \tag{5}$$

where w_j is the connection from modulated neuron j to the downstream unit. Notice that the weights and modulatory factors can be grouped together into a term that effectively behaves like a context-dependent synaptic connection. That is, consider the downstream response in context 1,

$$R(x, y = 1) = \sum_j u_j^{(1)} f_j(x) \tag{6}$$

where we have defined $u_j^{(1)} = w_j g_j(y = 1)$. The set of coefficients $u_j^{(1)}$ represents the effective synaptic weights that are active in context 1. Now comes the crucial result: because the modulatory factors g_j change as functions of context, those effective weights will be different in context 2, and so will be the downstream response,

$$R(x, y = 2) = \sum_j u_j^{(2)} f_j(x) \tag{7}$$

where $u_j^{(2)} = w_j g_j(y = 2)$. Thus, the effective network connections are $u_j^{(1)}$ in one context and $u_j^{(2)}$ in the other. The point is that the downstream response in the two contexts, given by (6) and (7), can produce completely different functions of x. That is exactly what needs to happen in the classification task of Fig. 4a and b – the downstream motor response to the same stimulus (x) must be able to vary arbitrarily from one context to another. The three expressions above show why the mechanism works: the multiplicative changes in the gain of the sensory responses act exactly as if the connectivity of the network changed from one context to the next, from a set of connections $u_j^{(1)}$ to a set $u_j^{(2)}$.

Figure 5 illustrates this with a model that performs the task in Fig. 4, classifying a stimulus set differently in three possible contexts. In contexts 1 and 2, the stimuli trigger left and right movements according, respectively, to their orientation or to whether they have a gap or not, as in Fig. 4a and b. In contrast, in context 3 the correct response is to make no movement at all; this is a no-go condition. The model network used to generate Fig. 5 included 160 gain-modulated sensory neurons in the first layer and 50 motor neurons in the second, or output layer. Each sensory neuron responded differently to a set of eight distinct stimuli like those shown in Fig. 4c. For each cell, the maximum suppression across the three contexts varied between 0 and 30%, with a mean of 15%. The figure shows the responses of all the

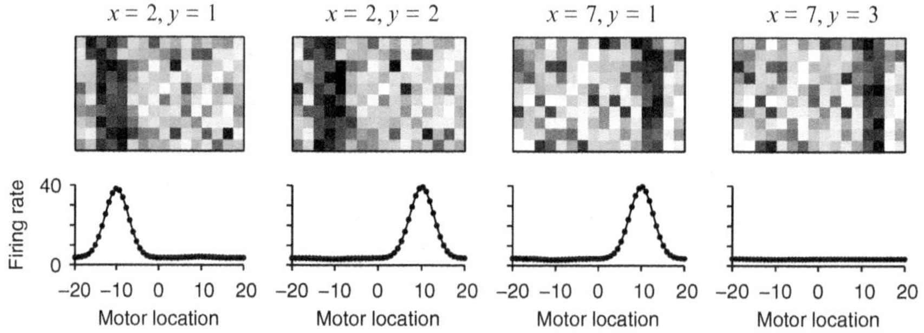

Fig. 5 A model network that performs the remapping task of Fig. 4. Gray-scale images represent the responses of 160 gain-modulated sensory neurons to the combinations of x (stimulus) and y (context) values shown above. White and black correspond to firing rates of 3 and 40 spikes/s, respectively. Plots at the bottom show the responses of 50 motor neurons driven by the gain-modulated units through synaptic connections. Identical stimuli lead to radically different motor responses depending on the current context

neurons in the network in four stimulus-context combinations. The firing rates of the gain-modulated neurons of the first layer are indicated by the gray-level maps. In these maps, the neurons are ordered such that their position along the x-axis roughly reflects their preferred stimulus, which goes from 1 to 8. The firing rates of the 50 motor units of the second layer are plotted below. For those cells, the location of the peak of activity encodes the location of a motor response.

There are a couple of notable things here. First, that the sensory activity changes quite markedly from one stimulus to another (compare the two conditions with $y = 1$), but much less so from one context to another (compare the two conditions with $x = 2$). This is because the modulatory factors do not need to change very much in order to produce a complete change in connectivity (more on this below). Second, what the model accomplishes is a drastic change in the motor responses of the second layer *to the same stimulus*, as a function of context. This includes the possibility of silencing the motor responses altogether or not, as can be seen by comparing the two conditions with $x = 7$.

In summary, in this type of network a set of sensory responses are nonlinearly modulated by contextual cues, but these do not need to be related in any way to the physical attributes of the stimuli. The model works extremely well, in that accuracy is limited simply by the variability (random fluctuations) in the gain-modulated responses and by the number of independent contexts [40]. In theory, it should work in any situation that involves choosing or switching between multiple functions or maps, as long as the modulation is nonlinear – as mentioned earlier, an exact multiplication as in (4) is not necessary, as long as x and y are combined nonlinearly. With the correct synaptic connections driving a group of downstream neurons, such a network can perform a variety of remapping tasks on the basis of a single transformation step between the two layers [40–42].

Switching as a Fundamental Operation

The model network discussed in the previous section can be thought of as implementing a context-dependent switch that changes the functional mapping between the two layers in the network. In fact, it can be shown that gain modulation has an effect that is equivalent to turning on and off different subsets of neurons as a function of context, or more generally, of the modulatory quantity y [41].

To illustrate this, consider the problem of routing information from a network A to two possible target networks B and C, such that either the connections $A \rightarrow B$ are active or the connections $A \rightarrow C$ are active, but not both. An example of this is a task in which the subject has to reach for an object that may appear at a variety of locations, but depending on the shape of the fixation point, or some other cue, he must reach with either the left or the right hand. In this situation, the contextual cue must act as a switch that activates one or the other downstream motor circuit, but it must do so for any stimulus value.

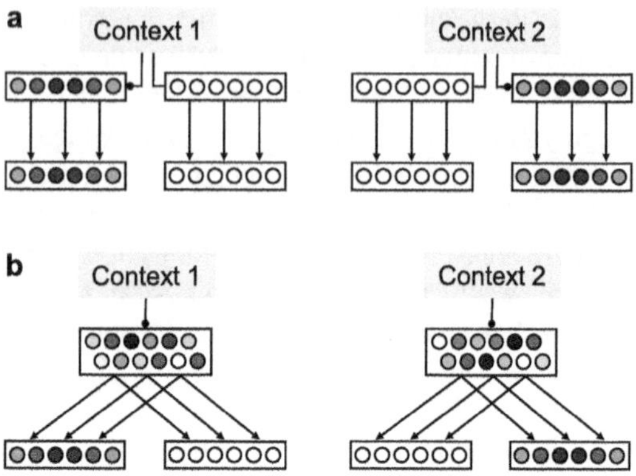

Fig. 6 Two possible schemes for routing information to either of two motor networks. One motor network should respond in context 1 and the other in context 2. (**a**) With two subpopulations of "switching neurons," which are 100% suppressed in their nonpreferred contexts. (**b**) With one population of partially modulated neurons, which show mild suppression in their nonpreferred contexts

Figure 6 schematizes two possible ways in which such a switch could be implemented. In the network of Fig. 6a, there are two separate sets of sensory responses that are switched on and off according to the context, and each subpopulation drives its own motor population downstream. This scheme requires true "switching neurons" that can be fully suppressed by the nonpreferred context. This type of neuron would need to respond as in Fig. 2a. This, admittedly, is a rather extreme and unrealistic possibility, but it nonetheless provides an interesting point of comparison: clearly, with such switching neurons it is trivial to direct the sensory information about target location to the correct motor network; and it is also clear that the two motor networks could respond in different ways to the same sensory input. In a way, the problem has simply been pushed back one step in the process.

An alternative network organization is shown in Fig. 6b. In this case, one single set of gain-modulated sensory neurons is connected to the two motor populations. What can be shown theoretically [41], is that any sensory-motor mapping that can be achieved with switching neurons (as in Fig. 6a), can also be implemented with standard, partially modulated neurons (as in Fig. 6b). Of course the necessary synaptic weights will be different in the two cases, but in principle, under very general conditions, if something can be done with 100% suppression, it can also be done with a much smaller amount of suppression (there may be a cost in terms of accuracy, but this depends on the level of noise in the responses and on the size of the network, among other factors [41]). In particular, a suppression of 100% is not necessary to produce a switch in the activated downstream population, as required by the task just discussed in which either the left or the right hand is used. This result offers a unique intuition as to why gain modulation is so powerful: modulating the activity of a population is equivalent to flipping a switch that turns on or off various sets of neurons.

This result is illustrated in Fig. 7, which shows the responses of a model network that has the architecture depicted in Fig. 6b. The x and y variables here are, respectively, the location of a stimulus, which takes a variety of values between -16 and 16, and the context, which takes two values. Figure 7a, c shows the activity of the 100 gain-modulated sensory neurons in the network in contexts 1 and 2,

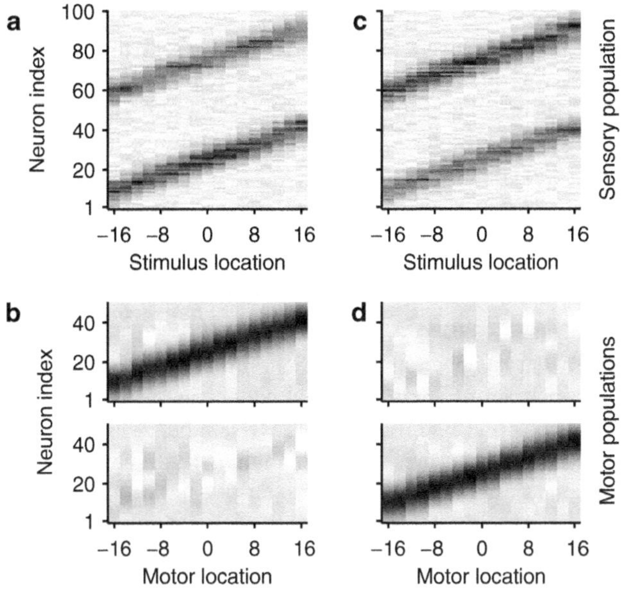

Fig. 7 Responses in a model network that routes information to two separate motor populations, as in Fig. ch07:fig6b. In context 1, one motor population is active and responds to all stimulus locations (x); in context 2, the other motor population is active. **a**, **c** Responses of the gain-modulated sensory neurons in contexts 1 and 2, respectively. **b**, **d** Responses of the motor neurons in contexts 1 and 2, respectively

respectively. These units are tuned to stimulus location, so they have peaked tuning curves like those in Fig. 2b, c. The 100 units have been split into two groups, those that have a higher gain or response amplitude in context 1 (units 1–50) and those that have a higher gain in context 2 (units 51–100). The neurons in each group are arranged according to their preferred stimulus location. This is why Fig. 7a and c show two bands: for each value of x, the neurons that fire most strongly in each group are those that have preferred values close to x. Notice that the activity patterns in Fig. 7a, c are very similar. This is because the differences in gain across the two contexts are not large: the percent suppression in this case ranged between 0 and 65%, with a mean of 32%.

What is interesting about this example is the behavior of the motor populations downstream. In context 1, one population responds and the other is silent (Fig. 7b), whereas in context 2 the roles are reversed, the second population responds and the first is silent (Fig. 7d). Importantly, this does not occur just for one value or a small number of values of x, but for any value. This switch, controlled by a contextual cue y, is generated on the basis of relatively small modulatory effects in the top layer. Thus, the modeling results reviewed here provide another interesting insight: small variations in the response gain of neurons in area A may be a signature of dramatically large variations in the activity of a downstream area B.

In the example of Fig. 7, the activated motor population simply mirrors the activation pattern in the sensory layer; that is, the sensory-motor map is one-to-one. This simple case was chosen to illustrate the switching mechanism without further complications, but in general, the relationship between stimulus location and the encoded motor variable (motor location) may be both much more complicated and different for the two contexts.

Flexible Responses to Complex Stimuli

To recapitulate what has been presented: coordinate transformations and arbitrary sensory-motor associations are types of mathematical problems that are akin to implementing a switch from one function of x to another, with the switch depending on a different variable y. Thus, in the mammalian brain, performing coordinate transformations and establishing arbitrary sensory-motor maps should lead to the deployment of similar neural mechanisms. Modeling work has shown that the nonlinear modulation of sensory activity can reduce the problem of implementing such a switch quite substantially, and this is consistent with experimental reports of nonlinear modulation by proprioceptive, task-dependent, and contextual cues in a variety of tasks ("Experimental Evidence for Gain Modulation").

There is, however, an important caveat. One potential limitation of gain modulation as a mechanism for performing transformations is that it may require an impossibly large number of neurons, particularly in problems such as invariant object recognition [18, 39, 44, 64] in which the dimensionality of the relevant space is high. The argument goes like this. Suppose that there are N stimuli to be classified according to M criteria, as a generalization of the task in Fig. 4. To perform this task, a neuronal network will need neurons with all possible combinations of stimulus and context sensitivities. Thus, there must be neurons that prefer stimuli $1, 2, \ldots N$ in context 1, neurons that prefer stimuli $1, 2, \ldots N$ in context 2, and so on. Therefore, the size of the network should grow as $N \times M$, which can be extremely large. This is sometimes referred to as "the curse of dimensionality."

Although this is certainly a challenging problem, its real severity is still unclear. In particular, O'Reilly and colleagues have argued [34, 35] that the combinatorial explosion may be much less of a problem than generally assumed. In essence, they say, this is because neurons are typically sensitive to a large range of input values, and the redundancy or overlap in this coarse code may allow a network to solve such problems accurately with many fewer than $N \times M$ units. At least in some model networks, it has indeed been demonstrated that accurate performance can be obtained with many fewer than the theoretical number of necessary units [10, 34, 35]. In practice, this means that it may be possible for real biological circuits to solve many remapping tasks with relatively few neurons, if they exhibit the appropriate set of broad responses to the stimuli and the contexts.

So, assuming that the curse of dimensionality does not impose a fundamental limitation, does gain modulation solve the flexibility problem discussed at the

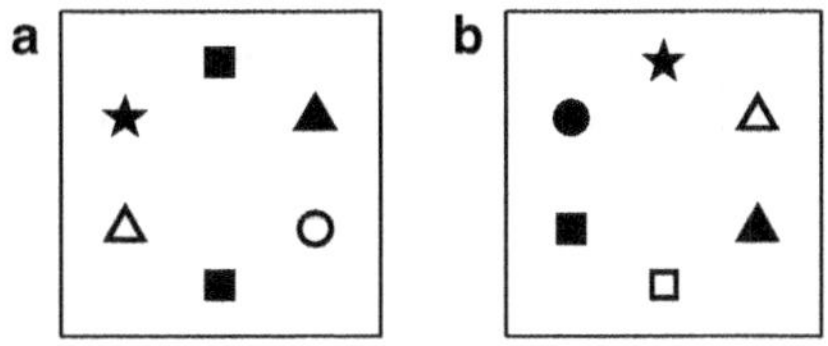

Fig. 8 Two examples of scenes with multiple objects. Each display can be considered as a single complex stimulus

beginning of the chapter? No, it does not, and this is why. All of the tasks and behaviors mentioned so far have been applied to single, isolated stimuli. If an organism always reacted to one stimulus at a time, then gain modulation would be enough for selecting the correct motor response under any particular context, assuming enough neurons or an appropriate coarse code. However, under natural conditions, a sensory scene in any modality contains multiple objects, and choices and behaviors are dictated by such crowded scenes. Other neural mechanisms are needed for (1) selecting one relevant object from the scene, or (2) integrating information across different locations or objects within a scene (note that similar problems arise whenever a choice needs to be made on the basis of a collection of objects, even if they appear one by one spread out over time). Thus, to produce a flexible, context-dependent response to a complex scene, gain modulation would need to be combined at least with other neural mechanisms for solving these two problems.

To put a simple example, consider the two displays shown in Fig. 8. If each display is taken as one stimulus, then there are many possible ways to map those two stimuli into "yes" or "no" responses. For instance, we can ask: Is there a triangle present? Is an open circle present on the right side of the display? Are there five objects in the scene? Or, are all objects different? Each one of these questions plays the role of a specific context that controls the association between the two stimuli and the two motor responses. The number of such potential questions (contexts) can be very large, but the key problem is that answering them requires an evaluation of the whole scene. Also, it is known from psychophysical experiments that subjects can change from one question to another extremely rapidly [61]. Gain modulation has the advantage of being very fast, and could indeed be used to switch these associations, but evidently the mechanisms for integrating information across a scene must be equally important.

As a first step in exploring this problem, we have constructed a model network that performs a variety of search tasks [8]. It analyzes a visual scene with up to eight objects and activates one of two output units to indicate whether a target is present or absent in the scene. As in the models described earlier, this network includes sensory neurons, which are now modulated by the identity of the search target, and output neurons that indicate the motor response. The target-dependent modulation allows the network to switch targets, that is, to respond to the question, is there a red vertical bar present, or to the question, is there a blue bar present, and so on. Not surprisingly, this model does not work unless an additional layer is included that integrates information across different parts of the scene. When this is done, gain modulation indeed still allows the network to switch targets, but the properties of the integration layer become crucial for performance too. There are other models for visual search that do not switch targets based on gain modulation; these are generally more abstract models [23, 60]. Our preliminary results [8] are encouraging because neurophysiological recordings in monkeys trained to perform search tasks have revealed sensory responses to individual objects in a search array that are indeed nonlinearly modulated according to the search target or the search instructions [9, 33]. Furthermore, theoretical work by Navalpakkam and Itti has shown that the gains of the sensory responses evoked by individual elements of a crowded scene can be set for optimizing the detection of a target [32]. This suggests that there is an optimal gain modulation strategy that depends on the demands of the task, and psychophysical results in their study indicate that human subjects do employ this optimal strategy [32].

In any case, the analysis of complex scenes should give a better idea of what can and cannot be done using gain modulation, and should provide ample material for future studies of the neural basis of flexible behavior.

Acknowledgments Research was partially supported by grant NS044894 from the National Institute of Neurological Disorders and Stroke.

References

1. Andersen RA, Bracewell RM, Barash S, Gnadt JW, Fogassi L (1990) Eye position effects on visual, memory, and saccade-related activity in areas LIP and 7a of macaque. J Neurosci 10:1176–1198
2. Andersen RA, Essick GK, Siegel RM (1985) Encoding of spatial location by posterior parietal neurons. Science 230:450–458
3. Andersen RA, Mountcastle VB (1983) The influence of the angle of gaze upon the excitability of light-sensitive neurons of the posterior parietal cortex. J Neurosci 3:532–548
4. Andersen RA, Snyder LH, Bradley DC, Xing J (1997) Multimodal representation of space in the posterior parietal cortex and its use in planning movements. Annu Rev Neurosci 20:303–330
5. Asaad WF, Rainer G, Miller EK (1998) Neural activity in the primate prefrontal cortex during associative learning. Neuron 21:1399–1407
6. Avillac M, Deneve S, Olivier E, Pouget A, Duhamel JR (2005) Reference frames for representing visual and tactile locations in parietal cortex. Nat Neurosci 8:941–949
7. Batista AP, Buneo CA, Snyder LH, Andersen RA (1999) Reach plans in eye-centered coordinates. Science 285:257–260
8. Bentley N, Salinas E (2007) A general, flexible decision model applied to visual search BMC Neuroscience 8(Suppl 2):P30, doi: 10.1186/1471-2202-8-S2-P30
9. Bichot NP, Rossi AF, Desimone R (2005) Parallel and serial neural mechanisms for visual search in macaque area V4. Science 308:529–534
10. Botvinick M, Watanabe T (2007) From numerosity to ordinal rank: a gain-field model of serial order representation in cortical working memory. J Neurosci 27:8636–8642
11. Britten KH, Heuer HW (1999) Spatial summation in the receptive fields of MT neurons. J Neurosci 19:5074–5084.
12. Brotchie PR, Andersen RA, Snyder LH, Goodman SJ (1995) Head position signals used by parietal neurons to encode locations of visual stimuli. Nature 375:232–235
13. Buneo CA, Jarvis MR, Batista AP, Andersen RA (2002) Direct visuomotor transformations for reaching. Nature 416:632–636

14. Connor CE, Preddie DC, Gallant JL and Van Essen DC (1997) Spatial attention effects in macaque area V4. J Neurosci 17:3201–3214
15. Cox DD, Meier P, Oertelt N, DiCarlo JJ (2005) 'Breaking' position-invariant object recognition. Nat Neurosci 8:1145–1147
16. Deneve S, Latham PE, Pouget A (2001) Efficient computation and cue integration with noisy population codes. Nat Neurosci 4:826–831
17. Desimone R, Albright TD, Gross CG, Bruce C (1984) Stimulus-selective properties of inferior temporal neurons in the macaque. J Neurosci 4:2051–2062
18. DiCarlo JJ, Cox DD (2007) Untangling invariant object recognition. Trends Cogn Sci 11:333–341
19. Duhamel J-R, Colby CL, Goldberg ME (1992) The updating of the representation of visual space in parietal cortex by intended eye movements. Science 255:90–92
20. Fitzpatrick D (2000) Seeing beyond the receptive field in primary visual cortex. Curr Opin Neurobiol 10:438–443.
21. Ghose GM, Maunsell JH (2008) Spatial summation can explain the attentional modulation of neuronal responses to multiple stimuli in area V4. J Neurosci 28:5115–5126
22. Graziano MSA, Hu TX, Gross CG (1997) Visuospatial properties of ventral premotor cortex. J Neurophysiol 77:2268–2292
23. Grossberg S, Mingolla E, Ross WD (1994) A neural theory of attentive visual search: interactions of boundary, surface, spatial, and object representations. Psychol Rev 101:470–489
24. Jay MF, Sparks DL (1984) Auditory receptive fields in primate superior colliculus shift with changes in eye position. Nature 309:345–347
25. Knierim JJ, Van Essen DC (1992) Neuronal responses to static texture patterns in area V1 of the alert macaque monkey. J Neurophysiol 67:961–980.
26. Koida K, Komatsu H (2007) Effects of task demands on the responses of color-selective neurons in the inferior temporal cortex. Nat Neurosci 10:108–116
27. Lauwereyns J, Sakagami M, Tsutsui K, Kobayashi S, Koizumi M, Hikosaka O (2001) Responses to task-irrelevant visual features by primate prefrontal neurons. J Neurophysiol 86:2001–2010
28. Levitt JB, Lund JS (1997) Contrast dependence of contextual effects in primate visual cortex. Nature 387:73–76
29. Mack A, Rock I (1998) Inattentional Blindness. MIT Press, Cambridge, Masachussetts.
30. Martínez-Trujillo JC, Treue S (2004) Feature-based attention increases the selectivity of population responses in primate visual cortex. Curr Biol 14:744–751
31. McAdams CJ, Maunsell JHR (1999) Effects of attention on orientation tuning functions of single neurons in macaque cortical area V4. J Neurosci 19:431–441
32. Navalpakkam V, Itti L (2007) Search goal tunes visual features optimally. Neuron 53:605–617
33. Ogawa T, Komatsu H (2004) Target selection in area V4 during a multidimensional visual search task. J Neurosci 24:6371–6382
34. O'Reilly RC, Busby RS (2002) Generalizable relational binding from coarse-coded distributed representations. In: Dietterich TG, Becker S, Ghahramani Z (eds) Advances in neural information processing systems (NIPS). MIT Press, Cambridge, MA
35. O'Reilly RC, Busby RS, Soto R (2003) Three forms of binding and their neural substrates: alternatives to temporal synchrony. In: Cleermans A (ed) The unity of consciousness: binding, integration and dissociation. OUP, Oxford, pp. 168–192
36. Platt ML, Glimcher PW (1999) Neural correlates of decision variables in parietal cortex. Nature 400:233–238
37. Pouget A, Sejnowski TJ (1997) Spatial tranformations in the parietal cortex using basis functions. J Cogn Neurosci 9:222–237
38. Pouget A, Snyder LH (2000) Computational approaches to sensorimotor transformations. Nat Neurosci 3:1192–1198
39. Riesenhuber M, Poggio T (1999) Hierarchical models of object recognition in cortex. Nat Neurosci 2:1019–1025
40. Salinas E (2004) Fast remapping of sensory stimuli onto motor actions on the basis of contextual modulation. J Neurosci 24:1113–1118

41. Salinas E (2004) Context-dependent selection of visuomotor maps. BMC Neurosci 5:47, doi: 10.1186/1471-2202-5-47.
42. Salinas E (2005) A model of target selection based on goal-dependent modulation. Neurocomputing 65-66C:161–166
43. Salinas E, Abbott LF (1995) Transfer of coded information from sensory to motor networks. J Neurosci 15:6461–6474
44. Salinas E, Abbott LF (1997) Invariant visual responses from attentional gain fields. J Neurophysiol 77:3267–3272
45. Salinas E, Abbott LF (1997) Attentional gain modulation as a basis for translation invariance. In: Bower J (ed) Computational neuroscience: trends in research 1997. Plenum, New York, pp. 807–812
46. Salinas E, Abbott LF (2001) Coordinate transformations in the visual system: how to generate gain fields and what to compute with them. Prog Brain Res 130:175–190
47. Salinas E, Sejnowski TJ (2001) Gain modulation in the central nervous system: where behavior, neurophysiology, and computation meet. Neuroscientist 7:430–40
48. Salinas E, Thier P (2000) Gain modulation: a major computational principle of the central nervous system. Neuron 27:15–21
49. Sato M, Hikosaka O (2002) Role of primate substantia nigra pars reticulata in reward-oriented saccadic eye movement. J Neurosci 22:2363–2373
50. Shenoy KV, Bradley DC, Andersen RA (1999) Influence of gaze rotation on the visual response of primate MSTd neurons. J Neurophysiol 81:2764–2786
51. Sillito AM, Grieve KL, Jones HE, Cudeiro J, Davis J (1995) Visual cortical mechanisms detecting focal orientation discontinuities. Nature 378:492–496
52. Simoncelli EP, Heeger DJ (1998) A model of neuronal responses in visual area MT. Vision Res 38:743–761
53. Schwartz O, Simoncelli EP (2001) Natural signal statistics and sensory gain control. Nat Neurosci 4:819–825
54. Sohn JW, Lee D (2007) Order-dependent modulation of directional signals in the supplementary and presupplementary motor areas. J Neurosci 27:13655–13666
55. Stricanne B, Andersen RA, Mazzoni P (1996) Eye-centered, head-centered, and intermediate coding of remembered sound locations in area LIP. J Neurophysiol 76:2071–2076
56. Tovee MJ, Rolls ET, Azzopardi P (1994) Translation invariance in the responses to faces of single neurons in the temporal visual cortical areas of the alert macaque. J Neurophysiol 72:1049–1060
57. Treue S, Martínez-Trujillo JC (1999) Feature-based attention influences motion processing gain in macaque visual cortex. Nature 399:575–579
58. Wallis JD, Miller EK (2003) From rule to response: neuronal processes in the premotor and prefrontal cortex. J Neurophysiol 90:1790–1806
59. White IM, Wise SP (1999) Rule-dependent neuronal activity in the prefrontal cortex. Exp Brain Res 126:315–335
60. Wolfe JM (1994) Guided Search 2.0 – a revised model of visual search. Psychonomic Bul Rev 1:202–238
61. Wolfe JM, Horowitz TS, Kenner N, Hyle M, Vasan N (2004) How fast can you change your mind? The speed of top-down guidance in visual search. Vision Res 44:1411–1426
62. Xing J, Andersen RA (2000) Models of the posterior parietal cortex which perform multimodal integration and represent space in several coordinate frames. J Cogn Neurosci 12:601–614
63. Zipser D, Andersen RA (1988) A back-propagation programmed network that simulates response properties of a subset of posterior parietal neurons. Nature 331:679–684
64. Zoccolan D, Kouh M, Poggio T, DiCarlo JJ (2007) Trade-off between object selectivity and tolerance in monkey inferotemporal cortex. J Neurosci 27:12292–12307

Far in Space and Yet in Synchrony: Neuronal Mechanisms for Zero-Lag Long-Range Synchronization

Raul Vicente, Leonardo L. Gollo, Claudio R. Mirasso, Ingo Fischer, and Gordon Pipa

Abstract Distant neuronal populations are observed to synchronize their activity patterns at zero-lag during certain stages of cognitive acts. This chapter provides an overview of the problem of large-scale synchrony and some of the solutions that have been proposed for attaining long-range coherence in the nervous system despite long conduction delays. We also review in detail the synchronizing properties of a canonical neuronal microcircuit that naturally enhances the isochronous discharge of remote neuronal resources. The basic idea behind this mechanism is that when two neuronal populations relay their activities onto a third mediating population, the redistribution of the dynamics performed by the latter leads to a self-organized and lag-free synchronization among the pools of neurons being relayed. Exploring the physiological relevance of this mechanism, we discuss the role of associative thalamic nuclei and their bidirectional interaction with the neocortex as a relevant physiological structure in which the network module under study is densely embedded. These results are further supported by the recently proposed role of thalamocortical interactions as a substrate for the trans-areal cortical coordination.

Introduction

The development of multi-electrode recordings was a major breakthrough in the history of systems neuroscience [1]. The simultaneous monitoring of the extracellular electrical activity of several neurons provided a solid experimental basis for electrophysiologists to test the emergence of neuronal assemblies [2]. Specifically, the parallel registration of spike events resulting from different cells permitted the evaluation of temporal relationships among their trains of action potentials, an eventual signature of assembly organization. Modern multielectrode techniques have now the capacity to simultaneously listen to a few hundreds of cells and, in contrast to

R. Vicente (✉)
Department of Neurophysiology, Max-Planck Institute for Brain Research, Deutschordenstrasse 46, 60528 Frankfurt, Germany
e-mail: raulvicente@mpih-frankfurt.mpg.de

K. Josić et al. (eds.), *Coherent Behavior in Neuronal Networks*, Springer Series in Computational Neuroscience 3, DOI 10.1007/978-1-4419-0389-1_8,

serial single cell recordings, to reveal temporally coordinated firing among different neurons that is not linked to any external stimulus but rather to internal neuronal interactions. Only equipped with such class of technology it was possible to unveil one of the most interesting scenarios of structured timing among neurons, namely the consistent and precise simultaneous firing of several nerve cells, a process referred to as neuronal synchrony [3].

Neuronal synchronization has been hypothesized to underly the emergence of cell assemblies and to provide an important mechanism for the large-scale integration of distributed brain activity [3,4]. One of the basic ideas in the field is called the binding by synchrony theory which exploits the dimension that temporal domain offers for coding [3, 5–8]. Essentially, it states that synchrony can be instrumental for temporally bringing together the processing output of different functionally specialized areas in order to give rise to coherent percepts and behavior. The differential effect that synchronous vs. temporally dispersed inputs can exert onto a downstream neuron indicates how the temporal coherence of a set of neurons can become a flexible and potentially information-carrying variable that can modulate subsequent stages of processing [3,9, 10]. Despite an ongoing debate about its functional role in neuronal processing is still open, the last two decades have seen the accumulation of large amount of data which show evidence, at least in a correlative manner, for a role of synchrony and the oscillatory activity that often accompanies it in a variety of cognitive processes ranging from perceptual grouping or stimulus saliency to selective attention or working memory [7, 11–14].

Interestingly, neuronal synchrony is not restricted to the local environment of a single cortical column or area. Rather, long-range synchrony across multiple brain regions, even across inter-hemispheric domains, has been reported in several species including the cat and primate cortex [15–20]. However, the zero-lag correlated activity of remote neuronal populations seems to challenge a basic intuition. Namely, one tends to tacitly assume that since the interaction among distant systems is retarded by the conduction delays (and therefore, that it is the past dynamics of one system what is influencing the other one at present) it is not possible that such interaction alone can induce the isochronous covariation of the dynamics of two remote systems. Actually, the latencies associated with conducting nerve impulses down axonal processes can amount to several tens of milliseconds for a typical long-range fiber in species with medium- or large-sized brains [21–23]. These ranges of conduction delays are comparable with the time-scale in which neuronal processing unfolds and therefore they cannot be simply discarded without further justification. Furthermore, profound effects in the structure and dynamics of the nervous system might have arisen just as a consequence of the communication conditions imposed by the time delays [24, 25]. As an example, several proposals of the origin of the lateralization of brain functions are based on the temporal penalty to maintaining information transferring across both hemispheres [26, 27].

The aim of this chapter is to illustrate that appropriate neuronal circuitries can circumvent the phase-shifts associated with conduction delays and give rise to isochronous oscillations even for remote locations. The chapter begins with a brief review of some theories that have been proposed to sustain long-range synchrony in the nervous system. Then we explore a novel and simple mechanism able to account for zero-lag neuronal synchronization for a wide range of conduction delays [28–31]. For that purpose, we shall investigate the synchronizing properties of a specific network motif which is highly expressed in the cortico-thalamo-cortical loop and in the cortex itself [32–34]. Such circuitry consists of the relaying of two pools of neurons onto a third mediating population which indirectly connects them. The chapter goes on by presenting the results of numerical simulations of the dynamics of this circuit with two classes of models: first using Hodgkin and Huxley (HH) type of cells and second building large-scale networks of Integrate and Fire (IF) neurons. Finally, and after a detailed characterization of the influence of long-conduction delays in the synchrony of this neural module, we discuss our results in the light of the current theories about coherent cortical interactions.

How can Zero-Lag Long-Range Synchrony Emerge Despite of Conduction Delays?

Before discussing different mechanisms proposed to cope with the long-range synchrony problem, it is first necessary to understand the origin of the delay that arises in neuronal interactions. As a rule, it is possible to dissect in at least five different contributions the latency in the communication between two neurons via a prototypical axo-dendritic chemical synapse. For illustration purposes here we follow the time excursion of an action potential generated in a presynaptic cell up to becoming a triggering source for a new spike in a postsynaptic cell.

- The first component is due to the propagation of an action potential from the axon hillock to the synaptic terminal. The limited axonal conduction velocity imposes a delay ranging from a few to tens of milliseconds depending on the caliber, myelination, internodal distance, length of the axonal process, and even the past history of impulse conduction along the axon [23, 35, 36].
- A second element of brief latency occurs due to the synaptic transmission. After the action potential has reached the presynaptic ending several processes contribute to different degree to the so-called synaptic delay. These include the exocytosis of neurotransmitters triggered by calcium influx, the diffusion of the transmitters across the synaptic cleft, and their binding to the postsynaptic specializations. Altogether the complete process from the release to the binding to specialized channels can typically span from 0.3 ms to even 4 ms [37].
- Another source of delay is the rise time of the postsynaptic potential. Different ionic channels show different time-scales in producing a change in the membrane conductance which eventually induces the building-up of a significant potential. For fast ionotropic AMPA or $GABA_A$ receptors it can take a time of the order of half a millisecond for such a process to rise a postsynaptic potential [38].

- Dendritic propagation toward the soma by either passive or active conduction is also a source of a small lag which value depends on the dendritic morphology.
- Finally, the postsynaptic neuron can exploit several mechanisms, such as membrane potential fluctuations, to control to some degree an intrinsic latency in triggering a new action potential [39].

For long-distance fibers the most important contribution of delay typically comes from the axonal conduction. In human, an averaged-sized callosal axon connecting the temporal lobes of both hemispheres is reported to accumulate a delay of 25 ms [26]. This is certainly not a negligible quantity, specially in the light of results showing an important role for precise temporal relations among neuron discharges.

Nevertheless, a fiber connecting two brain regions is inevitably composed of non-identical axons, which give rise to a broad spectrum of axonal delays rather than a single latency value [26, 40]. This is one of the possible substrates for the establishment long-range synchrony, i.e., the systematic presence (within such a spectrum) of very fast axons reciprocally interconnecting all possible areas susceptible of expressing synchrony. Within this framework the combination of a hypothetical extensive network of very fast conducting axons with the phase resetting properties of some class of neurons could in principle sustain an almost zero-lag long-range synchrony process. GABAergic neurons have been indicated to meet the second of such requirements. Via a powerful perisomatic control this type of cells can exert a strong shunting and hyperpolarizing inhibition which can result in the resetting of oscillations at their target cells [41–43]. Their critical role in generating several local rhythms has been well described [44, 45]. However, their implication in the establishment of long-distance synchrony is heavily compromised because the expression of fast long-range projections by interneurons is more the exception than the rule [43, 45]. Another important consideration is that long-range connections in a brain do not come for free. Even a small fraction of long-distance wiring can occupy a considerably portion of brain volume, an important factor that severely restricts the use of fast large-diameter fibers [26, 45].

Electrical synapses, and in special gap junctions, have also been involved in explaining spread neuronal synchrony [46]. Gap junctions consist of clusters of specialized membrane channels that interconnect the intracellular media of two cells and mediate a direct electrical coupling and the transferring of small molecules between them [47]. Evidence for gap junctions' role in giving rise to fast rhythmic activity has been put forward by observations that fast oscillations can be generated in conditions where chemical synaptic transmission was blocked [48]. Gap junctions also present two clear advantages over chemical synapses for the induction of zero-lag synchrony. First, they are not affected by synaptic delays since no neurotransmitters are used. Second, the electrotonic coupling between cells mainly acts via diffusion mechanisms and therefore, it tends to homogenize the membrane potential of the cells involved. Thus, gap junctions can be considered of synchronizing nature rather than excitatory or inhibitory class [46]. However, as we have pointed out before for long-distance fibers the axonal delay is the largest component of latency and the saving corresponding the elimination of the synaptic delay can just correspond to a small fraction of the total. In any case, electrical synapses are believed to underly homogenization of firing among neurons and to foster synchrony in moderately distributed networks [46, 49, 50].

Proposals for explaining the observed long-range synchronous fast dynamics in the cortex have also been inspired by the study of coupling distant oscillations. In this context, R. Traub and others investigated the effect of applying dual tetanic stimulation in hippocampal slices [51]. The authors of [51] observed that a strong simultaneous tetanic stimulation at two distant sites in a slice preparation induced gamma-frequency oscillations that were synchronous. The concomitant firing of spike doublets by some interneurons with such double stimulation condition plus modeling support, led the authors to infer that a causal relationship between the interneuron doublet and the establishment of long-range synchrony should hold [51, 52].

From other perspective, it is important to recall that neuronal plasticity is a key element in determining the structural skeleton upon which dynamical states such as synchrony can be built. Therefore, the experience-driven process of shaping neuronal connectivity can considerably impact the ability and characteristics of synchronization of a given neuronal structure. Interestingly, this interaction can go in both directions and correlated input activity can also influence the connectivity stabilization via certain plasticity processes [53]. With respect to the specific issue of the influence of axonal delays in long-range coherence, modeling studies have shown that spike-timing-dependent plasticity rules can stabilize synchronous gamma oscillations between distant cortical areas by reinforcing the connections the delay of which matches the period of the oscillatory activity [54, 55].

In summary, there are a number of factors and mechanisms that have been put forward to explain certain aspects of the long-range synchronization of nerve cells. Synchronization is a process or tendency toward the establishment of a dynamical order with many possible participating sources, and as a result it is not strange that several mechanisms can simultaneously contribute or influence it. Thus, neural systems might use distinct strategies for the emergence of coherent activity at different levels depending on the spatial scale (local or long-range), dynamical origin (intracortical or subcortical oscillations), and physiological state (sleep or awake), among others. Nevertheless, one should notice that a significant long-range synchronization is observed across different species with different brain sizes and at different stages of the developmental growth of brain structures. This point strongly suggests that any robust mechanism for generating zero time-lag long-distance cortical synchrony maintains its functionality for a wide range of axonal lengths. While it is possible that developmental mechanisms compensate for the resulting delay variations [56] it is still difficult to explain all the phenomenology of long-distance synchronization without a mechanism that inherently allows for zero-lag synchronization for a broad range of conduction delays and cell types. In the following parts of this chapter, we focus our attention on a recently proposed scheme named dynamical relaying which might contribute to such mechanism [28–31].

Zero-Lag Long-Range Neuronal Synchrony via Dynamical Relaying

In this section, we explore a simple network module that naturally accounts for the zero-lag synchrony among two arbitrarily separated neuronal populations. The basic idea that we shall further develop later, is that when two neuronal populations relay their activities onto a mediating population, the redistribution of the dynamics performed by this unit can lead to a robust and self-organized zero-lag synchrony among the outer populations [28–31].

At this point it is important to recall the separation of processes generating local rhythms or oscillations in a brain structure from the mechanisms responsible for their mutual synchronization. The model and simulations that are presented below provide a proof of principle for a synchronizing mechanism among remote neuronal resources despite long axonal delays. No particular brain structure or physiological condition is intended to be faithfully reproduced, rather the main objective is the demonstration that under quite general conditions an appropriate connectivity can circumvent the phase lags associated to conduction delays and induce a zero-lag long-range synchrony among remote neuronal populations. In any case, it is worth mentioning that the diffuse reciprocal connectivity, the dynamical consequences of which we study below, is characteristic of the interaction of the neocortex with several thalamic nuclei [32, 33]. Connectivity studies in primate cortex have also identified the pattern of connections investigated here as the most frequently repeated network motif at the level of cortico-cortical connections [34, 57, 58].

Illustration of Dynamical Relaying in a Module of Three HH Cells

The most simple configuration to illustrate the effects of dynamical relaying corresponds to the study of the activities of two neurons that interact by mutually relaying their dynamics onto a third one. We begin then by investigating a circuit composed of three HH cells with reciprocal delayed synaptic connections (see top panel in Fig. 1 for an schematic representation of the network architecture). We first consider a condition in which the isolated neurons already operate in an intrinsic spiking state and observe how the synaptic activity modifies the timing of their action potentials. To this end we add an intracellular constant current stimulation ($10\,\mu A/cm^2$) so that each isolated neuron develops a tonic firing mode with a natural period of 14.7 ms. The initial phase of the oscillations of each cell is randomly chosen to exclude any trivial coherent effect. Finally, we also set all axonal conduction delays in the communication between neurons to a considerably long value of 8 ms to mimic the long-range nature of the synaptic interactions. Further details about the methodology used in the following simulations can be found at the "Methods" section at the end of the chapter. In Fig. 1 we show the evolution of the membrane potentials under such conditions before and after an excitatory synaptic coupling among the cells is activated.

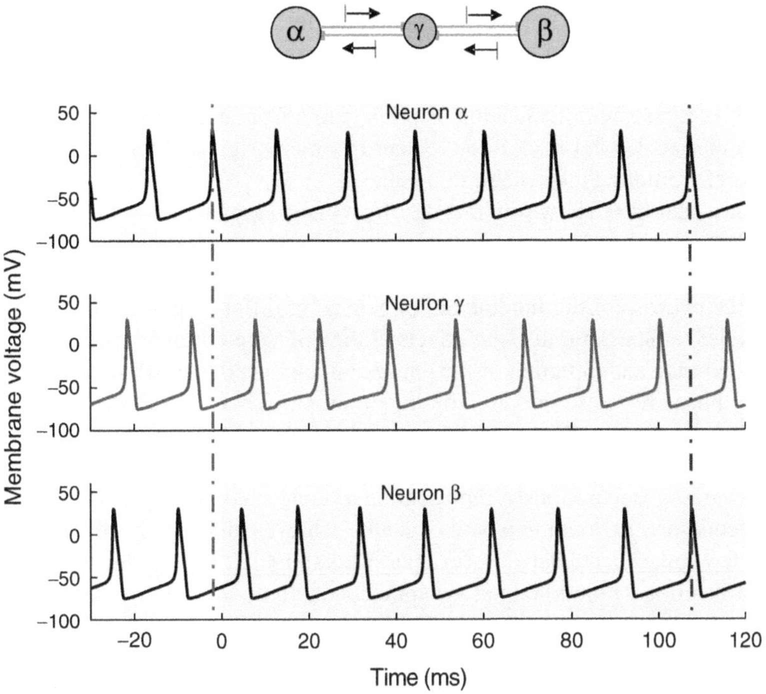

Fig. 1 Time series of the membrane voltage of three coupled HH cells $N_\alpha - N_\gamma - N_\beta$. At time $t = 0$ the excitatory synapses were activated. Conduction delay $\tau = 8$ ms. *Vertical lines* help the eye to compare the spike coherence before and after the interaction takes place

Previously to the switch-on of the synaptic coupling between the cells we can observe how the three neurons fire out of phase as indicated by the left vertical guide to the eye in Fig. 1. However, once the interaction becomes effective at $t = 0$ and synaptic activity is allowed to propagate, a self-organized process, in which the outer neurons synchronize their periodic spikes at zero-phase even in the presence of long conducting delays, is observed. It is important to notice that no external agent or influence is responsible for the setting of the synchronous state but this is entirely negotiated by the network itself. Furthermore, we checked that the present synchrony is not just a phase condition between purely periodic oscillators but a true temporal relationship. To that end, we added independent noisy membrane fluctuations to each neuron that resulted in a nonperfectly deterministic firing of the three neurons. In this case, the circuit maintained an approximated zero-lag synchrony between the outer neurons, reflecting both the robustness of the synchrony mechanism to moderate noise perturbations and showing that the synchrony process can be generalized beyond a phase relation.

The mechanism responsible for synchronization depends on the ability of an EPSP to modify the firing latencies of a postsynaptic neuron in a consistent manner. It further relies on the symmetric relay that the central neuron provides for the indirect communication between the outer neurons. The key idea is that the network motif under study allows for the outer neurons to exert an influence on each other via the intermediate relay cell. Thus, the reciprocal connections from the relay cell assure that the same influence that is propagating from one extreme of the network to the other is also fed-back into the neuron which originated the perturbation and therefore, promoting the synchronous state.

It must be noticed, however, that the effect of a postsynaptic potential on a neuron strongly depends on the internal state of the receiving cell, and more specifically on the phase of its spiking cycle at which a postsynaptic potential (PSP) arrives [59,60]. Since the neurons of the module are in general at different phases of their oscillatory cycles (at least initially) the effects of the PSPs are different for the three cells. The magnitude and direction of the phase-shifts induced by PSPs can be characterized by phase response curves. The important point here is that the accumulation of such corrections to the interspike intervals of the outer neurons is such that after receiving a few PSPs they compensate the initial phase difference and both cells end up discharging isochronously, representing a stable state. Simulations predict that a millisecond-precise locking of spikes can be achieved already after the exchange of only a few spikes in the network (in a period as short as 100 ms). This value is found to be a function of the maximal synaptic conductivity and can be even shorter for stronger synapses.

A key issue of the synchronization properties exhibited by such network architecture is whether the zero-lag correlation can be maintained for different axonal lengths or whether it is specific to a narrow range of axonal delays. To resolve this issue we need to test the robustness of the synchronous solution for other values of the conduction delays. In Fig. 2, we show the quality of the zero-lag synchronization for two HH cells as a function of the conduction delay. In that graph we plot the results for two different scenarios: one in which the neurons are directly coupled via

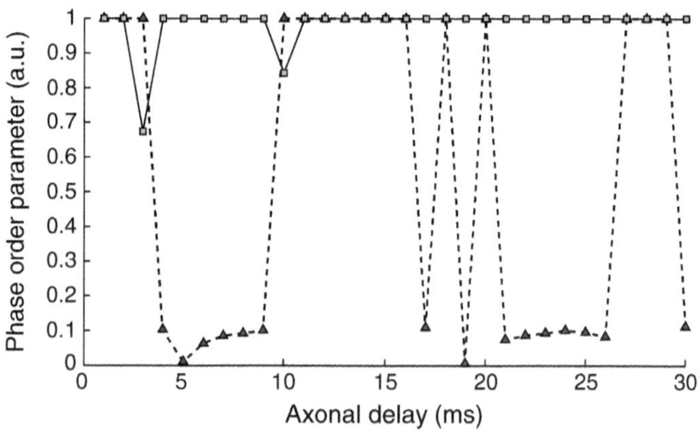

Fig. 2 Dependence of zero time-lag synchronization as a function of the axonal delay for a scheme of two coupled cells (*dashed line*) and three coupled cells (*solid line*). In the case of the three interacting cells only the synchrony between the outer neurons is plotted here

excitatory synapses (dashed line) and a second one in which the two neurons interact through a relay cell also in an excitatory manner (solid line). A quick comparison already reveals that while the direct excitatory coupling exhibits large regions of axonal conduction delays where the zero-lag synchrony is not achieved, the relay-mediated interaction leads to zero time-lag synchrony in 28 out of the 30 delay values explored, (1–30 ms). Only for the cases of $\tau = 3$ ms and $\tau = 10$ ms the network motif under study does not converge to the isochronous discharge for the outer neurons. For such latencies the three cells entered into a chaotic firing mode in which the neurons neither oscillate with a stable frequency nor exhibit a consistent relative lag between their respective spike trains.

Robust zero-lag synchrony among the outer neurons is also observed when the synaptic interaction between the cells is inhibitory instead of excitatory. Different synaptic rise and decay times within the typical range of fast AMPA and $GABA_A$ mediated transmission were tested with identical results as those reported above. These results indicate that the network motif of two neurons relaying their activities through a third neuron leads to a robust zero-lag synchrony almost independently of the delay times and type of synaptic interactions. We have also conducted simulations to test the robustness of this type of synchrony with respect to the nature of the relay cell. The results indicate that when a relay cell is operating in a parameter regime different from the outer ones (such as different firing rate or conductances), the zero-lag synchrony is not disturbed. Remarkably, even in the case where the relay cell is operating in a subthreshold regime, and thus only spiking due to the excitatory input from any of the outer neurons, the process of self-organization toward the zero-lag synchrony is still observed. It is also worth mentioning that in all cases such firing coherence is achieved through small shifts in the spiking latencies which leave the mean frequency of discharges (or rate) almost unchanged.

Effect of a Broad Distribution of Conduction Delays

Axons show a significant dispersion in properties such as diameter, myelin thickness, internodal distance, and past history of nerve conduction. Within a fiber bundle the variability from one axon to another of these characteristics is directly related to the speed of propagation of action potentials along them and eventually translates into the existence of a whole range of latencies in the neuronal communication between two separated brain areas. Thus, conduction times along fibers are more suitably considered as a spectrum or distribution rather than a single latency value [26, 40].

A crucial question is therefore whether the synchronization transition that we have described in the former section is restricted to single latency synaptic pathways or preserved also for broad distributions of axonal delays. To answer this issue we model the dispersion of axonal latencies by assuming that individual temporal delays of the arrivals of presynaptic potentials (i.e., latency times) are spread according to a given distribution. This intends to mimic the variability among the different

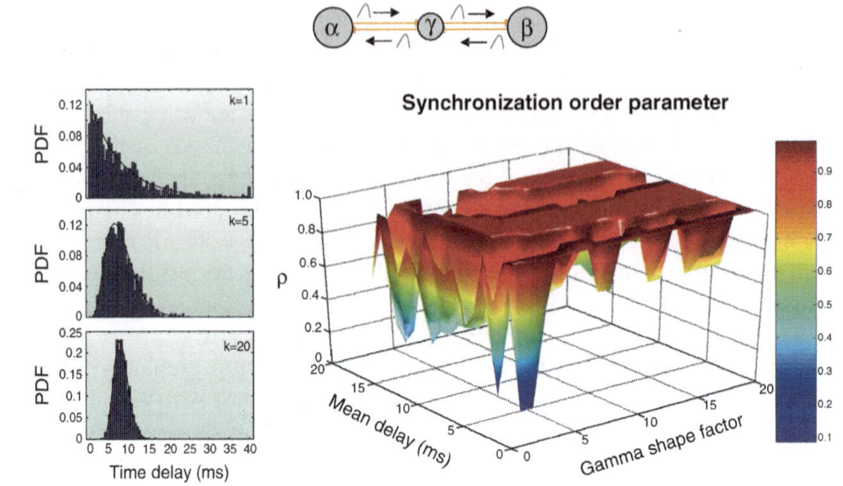

Fig. 3 *Left panels*: gamma distribution of delays with different shape factors ($k = 1, 5$, and 20) and the same mean ($\tau = 8$ ms). *Right panel*: synchronization index at zero-lag of the outer neurons as a function of the shape factor and mean of the distribution of delays.

axons within a fiber bundle connecting two neuronal resources. Since data about axonal distributions of conduction velocities in long-range fibers is limited, specially in the case of humans [26, 40], and there is probably not a unique prototypical form of such distributions we explore a whole family of gamma distributions with different shapes (see the "Methods" section). The left panels shown in Fig. 3 illustrate different gamma distributions of axonal delays for three different shape factors.

Our numerical simulations indicate that for a large region of mean delays (between 3 and 10 ms) the outer neurons synchronize independently of the shape of the distribution. These results can be observed in the right panel of Fig. 3 where we plot the zero-lag synchronization index of the outer neurons of the network motif as a function of the shape of the gamma distribution of axonal delays and its mean value. Only distributions with unrealistic small shape factor (i.e., exponentially decaying distributions) prevent synchrony irrespective of the average delay of the synaptic connections. For more realistic distributions, there is a large region of axonal delays that gives rise to the zero-lag synchrony among the outer neurons. As in the case of single latencies, we find a drop in the synchrony quality for distributions with a mean value around $\hat{\tau} \sim (10\text{–}12)$ ms, where chaotic firing is observed. The isochronous spiking coherence is in general recovered for larger mean delay values.

So far we have considered a rather symmetric situation in which similar distributions of axonal delays are present in each of the two branches that connect the relay neuron to the outer units. This assumption can only hold when the relay cell is approximately equidistant from the outer ones. In the final section of this chapter we refer to several results pointing to the thalamic nuclei and their circuitry as ideal relay centers of cortical communication which approximately satisfy this condition. It is nevertheless advisable to investigate the situation in which the axonal delays

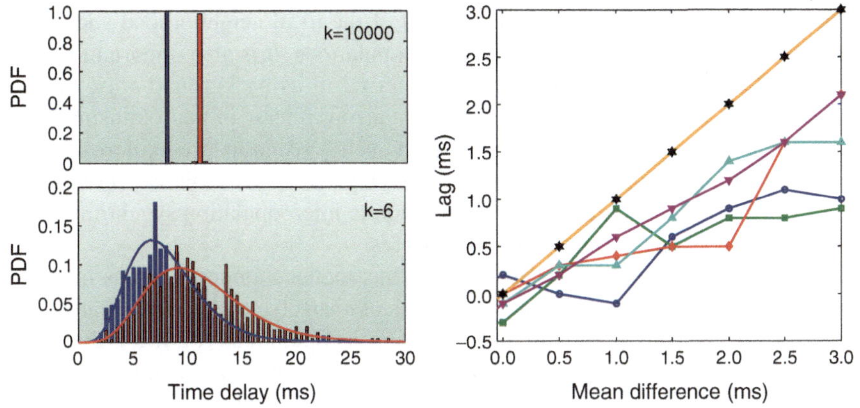

Fig. 4 *Left panels*: different gamma distributions of delays used for the two dissimilar branches of the network module. *Upper left panel* shows distributions with shape factor $k = 10,000$ (quasi-delta) and means of 8 and 11 ms. *Bottom left panel* shows distributions with shape factor $k = 6$ and means of 8 and 11 ms. *Right panel*: lag between the discharges of the outer neurons as a function of the difference in the mean of the distributions of delays for the two branches. Shape factors $k = 6$ (*squares*), $k = 8$ (*circles*), $k = 10$ (*diamonds*), $k = 12$ (*up-triangles*), $k = 14$ (*down-triangles*), and $k = 10,000$ (*stars*) were tested.

of each of the two pathways of the network motif are described by dissimilar distributions. In this case, we find that if the distributions of delays for each branch have different mean values then a nonzero phase-lag appears between the dynamics of the outer neurons. This effect is illustrated for gamma distributions of different shape factors in Fig. 4. For delta distributions of delays (which is equivalent to the single latency case) the lag amounts to the difference in mean values. Thus, if one of the pathways is described by a delta distribution of delays centered at $\tau_a = 5$ ms while the other is represented by a latency of $\tau_b = 7$ ms, then after some transient the neuron closer to the relay cell consistently fires 2 ms (i.e., $\tau_b - \tau_a$) in advance to the other outer neuron. It is worth to note that such value is still much smaller than the total delay accumulated to communicate both neurons ($\tau_a + \tau_b = 12$ ms). When studying the effect of broader distributions of delays, we observed that outer cells tend to fire with a lag even smaller than the difference in the mean values of the distributions. Thus, our results suggest that broader distributions of delays can help distant neurons to fire almost isochronously.

Dynamical Relaying in Large-Scale Neuronal Networks

A further key step in demonstrating the feasibility of synchronizing widely separated neurons via dynamical relaying is the extension of the previous results to the level of neuronal populations, the scale at which neuronal microcircuits develop their function [61]. Far from being independent, the dynamical response of any neuron is massively affected by the activity of the local neighborhood and by the long-range afferents originating in distant populations. It is also important to consider the random-like influences usually referred to as background noise, a term that collects a variety of processes from spontaneous release of neurotransmitters to fluctuations of unspecific inputs [62, 63]. In such a scenario, we explore whether long-range fibers supporting dynamical relaying, and thus indirectly connecting pools of neurons, are suitable to promote remote interpopulation synchrony in the presence of local interactions and noise sources.

To check if zero-lag correlated firing is thus induced among neurons in different populations we built three large networks of sparsely connected excitatory and inhibitory IF neurons. We interconnect the three populations following the topology of the network motif under study, i.e., the mutual relaying of activities of two external populations onto an intermediate pool of relay neurons. For details on the building of each network and their connectivity see the "Methods" section.

We first begin by initializing the three networks without the long-range interpopulation connections. Thus only the recurrent local connections and the Poissonian external background are active and then responsible for any dynamics in the stand-alone networks. Consequently, each population initially exhibits incoherent spiking of their neurons with respect to neurons belonging to any of the other populations. Once the long-range synapses are activated at $t = 100$ ms, we observe how the firing of the neurons organize toward the collective synchrony of the outer populations. Thus, the firing cycles of the outer networks of neurons occur with decreasing phase lags until both populations discharge near simultaneously and exhibit almost zero-phase synchrony. Figure 5 illustrates the typical raster plots, firing histograms, and cross-correlograms of neurons among the three inter-connected networks for a long conduction delay of 12 ms. Similar results are observed when other axonal delays in the range of 2–20 ms are explored.

The effective coupling of the networks modifies the relative timing among their spikes yielding populations 1 and 3 to rapidly synchronize. However, the qualitative dynamics of each single neuron seems to be not so much altered by the interaction and periodic firing of comparable characteristics is found in both the coupled and uncoupled case (compare the firing of the central population in Figs. 5 and 6 where in the latter the population 2 remains uncoupled from other populations). Indeed, the mean period of the coupled oscillatory activity ($\sim$32 ms) is found to be close to the local rhythm of an isolated network ($\sim$34 ms), and therefore the coupling has little effect on the frequency of oscillation. This indicates that zero-lag synchrony can be brought by this mechanism via small latency shifts without hardly affecting the nature of the neuronal dynamics. A different situation might appear when no prominent oscillatory activity is present in the isolated networks before they are functionally coupled. In that case (not illustrated here) we find that the reciprocal coupling among the networks can act as a generator of oscillations and zero-lag synchrony. In the latter case we find the period of the population oscillations strongly influenced by the conduction delay times and as a result by the coupling.

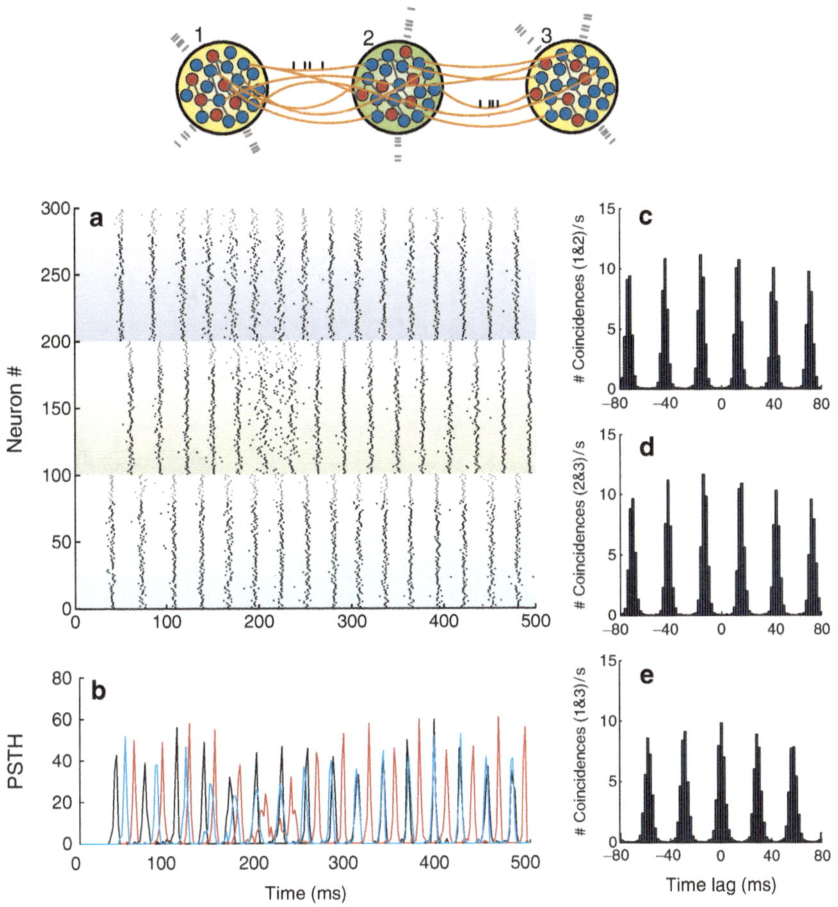

Fig. 5 Dynamics of three large-scale networks interacting through dynamical relaying. Panel (**a**) raster plot of 300 neurons randomly selected among the three populations (neurons 1–100 are from Pop. 1, 101–200 from Pop. 2, and 201–300 from Pop. 3). The top 20 neurons of each subpopulation (*plotted in gray*) are inhibitory, and the rest excitatory (*black*). Panel (**b**) firing histogram of each subpopulation of 100 randomly selected neurons (*black*, *red*, and *blue colors* code for populations 1, 2, and 3, respectively). Panel (**c**) averaged cross-correlogram between neurons of Pop. 1 and Pop. 2. Panel (**d**) averaged cross-correlogram between neurons of Pop. 2 and Pop. 3. Panel (**e**) averaged cross-correlogram between neurons of Pop. 1 and Pop. 3. At $t = 100$ ms the external interpopulation synapses become active. Bin sizes for the histogram and correlograms is set to 2 ms. Interpopulation axonal delays are set to 12 ms.

To better determine the role of the relay cells (Pop. 2) in shaping the synchronization among cells belonging to remote neuronal networks (Pop. 1 and Pop. 3), we designed the following control simulation. We investigated the neuronal dynamics obtained under exactly the same conditions as in the former approach with the only variation that this time the two outer networks interacted directly. The results

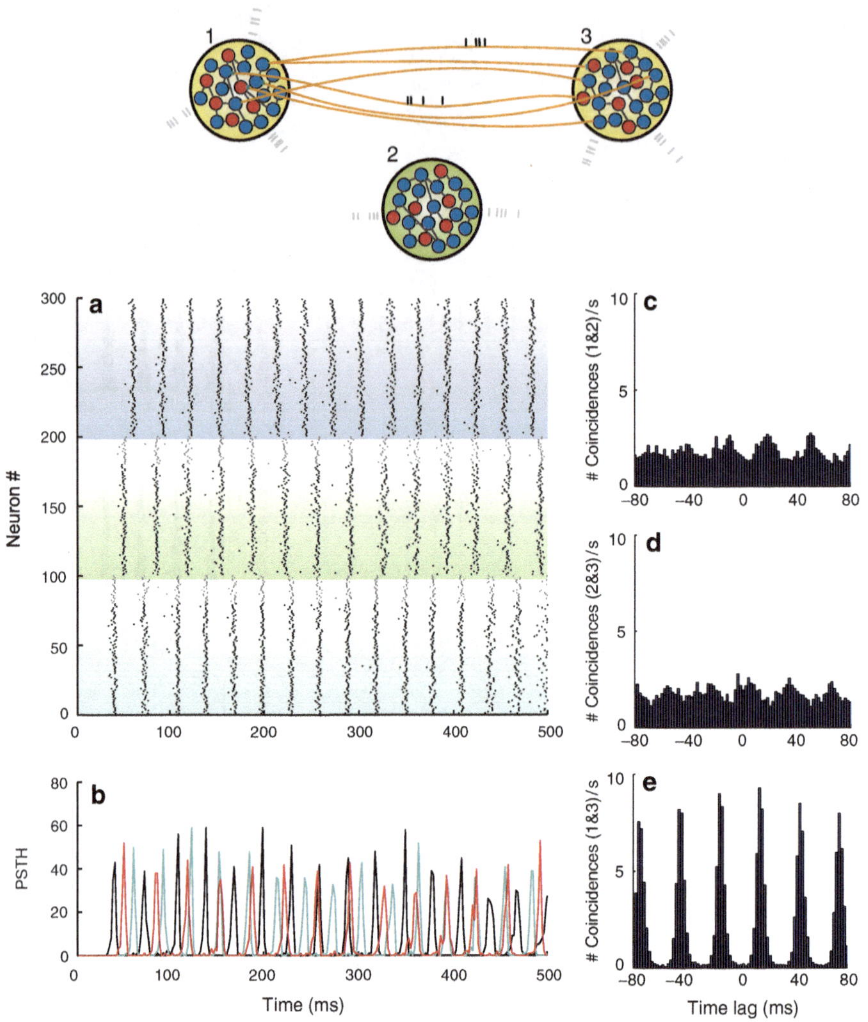

Fig. 6 Dynamics of two large-scale networks interacting directly. Population 2 is disconnected from other populations. Panel (**a**) raster plot of 300 neurons randomly selected among the three populations (neurons 1–100 are from Pop. 1, 101–200 from Pop. 2, and 201–300 from Pop. 3). The top 20 neurons of each subpopulation (*plotted in gray*) are inhibitory, and the rest excitatory (*black*). Panel (**b**) firing histogram of each subpopulation of 100 randomly selected neurons (*black, red*, and *blue* colors code for populations 1, 2, and 3, respectively). Panel (**c**) averaged cross-correlogram between neurons of Pop. 1 and Pop. 2. Panel (**d**) averaged cross-correlogram between neurons of Pop. 2 and Pop. 3. Panel (**e**) averaged cross-correlogram between neurons of Pop. 1 and Pop. 3. At $t = 100$ ms the external interpopulation synapses become active. Bin sizes for the histogram and correlograms is set to 2 ms. Interpopulation axonal delays are set to 12 ms.

are summarized in Fig. 6. Although only the topology of the connections has been changed, this is enough to eliminate the zero-lag synchronization of networks 1 and 3, highlighting the essential role of a relaying population.

So far we have focused on studying how a V-shaped network motif with reciprocal interactions determines the synchronization properties of the neurons composing it and compared them to the case of a direct reciprocal coupling between two populations. The results above indicate that the V-shaped structure can promote the zero-lag synchrony between their indirectly coupled outer populations for long delays, while just direct connections between two populations can sustain zero-lag synchrony only for limited amounts of axonal latencies. However, usually both situations are expected to occur simultaneously, this is neuronal populations that need to be coordinated might be linked by both direct (monosynaptic) and nondirect (multisynaptic) pathways. Therefore, we also conducted some numerical studies about how the addition of a direct bidirectional coupling between the populations 1 and 3 (and thus closing the open end of the sketch shown in top of Fig. 5 to form a ring) modified the synchronization properties formerly described, which were due only to their indirect communication via population 2. We observed that when the connectivity (or number of synapses) between the pools of neurons 1 and 3 is moderate and smaller than the connectivity between these populations and the relay population 2, then a zero-lag synchronous dynamics between the outer populations still emerges. This holds even for the case when the synapses linking Pop. 1 and Pop. 3 have a different delay than the ones linking these to the relay center. As expected, when the reciprocal connectivity between pools 1 and 3 is stronger then direct coupling dominates and, depending on the delay, it can impose a nonzero lag synchronous solution.

Among other relevant networks that might sustain the emergence zero-lag synchrony among some of its nodes stands the star topology where a central hub is reciprocally connected to other nodes. For such an arrangement, which in some sense can be understood as to be composed of several V-shaped motifs, numerical simulations show that the outer elements of the star that are bidirectionally connected to the central hub also tend to engage in zero-lag synchronous spiking.

General Discussion, Conclusions and Perspectives

In this chapter, we have dealt with the intriguing problem of explaining how long-range synchrony can emerge in the presence of extensive conduction delays. This challenging question that has attracted the attention of many researchers is still far from being fully clarified. Nevertheless, our main goal in the previous pages was to disseminate the idea that in addition to intrinsic cellular properties an appropriate neuronal circuitry can be essential in circumventing the phase shifts associated with conduction delays. In particular, here we have explored and shown how a simple network topology can naturally enhance the zero-lag synchronization of distant populations of neurons. The neuronal microcircuit that we have considered consists of the relaying of two pools of neurons onto a third mediating population which indirectly connects them. Simulations of Hodgkin and Huxley cells as well as large networks of integrate and fire neurons arranged in the mentioned configuration demonstrated a self-organized tendency toward the zero-lag synchronous state despite of large axonal delays. These results suggest that the presence of such connectivity pattern in neuronal circuits may contribute to the large-scale synchronization phenomena reported in a number of experiments in the last two decades [3, 4].

Is there in the brain any particular structure where such connectivity pattern is significantly common and one of its main building blocks? Within the brain complex network the thalamus and its bidirectional and radial connectivity with the neocortex form a key partnership. Several authors have indicated that the reciprocal coupling of cortical areas with the different thalamic nuclei may support mechanisms of distributed cortical processing and even form a substrate for the emergence of consciousness [64–67]. It has also been explicitly proposed that diffuse cortical projections of matrix cells in the dorsal thalamus together layer V corticothalamic projections are an ideal circuitry to extend thalamocortical activity and sustain the synchronization of widespread cortical and thalamic cells [32, 33]. The resemblance of such circuitry with the topology studied here is evident once the identification of the associative nuclei of the thalamus as our relay population for cortical activity is done. Altogether, the results described in this chapter point to the direction that long axonal latencies associated with cortico-thalamo-cortical loops are still perfectly compatible with the isochronous cortical synchronization across large distances. Within this scheme the most important requirement for the occurrence of zero-lag synchronization is that the relay population of cells occupies a temporally equidistant location from the pools of neurons to be synchronized. It is then highly significant that recent studies have identified a constant temporal latency between thalamic nuclei and almost any area in the mammalian neocortex [68]. Remarkably, this occurs irrespectively of the very different distances that separate the thalamus and the different cortex regions involved and relies on the adjustment of conduction velocity by myelination. Thus, thalamic nuclei occupy a central position for the mediation of zero-phase solutions.

Coherent dynamics between remote cortical populations could of course be generated also by reciprocally coupling these areas to yet another cortical area or other subcortical structures. It is important to remark that connectivity studies in primate cortex have identified the pattern of connections studied here as the most frequently repeated motif at the level of cortico-cortical connections in the visual and other cortical systems [34, 57, 58]. The functional relevance of this topology in cortical networks is unclear but according to our results is ideally suited to sustain coherent activity.

In general, it is quite possible that a variety of mechanisms are responsible for bringing synchrony at different levels (distinguishing for example, among local and long-distance synchrony) and different cerebral structures. The fact that each thalamus projects almost exclusively ipsilaterally (the massa intermedia is clearly inadequate for supporting the required interthalamic communication) is already an indication that the callosal commissure should play a prominent role in facilitating interhemispheric coherence. Lesion studies have since long confirmed this view [69]. However, within a single hemisphere the disruption of intracortical connectivity by a deep coronal cut through the suprasylvian gyrus in the cat cortex was observed to not disturb the synchrony of spindle oscillations across regions of cortex located at both sides of the lesion [70]. This suggests that subcortical, and in particular cortico-thalamic interactions, could be responsible not only for the generation of oscillations but also for maintaining both the long-range cortical and thalamic coherence found in such regimes. It is likely then that subcortical loops with widespread connectivity such as the associative or nonspecific cortico-thalamo-cortical circuits could run in parallel as an alternative pathway for the large-scale integration of cortical activity within a single hemisphere [33, 61, 67]. As we have proven here, with such connectivity pattern even large axonal conduction delays would not represent an impediment for the observation of zero time-lag coherence.

We would like to stress here that conduction delays are an important variable to consider not only in synchrony but in any temporal coding strategy. They contribute with an intrinsic temporal latency to neuronal communication that adds to the precise temporal dynamics of the neurons. Thus, they may have an important implication in gating mechanisms based in temporal relationships. For instance, when assisted by membrane oscillations neurons undergo repetitive periods of interleaved high and low excitability and it has been reported that the impact of a volley of spikes bombarding one of such oscillatory neuron is strongly influenced by the phase of the cycle (variable influenced by conduction delays) at which the action potentials reach the targeting neuron [39]. Conduction delays along with the frequency and phase difference of two respective oscillatory processes determine the timing of the arrival of inputs and therefore can control whether the incoming signal will be relatively ignored (when coinciding the trough of excitability) or processed further away (when reaching the neuron at the peak of the fluctuating depolarization) [10, 71]. By this mechanism it has been hypothesized that a dynamically changing coherent activity pattern may ride on top of the anatomical structure to provide flexible neuronal communication pathways [71]. Based on the properties formerly reviewed subcortical structures such as some thalamic nuclei might be in an excellent situation to play a role in regulating such coherence and contribute to the large-scale cortical communication.

In summary, the network motif highlighted here has the characteristic of naturally inducing zero-lag synchrony among the firing of two separated neuronal populations. Interestingly, such property is found to hold for a wide range of conduction delays, a highly convenient trait not easily reproduced by other proposed mechanisms, which have a more restricted functionality in terms of axonal latencies. Regarding its physiological substrate, the associative thalamic nuclei have the cortex as their main input and output sources and seem to represent active relay centers of cortical activity with properties well suitable for enhancing cortical coherence [33]. The advantage of this approach in terms of axonal economy, specially compared to an extensive network of fast long-range cortical links, is overwhelming. Ongoing research is being directed to a detailed modeling of the interaction between cortex and such nuclei with an emphasis in investigating the role of limited axonal conduction velocity. From the experimental side the relatively well controlled conditions of thalamocortical slice experiments, allowing for the identification of synaptically coupled neurons and cell class, might be a first step for testing whether the topology investigated here provides a significant substrate for coherent spiking activity. An important issue related to the physical substrate of synchrony is how the dynamic selection of the areas that engage and disengage into synchrony can be achieved but that is a subject beyond the scope of the present chapter.

Methods

Models

Two neuronal models were simulated to test the synchronization properties of the neuronal circuits investigated here.

In the most simplified version we focused on the dynamics of two single-compartment neurons that interact with each other via reciprocal synaptic connections with an intermediate third neuron of the same type (see top panel in Fig. 1). The dynamics of the membrane potential of each neuron was modeled by the classical Hodgkin–Huxley equations [72] plus the inclusion of appropriate synaptic currents that mimic the chemical interaction between nerve cells. The temporal evolution of the voltage across the membrane of each neuron is given by

$$C\frac{\mathrm{d}V}{\mathrm{d}t} = -g_{\mathrm{Na}}m^3h(V - E_{\mathrm{Na}}) - g_{\mathrm{K}}n^4(V - E_{\mathrm{K}})$$
$$-g_{\mathrm{L}}(V - E_{\mathrm{L}}) + I_{\mathrm{ext}} + I_{\mathrm{syn}}, \quad (1)$$

where $C = 1\ \mu\mathrm{F/cm}^2$ is the membrane capacitance, the constants $g_{\mathrm{Na}} = 120\,\mathrm{mS/cm}^2$, $g_{\mathrm{K}} = 36\,\mathrm{mS/cm}^2$, and $g_{\mathrm{L}} = 0.3\,\mathrm{mS/cm}^2$ are the maximal conductances of the sodium, potassium, and leakage channels, and $E_{\mathrm{Na}} = 50\,\mathrm{mV}$, $E_{\mathrm{K}} = -77\,\mathrm{mV}$, and $E_{\mathrm{L}} = -54.5\,\mathrm{mV}$ stand for the corresponding reversal potentials. According to Hodgkin and Huxley formulation the voltage-gated ion channels are described by the following set of differential equations

$$\frac{\mathrm{d}m}{\mathrm{d}t} = \alpha_m(V)(1 - m) - \beta_m(V)m, \quad (2)$$

$$\frac{\mathrm{d}h}{\mathrm{d}t} = \alpha_h(V)(1 - h) - \beta_h(V)h, \quad (3)$$

$$\frac{\mathrm{d}n}{\mathrm{d}t} = \alpha_n(V)(1 - n) - \beta_n(V)n, \quad (4)$$

where the gating variables $m(t)$, $h(t)$, and $n(t)$ represent the activation and inactivation of the sodium channels and the activation of the potassium channels, respectively. The experimentally fitted voltage-dependent transition rates are

$$\alpha_m(V) = \frac{0.1(V+40)}{1-\exp(-(V+40)/10)}, \tag{5}$$

$$\beta_m(V) = 4\exp(-(V+65)/18), \tag{6}$$

$$\alpha_h(V) = 0.07\exp(-(V+65)/20), \tag{7}$$

$$\beta_h(V) = [1+\exp(-(V+35)/10)]^{-1}, \tag{8}$$

$$\alpha_n(V) = \frac{(V+55)/10}{1-\exp(-0.1(V+55))}, \tag{9}$$

$$\beta_n(V) = 0.125\exp(-(V+65)/80). \tag{10}$$

The synaptic transmission between neurons is modeled by a postsynaptic conductance change with the form of an alpha-function

$$\alpha(t) = \frac{1}{\tau_{\mathrm{d}}-\tau_{\mathrm{r}}}\left(\exp(-t/\tau_{\mathrm{d}}) - \exp(-t/\tau_{\mathrm{r}})\right), \tag{11}$$

where the parameters τ_{d} and τ_{r} stand for the decay and rise time of the function and determine the duration of the response. Synaptic rise and decay times were set to $\tau_{\mathrm{r}} = 0.1$ and $\tau_{\mathrm{d}} = 3$ ms, respectively. Finally, the synaptic current takes the form

$$I_{\mathrm{syn}}(t) = -\frac{g_{\max}}{N}\sum_{\tau_{\mathrm{l}}}\sum_{\mathrm{spikes}} \alpha\left(t - t_{\mathrm{spike}} - \tau_{\mathrm{l}}\right)\left(V(t) - E_{\mathrm{syn}}\right), \tag{12}$$

where $g_{\max}$ (here fixed to $0.05\,\mathrm{mS/cm^2}$) describes the maximal synaptic conductance and the internal sum is extended over the train of presynaptic spikes occurring at t_{spike}. The delays arising from the finite conduction velocity of axons are taken into account through the latency time τ_{l} in the alpha-function. Thus, the external sum covers the N different latencies that arise from the conduction velocities that different axons may have in connecting two neuronal populations. N was typically set to 500 in the simulations. For the single-latency case, all τ_{l} were set to the same value, whereas when studying the effect of a distribution of delays we modeled such dispersion by a gamma distribution with a probability density of

$$f(\tau_{\mathrm{l}}) = \tau_{\mathrm{l}}^{k-1}\frac{\exp(-\tau_{\mathrm{l}}/\theta)}{\theta^k\Gamma(k)}, \tag{13}$$

where k and θ are shape and scale parameters of the gamma distribution. The mean time delay is given by $\hat{\tau}_{\mathrm{l}} = k\theta$.

Excitatory and inhibitory transmissions were differentiated by setting the synaptic reversal potential to be $E_{\mathrm{syn}} = 0$ mV or $E_{\mathrm{syn}} = -80$ mV, respectively. An external current stimulation I_{ext} was adjusted to a constant value of $10\,\mu\text{A/cm}^2$. Under such conditions a single HH type neuron enters into a periodic regime firing action potentials at a natural period of $T_{\text{nat}} = 14.66\,\text{ms}$.

The second class of models we have considered consists of three large balanced populations of integrate and fire neurons [73]. Top panel in Fig. 5 depicts a sketch of the connectivity. Each network consists of 4,175 neurons of which 80% are excitatory. The internal synaptic connectivity is chosen to be random, i.e., each neuron synapses with a 10% of randomly selected neurons within the same population, such that the total number of synapses in each network amounts to about 1,700,000. Additionally, to model background noise, each neuron is subjected to the influence of an external train of spikes with a Poissonian distribution as described below. The interpopulation synaptic links are arranged such that each neuron in any population receives input from 0.25% of the excitatory neurons in the neighboring population. This small number of interpopulation connections, compared to the much larger number of intrapopulation contacts, allows us to consider the system as three weakly interacting networks of neurons rather than a single homogeneous network. Intrapopulation axonal delays are set to 1.5 ms, whereas the fibers connecting different populations are assumed to involve much longer latencies in order to mimic the long-range character of such links.

The voltage dynamics of each neuron was then given by the following equation

$$\tau_{\text{m}} \frac{\mathrm{d}V_i}{\mathrm{d}t} = -V_i(t) + RI_i(t), \tag{14}$$

where τ_{m} stands for the membrane constant and $I(t)$ is a term collecting the currents arriving to the soma. The latter is decomposed in postsynaptic currents and external Poissonian noise

$$RI_i(t) = \tau_{\text{m}} \sum_j J_j \sum_k \delta(t - t_j^k - \tau_l) + A\xi_i, \tag{15}$$

where J_j is the postsynaptic potential amplitude, t_j^k is the emission time of the kth spike at neuron j, and τ_{l} is the transmission axonal delay. The external noise ξ_i is simulated by subjecting each neuron to the simultaneous input of 1,000 independent homogeneous Poissonian action potential trains with an individual rate of 5 Hz. Different cells were subjected to different realizations of the Poissonian processes to ensure the independence of noise sources for each neuron. J_{exc} and A amplitudes were set to 0.1 mV. The balance of the network was controlled by setting $J_{\text{inh}} = -3.5 J_{\text{exc}}$ to compensate the outnumber of excitatory units.

The dynamics of each neuron evolved from the reset potential of $V_{\text{r}} = 10\,\text{mV}$ by means of the synaptic currents up to the time when the potential of the ith neurons reached a threshold of 20 mV, value at which the neuron fires and its potential relaxes to V_{r}. The potential is clamped then to this quantity for a refractory period of 2 ms during which no event can perturb this neuron.

Simulations

The set of equations (1–12) was numerically integrated using the Heun method with a time step of 0.02 ms. For the first class of models we investigated, i.e., the three HH cells neuronal circuit, we proceeded as follows. Starting from random initial conditions each neuron was first simulated without any synaptic coupling for 200 ms after which frequency adaptation occurred and each neuron settled into a periodic firing regime with a well-defined frequency. The relation between the phases of the oscillatory activities of the neurons at the end of this warm up time was entirely determined by the initial conditions. Following this period and once the synaptic transmission was activated, a simulation time of 3 s was recorded. This allowed us to trace the change in the relative timing of the spikes induced by the synaptic coupling in this neural circuit.

The second class of model involving the interaction of heterogeneous large populations of neurons was built with the neuronal simulator package NEST [74]. The simulation of such networks uses a precise time-driven algorithm with the characteristic that the spike events are not constrained to the discrete time lattice. In a first stage of the simulation, the three populations were initialized being isolated from each other and let them to evolve just due to their internal local connectivity and external Poissonian noise. In a subsequent phase, the three populations were interconnected according to the motif investigated here and simulated during 1 s.

Data Analysis

The strength of the synchronization and the phase-difference between each individual pair of neurons (m, n) were derived for the first model of three HH neurons by the computation of the order parameter defined as

$$\rho(t) = \frac{1}{2}|\exp(\mathrm{i}\phi_m(t)) + \exp(\mathrm{i}\phi_n(t))|, \tag{16}$$

which takes the value of 1 when two systems oscillate in-phase and 0 when they oscillate in an antiphase or in an uncorrelated fashion. To compute this quantifier it is only necessary to estimate the phases of the individual neural oscillators. An advantage of this method is that one can easily reconstruct the phase of a neuronal oscillation from the train of spikes without the need of recording the full membrane potential time series [75]. The idea behind is that the time interval between two well-defined events (such as action potentials) define a complete cycle and the phase increase during this time amounts to 2π. Then, linear interpolation is used to assign a value to the phase between the spike events.

The synchrony among the large populations of neurons of the second model described here was assessed by the computation of averaged cross-correlograms. For that purpose, we randomly selected three neurons (one from each of the three populations) and computed for each pair of neurons belonging to different populations the histogram of coincidences (bin size of 2 ms) as a function of the time shift of one of the spike trains. We computed the cross-correlograms within the time window ranging from 500 to 1,000 ms to avoid the transients toward the synchronous state. The procedure was repeated 300 times to give rise to the estimated averaged distributions of coincidences exhibited in Figs. 5 and 6.

Acknowledgments The authors would like to thank Wolf Singer, Carl van Vreeswijk, Christopher J. Honey, and Nancy Kopell for fruitful discussions. This work was partially supported by the Hertie Foundation, the European Commission Project GABA (FP6-NEST contract 043309), and the Spanish MCyT and Feder under Project FISICO (FIS-2004-00953). R.V. and G.P. are also with the Frankfurt Institute for Advanced Studies (FIAS).

References

1. Nicolelis M, Ribeiro S (2002) Multielectrode recordings: the next steps. Curr. Opin. Neurobio. 12:602–606
2. Singer W, Engel AK, Kreiter AK, Munk MHJ, Neuenschwander S, Roelfsema PR (1997) Neuronal assemblies: necessity, signature and detectability. Trends Cogn. Sci. 1:252–260
3. Singer W (1999) Neuronal Synchrony: A Versatile Code for the Definition of Relations. Neuron 24:49–65
4. Varela FJ, Lachaux JP, Rodriguez E, Martinerie J (2001) The brainweb: phase synchronization and large-scale integration. Nat. Rev. Neurosci. 2:299–230
5. Milner PM (1974) A model for visual shape recognition. Psychol. Rev. 81:521–535
6. von der Malsburg, C (1981) The correlation theory of brain function. Intern. Rep. 81-2, Dept. of Neurobiology, Max-Planck-Institute for Biophysical Chemistry, Gottingen, Germany
7. Gray CM, Konig P, Engel AK, Singer W (1989) Oscillatory responses in cat visual cortex exhibit inter-columnar synchronization which reflects global stimulus properties. Nature 338:334–337
8. Gray CM (1999) The temporal correlation hypothesis of visual feature integration. Neuron 24:31–47
9. Salinas E, Sejnowski TJ (2000) Impact of correlated synaptic input on output firing rate and variability in simple neuronal models. J. Neurosci. 20:6193–6209
10. Salinas E, Sejnowski TJ (2001) Correlated neuronal activity and the flow of neuronal information. Nat. Rev. Neurosci. 2:539–550
11. Castelo-Branco M, Goebel R, Neuenschwander S, Singer W (2000) Neuronal synchrony correlates with surface segregation rules. Nature 405:685–689
12. Fries P, Roelfsema PR, Engel AK, Konig P, Singer W (1997) Synchronization of oscillatory responses in visual cortex correlates with perception in interocular rivalry. Proc. Natl. Acad. Sci. 94:12699–12704
13. Fries P, Reynolds JH, Rorie AE, Desimone R (2001) Modulation of oscillatory neuronal synchronization by selective visual attention. Science 291:1560–1563
14. Sarnthein J, Petsche H, Rappelsberger P, Shaw GL, von Stein A (1998) Synchronization between prefrontal and posterior association cortex during human working memory. Proc. Natl. Acad. Sci. 95:7092–7096
15. Roelfsema PR, Engel AK, Konig P, Singer W (1997) Visuomotor integration is associated with zero time-lag synchronization among cortical areas. Nature 385:157–161
16. Rodriguez E et al. (1999) Perception's shadow: long-distance synchronization of human brain activity. Nature 397:430–433
17. Mima T, Oluwatimilehin T, Hiraoka T, Hallett M (2001) Transient Interhemispheric Neuronal Synchrony Correlates with Object Recognition. J. Neurosci. 21:3942–3948

18. Uhlhaas PJ et al. (2006) Dysfunctional long-range coordination of neural activity during Gestalt perception in schizofrenia. J. Neurosci. 26:8168–8175
19. Soteropoulus DS, Baker S (2006) Cortico-cerebellar coherence during a precision grip task in the monkey. J. Neurophysiol. 95:1194–1206
20. Witham CL, Wang M, Baker S (2007) Cells in somatosensory areas show synchrony with beta oscillations in monkey motor cortex. Eur. J. Neurosci. 26:2677–2686
21. Swadlow HA, Rosene DL, Waxman SG (1978) Characteristics of interhemispheric impulse conduction between the prelunate gyri of the rhesus monkey. Exp. Brain Res. 33:455–467
22. Swadlow HA (1985) Physiological properties of individual cerebral axons studied in vivo for as long as one year. J. Neurophysiol. 54:1346–1362
23. Swadlow HA (1994) Efferent neurons and suspected interneurons in motor cortex of the awake rabbit: axonal properties, sensory receptive fields, and subthreshold synaptic inputs. J. Neurophysiol. 71:437–453
24. Miller R (2000) Time and the brain. Harwood Press, Switzerland
25. Wen Q, Chkolvskii DB (2005) Seggregation of the brain into Gray and White matter: a design minimiying conduction delays. PLoS Comput. Biol. 1:e78
26. Ringo JL, Doty RW, Demeter S, Simard, PY (1994) Time is the essence: A conjecture that hemispheric specialization arises from interhemispheric conduction delay. Cereb. Cortex 4:331–343
27. Miller R (1996) Axonal conduction time and human cerebal laterality: a psychobiological theory, 1st edn. Harwood Academics Publisher, Amsterdam
28. Vicente R, Gollo LL, Mirasso CR, Fischer I, Pipa G (2008) Dynamical relaying can yield zero time lag neuronal synchrony despite long conduction delays. Proc. Natl. Acad. Sci. 105:17157–17162
29. Fischer I, Vicente R, Buldu JM, Peil M, Mirasso CR, Torrent MC, Garcia-Ojalvo J (2006) Zero-lag long-range synchronization via dynamical relaying. Phys. Rev. Lett. 97:123902
30. Vicente R, Pipa G, Fischer I, Mirasso CR (2007) Zero-lag long range synchronization of neurons is enhanced by dynamical relaying. Lect. Notes Comp. Sci. 4688:904–913
31. D'Huys O, Vicente R, Erneux T, Danckaert J, Fischer I (2008) Synchronization properties of network motifs: Influence of coupling delay and symmetry. Chaos 18:037116
32. Jones EG (2002) Thalamic circuitry and thalamocortical synchrony. Phil. Trans. R. Soc. Lond. B 357:1659–1673
33. Shipp S (2003) The functional logic of cortico-pulvinar connections. Phil. Trans. R. Soc. Lond. B 358:1605–1624
34. Honey CJ, Kotter R, Breakspear M, Sporns O (2007) Network structure of cerebral cortex shapes functional connectivity on multiple time scales. Proc. Natl. Acad. Sci. 104:10240–10245
35. Soleng AF, Raastad M, Andersen P (1998) Conduction latency along CA3 hippocampal axons from the rat. Hippocampus 13:953–961
36. Swadlow HA, Waxman SG (1975) Observations on impulse conduction along central axons. Proc. Natl. Acad. Sci. 72:5156–5159
37. Katz B, Miledi R (1965) The measurement of synaptic delay, and the time course of acetylcholine release at the neuromuscular junction. Proc. R. Soc. Lond. Series B, Biol. Sci. 161:483–495
38. Shepherd GM (2004) The synaptic organization of the brain. Oxford University Press
39. Volgushev M, Chistiakova M, Singer W (1998) Modification of discharge patterns of neocortical neurons by induced oscillations of the membrane potential. Neuroscience 83:15–25
40. Aboitiz F, Scheibel AB, Fisher RS, Zaidel E (1992) Fiber composition of the human corpus callosum. Brain Behav. Evol. 598:143–153
41. Dickson CT, Biella G, de Curtis M (2003) Slow periodic events and their transition to gamma oscillations in the entorhinal cortex of the isolated guinea pig brain. J. Neurophysiol. 900:39–46
42. Rizzuto DS, Madsen JR, Bromfield EB, Schulze-Bonhage A, Seelig D, Aschenbrenner-Scheibe R, Kahana MJ (2003) Reset of human neocortical oscillations during a working memory task. Proc. Natl. Acad. Sci. 100:7931–7936

43. Mann EO, Paulsen O (2007) Role of GABAergic inhibition in hippocampal network oscillations. Trends Neurosci. 30:343–349
44. Whittington MA, Doheny HC, Traub RD, LeBeau FEN, Buhl EH (2001) Differential expression of synaptic and nonsynaptic mechanisms underlying stimulus-induced gamma oscillations in vitro. J. Neurosci. 21:1727–1738
45. Buzsaki G (2006) Rhythms of the brain. Oxford University Press
46. Bennet MVL, Zukin RS (2004) Electrical coupling and neuronal synchronization in the mammalian brain. Neuron 41:495–511
47. Caspar DLD, Goddenough DA, Makowski L, Phillips WC (1977) Gap junction structures. J. Cell Biol. 74:605–628
48. Draghun A, Traub RD, Schmitz D, Jefferys JGR (1998) Electrical coupling underlies high-frequency oscillations in the hippocampus in vitro. Nature 394:189–192
49. Traub RD, Kopell N, Bibbig A, Buhl EH, Lebeau FEN, Whittington MA (2001) Gap junctions between interneuron dendrites can enhance synchrony of gamma oscillations in distributed networks. J. Neurosci. 21:9478–9486
50. Kopell N, Ermentrout GB (2004) Chemical and electrical synapses perform complementary roles in the synchronization of interneuronal networks. Proc. Natl. Acad. Sci. 101:15482–15487
51. Traub RD, Whittington MA, Stanford IM, Jefferys JGR (1996) A mechanism for generation of long-range synchronous fast oscillations in the cortex. Nature 383:621–624
52. Bibbig A, Traub RD, Whittington MA (2002) Long-range synchronization of gamma and beta oscillations and the plasticity of excitatory and inhibitory synapses: a network model. J. Neurophysiol. 88:1634–1654
53. Lowel S, Singer W (1992) Selection of intrinsic horizontal connections in the visual cortex by correlated neuronal activity. Science 255:209–212
54. Knoblauch A, Sommer FT (2003) Synaptic plasticity, conduction delays, and inter-areal phase relations of spike activity in a model of reciprocally connected areas. Neurocomputing 52–54:301–306
55. Izhikevich E (2006) Polychronization: computation with spikes. Neural Comput. 18:245–282
56. Swindale NV (2003) Neural synchrony, axonal path lengths, and general anesthesia: a hipothesis. Neuroscientist 9:440–445
57. Sporns O, Kotter R (2004) Motifs in brain networks. PLoS Biol. 2:e369
58. Sporns O, Chialvo D, Kaiser M, Hiltetag CC (2004) Organization, development and function of complex brain networks. Trends Cogn. Sci. 8:418–425
59. Ermentrout, JB (1996) Type I membranes, phase resetting curves, and synchrony. Neural Comp. 8:979–1001
60. Reyes AD, Fetz EE (1993) Two modes of interspike interval shortening by brief transient depolarizations in cat neocortical neurons. J. Neurophysiol. 69:1661–1672
61. Douglas RJ, Martin KAC (2004) Neuronal circuits of the neocortex. Annu. Rev. Neurosci. 27:419–451
62. Pare D, Shink E, Gaudreau H, Destexhe A, Lang EJ (1998) Impact of spontaneous synaptic activity on the resting properties of cat neocortical pyramidal neurons in vivo. J. Neurophysiol. 78:1450–1460
63. Arieli A, Sterkin A, Grinvald A, Aersten A (1996) Dynamics of ongoing activity: explanation of the large variability in evoked cortical responses. Science 273:1868–1871
64. Llinas R, Pare D (1997) Coherent oscillations in specific and nonspecific thalamocortical networks and their role in cognition. In: Steriade M, Jones EG, McCormick DA (eds) Thalamus. Pergamon, New York.
65. Llinas R, Ribary U, Contreras D, Pedroarena C (1998) The neuronal basis for conciousness. Phil. Trans. R. Soc. Lond. B 353:1841–1849
66. Ribary U, Ioannides AA, Singh KD, Hasson R, Bolton JPR, Lado F, Mogilner A, Llinas R (1991) Magnetic field tomography of coherent thalamocortical 40-Hz oscillations in humans. Proc. Natl. Acad. Sci. 88:11037–11041
67. Sherman SM, Guillery, RW (2002) The role of the thalamus in the flow of information to the cortex. Phil. Trans. R. Soc. Lond. B 357:1695–1708

68. Salami M, Itami C, Tsumoto T, Kimura F (2003) Change of conduction velocity by regional myelination yields to constant latency irrespective of distance between thalamus to cortex. Proc. Natl. Acad. Sci. 100:6174–6179
69. Engel AK, Kreiter AK, Koenig P, Singer W (1991) Synchronization of oscillatory neuronal responses between striate and extrastriate visual cortical areas of the cat. Proc. Natl. Acad. Sci. 88:6048–6052
70. Contreras D, Destexhe A, Sejnowski TJ, Steriade M (1996) Control of spatiotemporal coherence of a thalamic oscillation by corticothalamic feedback. Science 274:771–774
71. Fries P (2005) Neuronal communication through neuronal coherence. Trends Cogn. Sci. 9:474–480
72. Hodgkin AL, Huxley AF (1952) A quantitative description of the membrane current and its application to conduction and excitation in nerve. J. Physiol. 117:500–544
73. Brunel N (2000) Dynamics of Sparsely Connected Networks of Excitatory and Inhibitory Spiking Neurons. J. Comput. Neurosci. 8:183–208
74. Brette R, et al (2007) Simulation of networks of spiking neurons: A review of tools and strategies. J. Comput. Neurosci. 23:349–398
75. Pikovsky A, Rosenblum M, Kurths J (2002) Synchronization: A universal Concept in Nonlinear Science. Cambridge University Press

Characterizing Oscillatory Cortical Networks with Granger Causality

Anil Bollimunta, Yonghong Chen, Charles E. Schroeder, and Mingzhou Ding

Abstract Multivariate neural recordings are becoming commonplace. Statistical techniques such as Granger causality promise to reveal the patterns of neural interactions and their functional significance in these data. In this chapter, we start by reviewing the essential mathematical elements of Granger causality with special emphasis on its spectral representation. Practical issues concerning the estimation of such measures from time series data via autoregressive models are discussed. Simulation examples are used to illustrate the technique. Finally, we analyze local field potential recordings from the visual cortex of behaving monkeys to address the neuronal mechanisms of the alpha oscillation.

Introduction

Oscillatory activities are ubiquitous in the cerebral cortex. Based on the frequency of signal rhythmicity, neural oscillations are classified according to the following approximate taxonomy: delta (1–3 Hz), theta (4–7 Hz), alpha (8–12 Hz), beta (13–30 Hz) and gamma (31–90 Hz). A number of mechanisms have been identified that contribute to the generation of neural oscillations. At the single cell level specific combinations of ionic conductances can lead to rhythmic discharge through burst firing [5, 11, 14, 21, 23]. This rhythmicity is then amplified by ensembles of neurons with similar physiological properties. Oscillation can also occur as an emergent phenomenon in an interconnected network of neurons [18]. In this case, no single neuron is capable of discharging rhythmically in isolation, but a network of neurons with reciprocal synaptic activations are the source of the oscillatory activity. While the physiological generating mechanisms and functions of brain rhythms remain a subject of debate, recent advances in experimental technology make it possible to

M. Ding (✉)
J. Crayton Pruitt Family Department of Biomedical Engineering, University of Florida, Gainesville, FL 32611, USA
e-mail: mding@bme.ufl.edu

K. Josić et al. (eds.), *Coherent Behavior in Neuronal Networks*, Springer Series in Computational Neuroscience 3, DOI 10.1007/978-1-4419-0389-1_9,

record neural activity from multiple sites simultaneously in the intact cortex, paving the way for understanding neuronal oscillations from a network perspective.

Multisite neural recordings produce massive quantities of data and these data form the basis for unraveling the patterns of neural interactions in oscillatory cortical networks. It has long been recognized that neural interactions are directional. Being able to infer directions of neural interactions from data is an important capability for fully realizing the potential of multisite data. Traditional interdependence measures include cross correlation and spectral coherence. These techniques do not yield directional information reliably. Granger causality has emerged in recent years as a statistically principled method for accomplishing that goal. The basis of Granger causality estimation is the autoregressive models of time series. Recent work has explored its application to multisite neural recordings [2, 3, 7, 13]. In this chapter, we start with a brief summary of the basics of Granger causality with emphasis on its spectral representation. The method is then demonstrated on simulation examples where the network connectivity is known a priori. Finally, we address the neuronal mechanisms underlying cortical alpha rhythm by applying the technique to laminar local field potentials and multiunit activities recorded from an awake and behaving monkey.

Granger Causality Analysis

The development below follows that of Geweke [8]. Also see Ding et al. [7] for more details. Consider two jointly stationary stochastic processes X_t and Y_t. Individually, X_t and Y_t are described by the following two autoregressive (AR) models [7]

$$X_t = \sum_{j=1}^{\infty} a_{1j} X_{t-j} + \varepsilon_{1t}, \tag{1}$$

$$Y_t = \sum_{j=1}^{\infty} d_{1j} Y_{t-j} + \eta_{1t}, \tag{2}$$

where the noise terms are uncorrelated over time with variances $\mathrm{var}(\varepsilon_{1t}) = \Sigma_1$ and $\mathrm{var}(\eta_{1t}) = \Gamma_1$. Together, their joint autoregressive representation is

$$X_t = \sum_{j=1}^{\infty} a_{2j} X_{t-j} + \sum_{j=1}^{\infty} b_{2j} Y_{t-j} + \varepsilon_{2t}, \tag{3}$$

$$Y_t = \sum_{j=1}^{\infty} c_{2j} X_{t-j} + \sum_{j=1}^{\infty} d_{2j} Y_{t-j} + \eta_{2t}, \tag{4}$$

where the noise vector is again uncorrelated over time and their contemporaneous covariance matrix is

$$\Sigma = \begin{pmatrix} \Sigma_2 & \Upsilon_2 \\ \Upsilon_2 & \Gamma_2 \end{pmatrix} \tag{5}$$

Here $\Sigma_2 = \text{var}(\varepsilon_{2t}), \Gamma_2 = \text{var}(\eta_{2t})$, and $\Upsilon_2 = \text{cov}(\varepsilon_{2t}, \eta_{2t})$. If X_t and Y_t are independent, then $\{b_{2j}\}$ and $\{c_{2j}\}$ are uniformly zero, $\Upsilon_2 = 0$, $\Sigma_1 = \Sigma_2$, and $\Gamma_1 = \Gamma_2$. This observation motivates the definition of total interdependence between X_t and Y_t as

$$F_{X,Y} = \ln \frac{\Sigma_1 \Gamma_1}{|\Sigma|} \tag{6}$$

where $|\cdot|$ is the symbol for determinant. Clearly, $F_{X,Y} = 0$ when the two time series are independent, and $F_{X,Y} > 0$ when they are not.

Consider (1) and (3). The value of Σ_1 measures the accuracy of the autoregressive prediction of X_t based on its previous values, whereas the value of Σ_2 represents the accuracy of predicting the present value of X_t based on the previous values of both X_t and Y_t. According to Wiener [24] and Granger [10], if Σ_2 is less than Σ_1 in some suitable statistical sense, then Y_t is said to have a causal influence on X_t. We quantify this causal influence by

$$F_{Y \to X} = \ln \frac{\Sigma_1}{\Sigma_2}. \tag{7}$$

It is clear that $F_{Y \to X} = 0$ when there is no causal influence from Y to X and $F_{Y \to X} > 0$ when there is. One can define causal influence from X to Y as

$$F_{X \to Y} = \ln \frac{\Gamma_1}{\Gamma_2}. \tag{8}$$

The value of this quantity can be similarly interpreted.

It is possible that the interdependence between X_t and Y_t cannot be fully explained by their interactions. The remaining interdependence is captured by Υ_2, the covariance between ε_{2t} and η_{2t}. This interdependence is referred to as instantaneous causality and is characterized by

$$F_{X \cdot Y} = \ln \frac{\Sigma_2 \Gamma_2}{|\Sigma|}. \tag{9}$$

When Υ_2 is zero, $F_{X \cdot Y}$ is also zero. When Υ_2 is not zero, $F_{X \cdot Y} > 0$. From (6)–(9) one concludes that

$$F_{X,Y} = F_{X \to Y} + F_{Y \to X} + F_{X \cdot Y} \tag{10}$$

This formula demonstrates that the total interdependence between two time series X_t and Y_t can be decomposed into three components: two directional causal influences due to their interaction patterns, and the instantaneous causality due to factors possibly exogenous to the (X, Y) system (e.g., a common driving input).

To develop the spectral representation of Granger causality, we introduce the lag operator L: $LX_t = X_{t-1}$. Equations (3) and (4) can be rewritten as

$$\begin{pmatrix} A_2(L) & B_2(L) \\ C_2(L) & D_2(L) \end{pmatrix} \begin{pmatrix} X_t \\ Y_t \end{pmatrix} = \begin{pmatrix} \varepsilon_{2t} \\ \eta_{2t} \end{pmatrix}, \tag{11}$$

where $A_2(L)$, $B_2(L)$, $C_2(L)$, and $D_2(L)$ are power series in L with $A_2(0) = 1$, $B_2(0) = 0$, $C_2(0) = 0$, and $D_2(0) = 1$. A Fourier transform of (11) yields

$$\begin{pmatrix} a_2(\omega) & b_2(\omega) \\ c_2(\omega) & d_2(\omega) \end{pmatrix} \begin{pmatrix} X(\omega) \\ Y(\omega) \end{pmatrix} = \begin{pmatrix} E_x(\omega) \\ E_y(\omega) \end{pmatrix}, \tag{12}$$

where $\omega = 2\pi f$ and the components of the coefficient matrix $\mathbf{A}(\omega)$ are

$$a_2(\omega) = 1 - \sum_{j=1}^{\infty} a_{2j} \mathrm{e}^{-\mathrm{i}\omega j},\ b_2(\omega) = -\sum_{j=1}^{\infty} b_{2j} \mathrm{e}^{-\mathrm{i}\omega j},$$

$$c_2(\omega) = -\sum_{j=1}^{\infty} c_{2j} \mathrm{e}^{-\mathrm{i}\omega j},\ d_2(\omega) = 1 - \sum_{j=1}^{\infty} d_{2j} \mathrm{e}^{-\mathrm{i}\omega j},$$

In terms of transfer functions, (12) becomes

$$\begin{pmatrix} X(\omega) \\ Y(\omega) \end{pmatrix} = \begin{pmatrix} H_{xx}(\omega) & H_{xy}(\omega) \\ H_{yx}(\omega) & H_{yy}(\omega) \end{pmatrix} \begin{pmatrix} E_x(\omega) \\ E_y(\omega) \end{pmatrix}, \tag{13}$$

where $\mathbf{H}(\omega) = \mathbf{A}^{-1}(\omega)$ is the transfer function whose components are

$$\begin{aligned} H_{xx}(\omega) &= \frac{1}{\det\mathbf{A}} d_2(\omega),\ H_{xy}(\omega) = -\frac{1}{\det\mathbf{A}} b_2(\omega), \\ H_{yx}(\omega) &= -\frac{1}{\det\mathbf{A}} c_2(\omega),\ H_{yy}(\omega) = \frac{1}{\det\mathbf{A}} a_2(\omega). \end{aligned} \tag{14}$$

After proper ensemble averaging the spectral matrix is obtained according to

$$\mathbf{S}(\omega) = \mathbf{H}(\omega)\Sigma\mathbf{H}^*(\omega) \tag{15}$$

where $*$ denotes complex conjugate and matrix transpose and Σ is defined in (5).

The spectral matrix contains cross spectra (off-diagonal terms) and auto spectra (diagonal terms). If X_t and Y_t are independent, then the cross spectra are zero and $|\mathbf{S}(\omega)|$ equals the product of two auto spectra. This observation, analogous to that leading to the definition of total interdependence in the time domain in (6), motivates the spectral domain representation of total interdependence between X_t and Y_t as

$$f_{X,Y}(\omega) = \ln \frac{S_{xx}(\omega) S_{yy}(\omega)}{|\mathbf{S}(\omega)|}, \tag{16}$$

where $|\mathbf{S}(\omega)| = S_{xx}(\omega)S_{yy}(\omega) - S_{xy}(\omega)S_{yx}(\omega)$ and $S_{yx}(\omega) = S_{xy}^*(\omega)$. It is easy to see that this decomposition of interdependence is related to coherence by the following relation:

$$f_{X,Y}(\omega) = -\ln(1 - C(\omega)), \tag{17}$$

where coherence is defined as

$$C(\omega) = \frac{|S_{xy}(\omega)|^2}{S_{xx}(\omega)S_{yy}(\omega)}. \tag{18}$$

Coherence is a normalized quantity, with values ranging between 0 and 1, with 1 indicating maximum interdependence between the two time series at frequency ω and 0 indicating independence.

From (15), the auto spectrum of X_t is:

$$S_{xx}(\omega) = H_{xx}(\omega)\Sigma_2 H_{xx}^*(\omega) + 2\Upsilon_2 \mathrm{Re}(H_{xx}(\omega)H_{xy}^*(\omega)) + H_{xy}(\omega)\Gamma_2 H_{xy}^*(\omega). \tag{19}$$

To fix ideas, let us start with $\Upsilon_2 = 0$. In this case there is no instantaneous causality and the interdependence between X_t and Y_t is entirely due to their interactions through the regression terms on the right-hand sides of (3) and (4). The spectrum has two terms. The first term, involving only the variance of ε_{2t} which is the noise term that drives the X_t time series, can be viewed as the intrinsic contribution to the power of X_t. The second term, involving only the variance of η_{2t} which is the noise term that drives the Y_t time series, can be viewed as the causal contribution to the power of X_t from Y_t. This decomposition of power into an intrinsic part and a causal part forms the basis for defining spectral domain causality measures.

When Υ_2 is not zero, Geweke [8] introduced the following transformation to remove the cross term and make the identification of an intrinsic power term and a causal power term possible. The procedure is called normalization and it consists of left-multiplying

$$\mathbf{P} = \begin{pmatrix} 1 & 0 \\ -\frac{\Upsilon_2}{\Sigma_2} & 1 \end{pmatrix} \tag{20}$$

on both sides of (12). The result is

$$\begin{pmatrix} a_2(\omega) & b_2(\omega) \\ c_3(\omega) & d_3(\omega) \end{pmatrix} \begin{pmatrix} X(\omega) \\ Y(\omega) \end{pmatrix} = \begin{pmatrix} E_x(\omega) \\ \tilde{E}_y(\omega) \end{pmatrix}, \tag{21}$$

where $c_3(\omega) = c_2(\omega) - \frac{\Upsilon_2}{\Sigma_2}a_2(\omega)$, $d_3(\omega) = d_2(\omega) - \frac{\Upsilon_2}{\Sigma_2}b_2(\omega)$, $\tilde{E}_y(\omega) = E_y(\omega) - \frac{\Upsilon_2}{\Sigma_2}E_x(\omega)$. The new transfer function $\tilde{\mathbf{H}}(\omega)$ for (21) is the inverse of the new coefficient matrix $\tilde{\mathbf{A}}(\omega)$:

$$\tilde{\mathbf{H}}(\omega) = \begin{pmatrix} \tilde{H}_{xx}(\omega) & \tilde{H}_{xy}(\omega) \\ \tilde{H}_{yx}(\omega) & \tilde{H}_{yy}(\omega) \end{pmatrix} = \frac{1}{\det \tilde{\mathbf{A}}} \begin{pmatrix} d_3(\omega) & -b_2(\omega) \\ -c_3(\omega) & a_2(\omega) \end{pmatrix}. \tag{22}$$

Since $\det \tilde{\mathbf{A}} = \det \mathbf{A}$ we have

$$\tilde{H}_{xx}(\omega) = H_{xx}(\omega) + \frac{\Upsilon_2}{\Sigma_2} H_{xy}(\omega), \, \tilde{H}_{xy}(\omega) = H_{xy}(\omega),$$

$$\tilde{H}_{yx}(\omega) = H_{yx}(\omega) + \frac{\Upsilon_2}{\Sigma_2} H_{xx}(\omega), \, \tilde{H}_{yy}(\omega) = H_{yy}(\omega). \tag{23}$$

From (21), following the same steps that lead to (19), the spectrum of X_t is found to be:

$$S_{xx}(\omega) = \tilde{H}_{xx}(\omega)\Sigma_2 \tilde{H}^*_{xx}(\omega) + H_{xy}(\omega)\tilde{\Gamma}_2 H^*_{xy}(\omega). \tag{24}$$

Here the first term is interpreted as the intrinsic power and the second term as the causal power of X_t due to Y_t. Based on this interpretation, we define the causal influence from Y_t to X_t at frequency ω as

$$f_{Y \to X}(\omega) = \ln \frac{S_{xx}(\omega)}{\tilde{H}_{xx}(\omega)\Sigma_2 \tilde{H}^*_{xx}(\omega)}. \tag{25}$$

According to this definition the causal influence is zero when the causal power is zero (i.e., the intrinsic power equals the total power), and it increases as the causal power increases.

By taking the transformation matrix as $\begin{pmatrix} 1 & -\frac{\Upsilon_2}{\Gamma_2} \\ 0 & 1 \end{pmatrix}$ and performing the same analysis, we get the causal influence from X_t to Y_t:

$$f_{X \to Y}(\omega) = \ln \frac{S_{yy}(\omega)}{\hat{H}_{yy}(\omega)\Gamma_2 \hat{H}^*_{yy}(\omega)}, \tag{26}$$

where $\hat{H}_{yy}(\omega) = H_{yy}(\omega) + \frac{\Upsilon_2}{\Gamma_2} H_{yx}(\omega)$.

Letting the spectral decomposition of instantaneous causality be

$$f_{Y.X}(\omega) = \ln \frac{(\tilde{H}_{xx}(\omega)\Sigma_2 \tilde{H}^*_{xx}(\omega))(\hat{H}_{yy}(\omega)\Gamma_2 \hat{H}^*_{yy}(\omega))}{|\mathbf{S}(\omega)|}, \tag{27}$$

we obtain a spectral domain expression for the total interdependence that is analogous to (10) in the time domain:

$$f_{X,Y}(\omega) = f_{X \to Y}(\omega) + f_{Y \to X}(\omega) + f_{X.Y}(\omega). \tag{28}$$

It is important to note that the spectral instantaneous causality may become negative for some frequencies in certain situations and may not have a readily interpretable physical meaning.

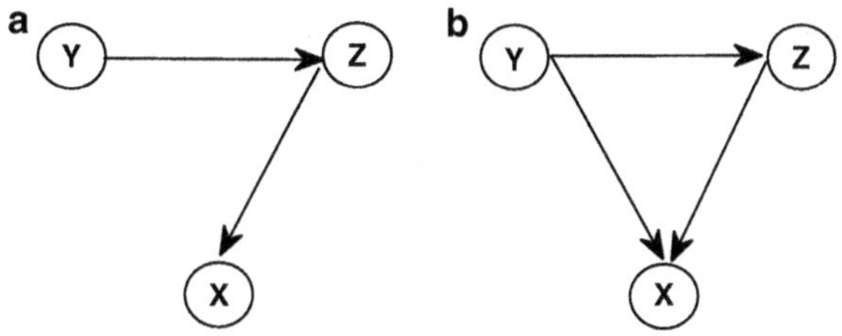

Fig. 1 Two possible coupling schemes for three time series. A pairwise causality analysis cannot distinguish these two connectivity patterns

Geweke proved that [8], under general conditions, the above spectral causality measures relate to the time domain measures through:

$$\begin{aligned}
F_{X,Y} &= \frac{1}{2\pi}\int_{-\pi}^{\pi} f_{X,Y}(\omega)\mathrm{d}\omega,\\
F_{X\rightarrow Y} &= \frac{1}{2\pi}\int_{-\pi}^{\pi} f_{X\rightarrow Y}(\omega)\mathrm{d}\omega,\\
F_{Y\rightarrow X} &= \frac{1}{2\pi}\int_{-\pi}^{\pi} f_{Y\rightarrow X}(\omega)\mathrm{d}\omega,\\
F_{Y,X} &= \frac{1}{2\pi}\int_{-\pi}^{\pi} f_{X,Y}(\omega)\mathrm{d}\omega.
\end{aligned}$$

If those conditions are not met, these equalities become inequalities.

When there are more than two time series a pairwise analysis may not fully resolve the connectivity pattern. Figure 1 shows two connectivity schemes among three time series. A pairwise analysis will conclude that the connectivity pattern in Fig. 1b applies to both cases. In other words, pairwise analysis cannot distinguish whether the drive from Y to X has a direct component (Fig. 1b) or is mediated entirely by Z (Fig. 1a). In addition, for three processes, if one process drives the other two with differential time delays, a pairwise analysis would indicate a causal influence from the process that receives an early input to the process that receives a late input. To overcome these problems, conditional Granger causality [4, 9] has been proposed in both the time as well as the frequency domain (see [7] for a more detailed development of this measure).

Estimation of Autoregressive Models

The estimation of Granger causality involves fitting autoregressive models to time series data. The basic steps are discussed below for the general case of p recording channels. One emphasis is the incorporation of multiple time series segments into the estimation procedure [6]. This consideration is motivated by the goal of applying autoregressive modeling to neuroscience problems. It is typical in behavioral and cognitive sciences that the same experiment be repeated on many successive trials.

Under appropriate conditions, physiological data recorded from these repeated trials may be viewed as realizations of a common underlying stochastic process.

Let $\mathbf{X}(t) = [X_1(t), X_2(t), \cdots, X_p(t)]^{\mathrm{T}}$ be a p dimensional random process. Here T denotes matrix transposition. For multivariate neural data, p stands for the total number of recording channels. To avoid confusion with the channel designation in the subscript, the time variable t is written as the argument of the process. Assume that $\mathbf{X}(t)$ is stationary and can be described by the following mth order autoregressive equation

$$\mathbf{X}(t) + \mathbf{A}(1)\mathbf{X}(t-1) + \cdots + \mathbf{A}(m)\mathbf{X}(t-m) = \mathbf{E}(t) \tag{29}$$

where $\mathbf{A}(i)$ are $p \times p$ coefficient matrices and $\mathbf{E}(t) = [E_1(t), E_2(t), \ldots, E_p(t)]^{\mathrm{T}}$ is a zero mean uncorrelated noise vector with covariance matrix Σ.

To estimate $\mathbf{A}(i)$ and Σ, (29) is multiplied from the right by $\mathbf{X}^{\mathrm{T}}(t-k)$, where $k = 1, 2, \ldots, m$. Taking expectations, we obtain the Yule-Walker equations

$$\mathbf{R}(-k) + \mathbf{A}(1)\mathbf{R}(-k+1) + \cdots + \mathbf{A}(m)\mathbf{R}(-k+m) = 0, \tag{30}$$

where $\mathbf{R}(n) =< \mathbf{X}(t)\mathbf{X}^{\mathrm{T}}(t+n) >$ is $\mathbf{X}(t)$'s covariance matrix of lag n and $\mathbf{R}(-n) = \mathbf{R}^{\mathrm{T}}(n)$. Here $< \mathbf{E}(t)\mathbf{X}^{\mathrm{T}}(t-k) >= 0$ since $\mathbf{E}(t)$ is an uncorrelated process.

Assume that L realizations of the $\mathbf{X}$ process are available, $\{\mathbf{x}_l(i)\}_{i=1}^{N}$, where $l = 1, 2, 3, \ldots, L$. The ensemble mean is estimated and removed from each individual realization. The covariance matrix in (30) is estimated by averaging the following matrix over l:

$$\tilde{\mathbf{R}}_l(n) = \frac{1}{N-n} \sum_{i=1}^{N-n} \mathbf{x}_l(i)\mathbf{x}_l^{\mathrm{T}}(i+n). \tag{31}$$

For neural data, each trial is considered a realization.

Equation (29) contains a total of mp^2 unknown model coefficients. In (30) there is exactly the same number of simultaneous linear equations. One can simply solve these equations to obtain the model coefficients. An alternative approach is to use the Levinson, Wiggins, Robinson (LWR) algorithm, which is a more robust solution procedure based on the ideas of maximum entropy [6]. This algorithm was implemented in the analysis of numerical examples and neural data described in the following sections. The noise covariance matrix Σ may be obtained as part of the LWR algorithm. Otherwise one may obtain Σ through

$$\Sigma = \mathbf{R}(0) + \sum_{i=1}^{m} \mathbf{A}(i)\mathbf{R}(i). \tag{32}$$

The above estimation procedure can be carried out for any model order m. The correct m, representing the tradeoff between sufficient spectral resolution and over-parametrization, is usually determined by minimizing the Akaike Information Criterion (AIC) defined as

$$\mathrm{AIC}(m) = -2\log[\det(\Sigma)] + \frac{2p^2m}{N_{\text{total}}} \tag{33}$$

where N_{total} is the total number of data points from all the trials. Plotted as a function of m the proper model order corresponds to the minimum of this function. It is often the case that for neurobiological data N_{total} is very large. Consequently, for a reasonable range of m, the AIC function does not achieve a minimum. An alternative criterion is the Bayesian Information Criterion (BIC), which is defined as

$$\mathrm{BIC}(m) = -2\log[\det(\Sigma)] + \frac{2p^2m\log N_{\text{total}}}{N_{\text{total}}}. \tag{34}$$

This criterion can compensate for the large number of data points and may perform better in neural applications. A final step, necessary for determining whether the autoregressive time series model is suited for a given data set, is to check whether the residual noise is white. Here the residual noise is obtained by computing the difference between the value predicted by the model, $-(\mathbf{A}(1)\mathbf{X}(t-1)+\cdots+\mathbf{A}(m)\mathbf{X}(t-m))$, and the actually measured value, $\mathbf{X}(t)$.

Once an autoregressive model is adequately estimated, it becomes the basis for both time domain and spectral domain Granger causality analysis. Specifically, in the spectral domain, (29) can be written as

$$\mathbf{X}(\omega) = \mathbf{H}(\omega)\mathbf{E}(\omega) \tag{35}$$

where

$$\mathbf{H}(\omega) = \Big(\sum_{j=0}^{m}\mathbf{A}(j)\mathrm{e}^{-\mathrm{i}\omega j}\Big)^{-1} \tag{36}$$

is the transfer function with $\mathbf{A}(0)$ being the identity matrix. From (35), after proper ensemble averaging, we obtain the spectral matrix

$$\mathbf{S}(\omega) = \mathbf{H}(\omega)\Sigma\mathbf{H}^*(\omega) \tag{37}$$

According to the procedures outlined in the previous section, the transfer function, the noise covariance, and the spectral matrix constitute the basis for carrying out Granger causality analysis.

Numerical Simulations

In this section, we use three examples to illustrate various aspects of the approach given earlier. Two of the examples involve coupled autoregressive models. Another example is based on equations derived from neuronal population dynamics.

Example 1. Consider the following two variable model:

$$
\begin{aligned}
X_t &= \varepsilon_t, \\
Y_t &= 0.5Y_{t-1} + X_{t-1} + \eta_t,
\end{aligned}
\tag{38}
$$

where ε_t, η_t are independent Gaussian white noise processes with zero means and variances $\text{var}(\varepsilon_t) = 1$, $\text{var}(\eta_t) = 0.09$, respectively. Assume that each time step is 5 ms. The sampling rate is 200 Hz. For such a simple model, it is not hard to derive the theoretical coherence between X_t and Y_t, which is 0.92 for all frequencies. Also, from the construction of the model, it can be seen that there is only a unidirectional causal influence from X_t to Y_t; the feedback from Y_t to X_t is zero. In addition, there is no instantaneous causality since the two white noise processes are independent. Based on (17), the unidirectional Granger causality from X_t to Y_t is analytically determined to be: $f_{X\to Y} = -\ln(1-0.92) = 2.49$.

Equation (38) was simulated to generate a data set of 500 realizations with each realization consisting of 100 time points. Assuming no knowledge of (38), we fitted an AR model to the simulation data set and computed coherence and Granger causality spectra, which are shown in Fig. 2. The agreement between theoretical and simulated values is excellent.

Example 2. A simple neural model is considered [12]. An excitatory and an inhibitory neuronal population are coupled to form a cortical column. The columns are then coupled through mutually excitatory interactions to form a network (Fig. 3a):

$$
\begin{aligned}
\frac{\mathrm{d}^2 x_n(t)}{\mathrm{d}t^2} + (a+b)\frac{\mathrm{d}x_n(t)}{\mathrm{d}t} + abx_n(t) &= -k_{\mathrm{ei}} Q(y_n(t), Q_m) + \\
&\quad + \frac{1}{N}\sum_{p=1}^{N} c_{np} Q(x_p(t-\tau_{np}), Q_m) + \xi_{x_n}(t) + I_n, \\
\frac{\mathrm{d}^2 y_n(t)}{\mathrm{d}t^2} + (a+b)\frac{\mathrm{d}y_n(t)}{\mathrm{d}t} + aby_n(t) &= k_{\mathrm{ie}} Q(x_n(t), Q_m) + \xi_{y_n}(t).
\end{aligned}
\tag{39}
$$

Here x_n and y_n represent the local field potentials of the excitatory and inhibitory populations in the nth column, ξ_{x_n} and ξ_{y_n} are local white noise, and I_n is external input. The constants $a, b > 0$ are parameters describing the intrinsic properties of

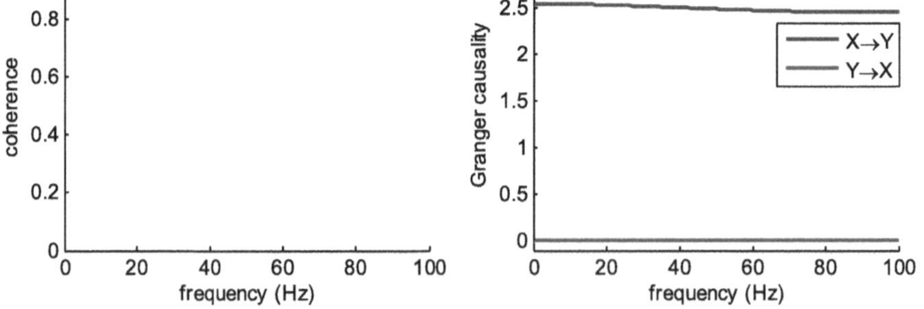

Fig. 2 Coherence and Granger causality spectra for simulation example 1.

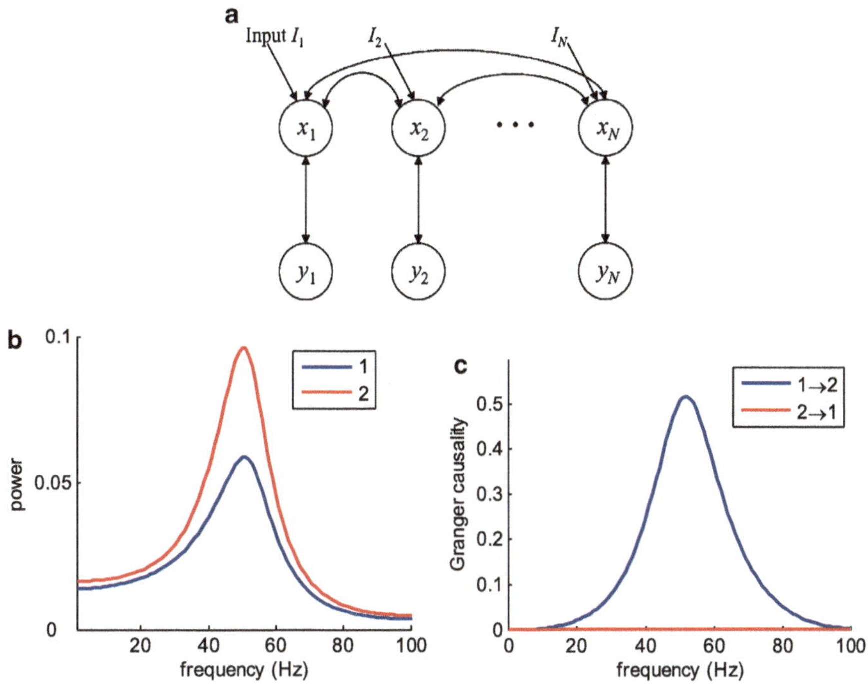

Fig. 3 (**a**) General coupling topology of the neural population model in simulation example 2. (**b**) and (**c**) Power and Granger causality spectra of two coupled columns.

each population. The parameter $k_{\text{ie}} > 0$ gives the coupling gain from the excitatory (x) to the inhibitory (y) population, whereas $k_{\text{ei}} > 0$ represents the strength of the reciprocal coupling. The coupling strength c_{np} is the gain from the excitatory population of column p to the excitatory population of column n. The sigmoid coupling function Q can be found in [12].

The values of the parameters used in the simulation are: $N = 2, a = 0.22, b = 0.72, k_{\text{ei}} = 0.4, k_{\text{ie}} = 0.1, c_{11} = c_{21} = c_{22} = 0, c_{12} = 0.5, \tau_{12} = 15\,ms, Q_m = 5$, $I_1 = I_2 = 0$, and the variances of the white noise inputs are 0.04. In other words, two columns were coupled together, where column 1 unidirectionally drives column 2 with no feedback from column 2 to column 1. The delayed differential equations were solved using a fourth order Runge-Kutta method with a fixed step of 0.1 ms. 1,01,000 points were generated and later down sampled to 200 Hz after discarding the first 1,000 transient points. The data set analyzed consisted of 2,000 data points.

An autoregressive model of order 5 was fitted to the data. Power, coherence, and Granger causality spectra were computed based on the fitted model. The results for power and Granger causality spectra are shown in Fig. 3b and c, respectively. It is clear that the network connectivity is correctly identified in Fig. 3c. Interestingly, the power in the driven column (column 2) is actually higher than that in the driving column (column 1). This indicates that one cannot easily infer causal relationships in a multinode network by using the magnitude of power as the sole indicator.

Example 3. In this example we illustrate the importance of conditional causality analysis in revealing the true connectivity pattern in a network of three coupled AR models. Consider the following AR(2) processes:

$$\begin{aligned} x_1(t) &= 0.55x_1(t-1) - 0.7x_1(t-2) + 0.4x_3(t-1) + \eta_1(t), \\ x_2(t) &= 0.56x_2(t-1) - 0.8x_2(t-2) + \eta_2(t), \\ x_3(t) &= 0.58x_3(t-1) - 0.9x_3(t-2) + 0.4x_2(t-1) + \eta_3(t), \end{aligned} \tag{40}$$

where $\eta_1(t)$, $\eta_2(t)$, and $\eta_3(t)$ are independent white noise processes with zero mean and unit variance. From model construction, there are causal influences from x_3 to x_1 and from x_2 to x_3, but there is no direct causal influence between x_1 and x_2. The coupling scheme here corresponds to Fig. 1a.

Simulating this model, we created a data set of 100 trials where each trial contained 1,024 time points. Assuming no knowledge of the model, a pairwise Ganger causality analysis was performed using a model order of 3. The results are shown in Fig. 4a. The causal influences from x_3 to x_1 and from x_2 to x_3 are both correctly identified. However, this analysis also revealed a causal influence from x_2 to x_1. This influence is not part of the model and is thus an artifact of the pairwise analysis. After applying conditional causality analysis, this artifact disappeared, as shown in Fig. 4b.

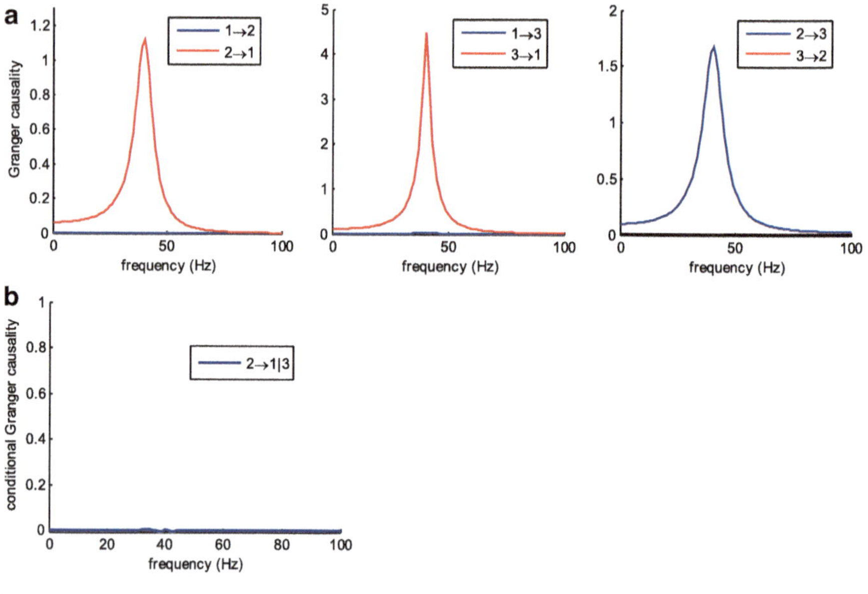

Fig. 4 (**a**) Pairwise analysis results for simulation example 3. (**b**) Conditional Granger causality analysis result.

Laminar Organization of the Cortical Alpha Rhythm

Oscillatory activity in the brain can appear in a number of frequency bands. Among them, the alpha rhythm (8–12 Hz) is prominent in human EEG recordings over the occipital and parietal areas during wakefulness. Nearly 80 years after its discovery, its genesis, cellular mechanisms, and functions remain unclear [20]. Early work on the genesis of the cortical alpha rhythm emphasized the pacemaking role of the thalamus [1]. A series of in vivo studies in dogs suggested that the alpha rhythm could be of a cortical origin with large layer 5 pyramidal neurons acting as pacemakers [14]. This hypothesis has found support in in vitro studies on slices from sensory cortices [21]. While in vitro preparations have proven an invaluable tool for understanding the physiology of cortical oscillations, recent writings have cautioned about the applicability of the findings made in these preparations to the intact brain [22]. Full anatomical connectivity brings the influence of various neuromodulatory systems on cell groups, resulting in changes in membrane potential and firing properties [22], the impact of which on the laminar organization of cortical oscillations remains unclear. Moreover, some of the powerful in vitro techniques such as trisection are not possible in behaving animals. Advanced computational methods in conjunction with properly recorded neural data hold the key to future progress in this area. Below we demonstrate the effectiveness of the method outlined earlier by applying it to characterize the "spontaneous" alpha rhythm in the visual cortex in the alert monkey. A more thorough study has been carried out in [2].

As part of an experiment involving switching attention between auditory to visual input streams, a macaque monkey was trained to perform an auditory oddball discrimination task [16]. Pure tones of 100 ms duration were presented at approximately 1.5 Hz. The stream of these standard stimuli was randomly interrupted by tones that differed in frequency (deviants). The monkey was required to respond to these deviant stimuli immediately following its onset. Local field potential (LFP) and multiunit activity (MUA) were sampled (2 kHz) with a linear array electrode with 14 contacts spanning all six cortical layers in visual area V4. The intercontact spacing was 200 m. The reason for analyzing activity in visual cortices during auditory discrimination was that the discrimination kept the monkey verifiably alert without using visual stimuli, so that we could study spontaneous neural activity.

To characterize the laminar organization of the cortical alpha rhythm we followed a three-step analysis protocol. First, laminar generators of LFP oscillations at the alpha frequency are identified by calculating the transmembrane current flow profile using the current source density (CSD) method. While the CSD analysis has been performed extensively on local field potentials with respect to the onset of a repetitive sensory stimulus [17, 19], its extension to ongoing neural activity is more difficult to ascertain. Single-trial CSD estimates tend to be noisy, and as there is no stimulus-related trigger, LFP averaging requires an alternate procedure for the alignment of trials. Here we use the phase of the alpha oscillation in a short epoch (described below) as a trigger for averaging LFPs. Second, alpha current generators that have the potential of pacemaking are identified with CSD-MUA coherence. In the context of studying evoked potentials, a source or sink is considered active if simultaneously recorded MUA is depressed or enhanced, indexing net local hyperpolarization or depolarization, respectively [17, 19]. For ongoing oscillatory activity, the membrane undergoes rhythmic transition between hyperpolarization and depolarization. In particular, during the depolarizing phase of the oscillation, the pacemaker cells may fire bursts of action potentials, which, via synaptic transmission, entrain neural activity in other laminae and cortical areas. For the present work, significant phase coherence between CSD and MUA is taken to indicate that a current generator is accompanied by rhythmic firing and thus has the potential of pacemaking. Third, the primary pacemaking generator is identified with the Granger causality analysis. For a cortical column with multiple alpha current generators distributed across different layers, the relationship among these generators needs to be further delineated. This is particularly so if the second step reveals that more than one generator has the potential of being the pacemaker. Granger causality analysis is used to further disambiguate the roles of different current generators, as the primary pacemaking generator is expected to exert unidirectional causal influence on other neural ensembles.

Contiguous LFP data of 30 s in duration was high-pass filtered (3 Hz, zero phase-shift), down-sampled to 200 Hz, and divided into 200 ms epochs. Each epoch, also referred to as a trial, was treated as a realization of an underlying stochastic process. The power spectrum of each of the 14 recording contacts was estimated and the contact showing the highest power spectral density at the alpha frequency was chosen as the "phase index" contact. Figure 5b shows the laminar distribution of the peak (10 Hz) LFP power. It can be seen that infragranular (IG) layers (electrode contacts 10–14) have higher alpha power than the granular (G) (electrode contacts 8 and 9) as well as the supragranular (SG) layers (electrode contacts 1–7). Contact 13 was chosen as the "phase index" contact. A sinusoid of the same frequency (10 Hz) was then fitted to the data from the phase index contact for each epoch to obtain the phase at that frequency with respect to the beginning of the epoch. The LFP data from all the contacts were shifted according to this estimated phase to realign all the trials, and the realigned signals were then averaged across epochs (trials) to obtain the averaged LFP for each contact. The current source density (CSD) profile was derived by taking the second spatial derivative. From the CSD profile the current sources (blue) and sinks (red) underlying the generation of the oscillatory alpha field activity are readily identified in G, IG as well as SG layers (Fig. 5a).

To assess the pacemaking potential of each current generator, CSD–MUA coherence was computed. The MUA data were epoched the same way as the LFP data and down-sampled from 2 kHz by taking a temporal average in nonoverlapping windows of 5 ms duration to achieve effectively the same sampling resolution of 200 Hz as the down-sampled LFPs. The coherence between single trial CSDs around alpha current generators and the corresponding mean-centered single-trial MUAs was calculated and the coherence spectra have clear peaks at around 10 Hz in the IG and G layers as shown in Fig. 5c. The peak coherence is 0.53 ($p < 0.01$) in IG layers, and 0.35 ($p < 0.01$) in G layer, suggesting that the neuronal firing at these generators is phase-locked to the oscillatory current. In contrast, the CSD–MUA coherence for the SG layer did not show an alpha peak (Fig. 5c) and the coherence value at

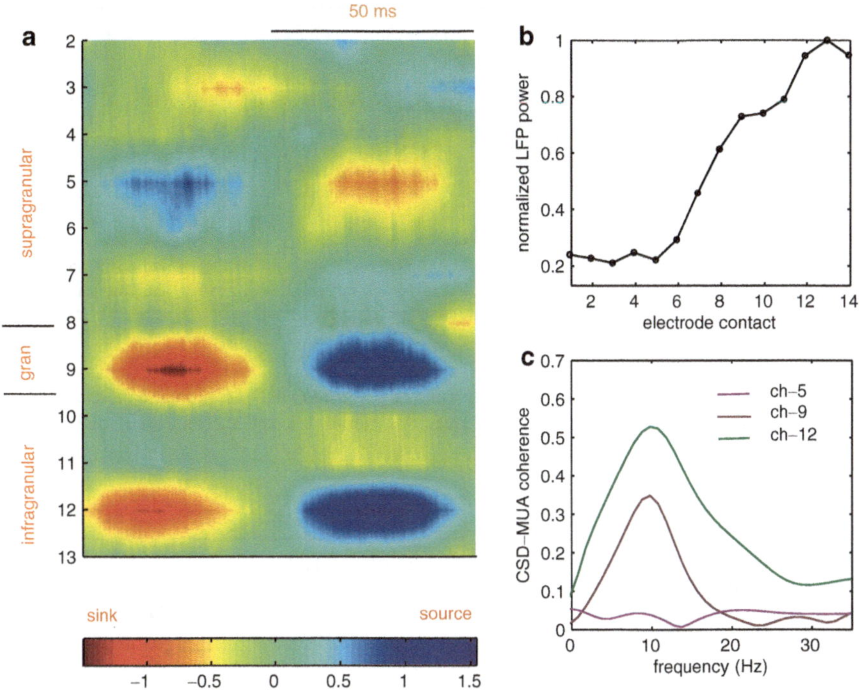

Fig. 5 (**a**) Current source density displayed as a color coded plot. (**b**) Laminar distribution of the LFP power at 10 Hz (normalized). (**c**) CSD–MUA coherence spectra at different recording contacts.

10 Hz was not significant. Here the significance level is determined using a random permutation procedure [2]. Note that the SG current generator is out of phase with that in G and IG layers. A plausible explanation for the lack of significant CSD–MUA coherence in the SG layers is dampening due to inhibition. Thus, the biasing of the CSD–MUA coherence toward the G and the IG layers (Fig. 5c), together with the laminar distribution of alpha power in Fig. 5b, strongly suggest that the neural ensembles in the G and IG layers are potential alpha pacemakers.

The more precise relationship between these potential pacemakers is examined by a Granger causality analysis using signals that represent local neural activity around each current generator. Typically, LFPs are recorded against a distant reference, making them susceptible to volume conduction of potentials from other sites This can affect interdependence analysis (see next section). The first derivative used for generating the bipolar LFPs and the second derivative approximation used for the current source density analysis help to eliminate this problem. For the data set shown in Fig. 5 the three bipolar signals are: SG = LFP(contact 5) - LFP(contact 3), G = LFP(contact 9) - LFP(contact 7), and IG = LFP(contact 13) - LFP(contact 11).

Bipolar LFPs representing local neural activity around each current generator were subjected to parametric spectral and Granger causality analysis. The AR model

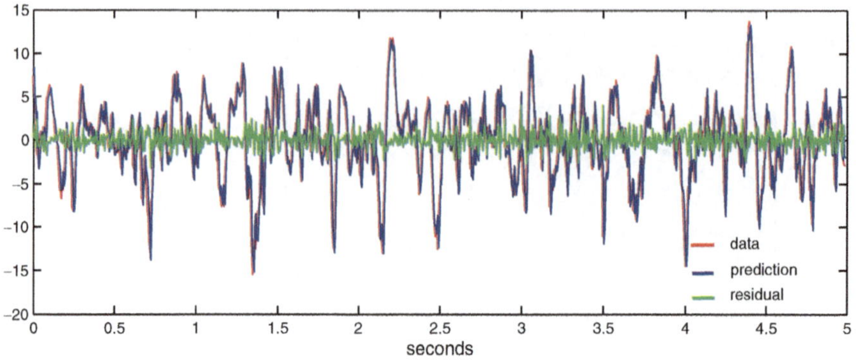

Fig. 6 Autoregressive model estimation performance. Overlaid are 5 s of bipolar LFP data (red) in the unit of microvolt from the infragranular layer, AR model based prediction (blue), and residual (green).

Fig. 7 Fourier based power spectra of the bipolar LFP data, AR model based prediction and residual process in Fig. 6.

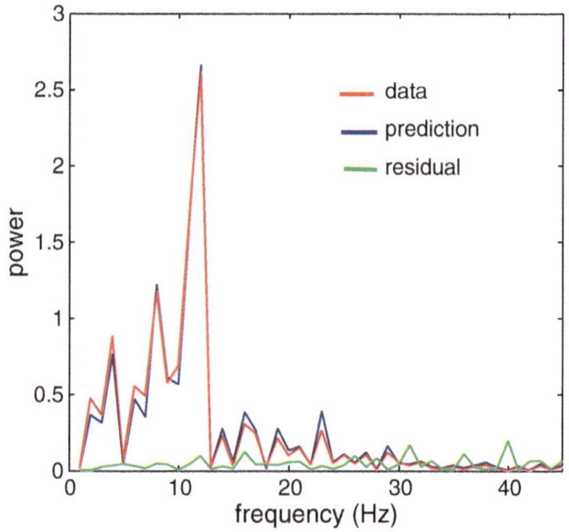

order of $m = 10$ (50 ms) was chosen as a tradeoff between sufficient spectral resolution and over-parameterization. Before proceeding with the result presentation, we consider the adequacy of using autoregressive models to represent neural data. Figure 6 shows the performance of the AR model on 5 s of contiguous bipolar LFP data from the IG layer. The model based one-step prediction data (blue curve in Fig. 6) closely follows the bipolar LFP data (red curve). The difference between the one-step prediction and the actual data, called the residual process, is overlaid (green curve). Figure 7 shows the Fourier based power spectra of the data, AR model prediction and the residual process in Fig. 6. An adequate parametric model fit of the data means that the residual noise process must be temporally uncorrelated (white). The power spectrum of the residual process (green curve in Fig. 7) does not have

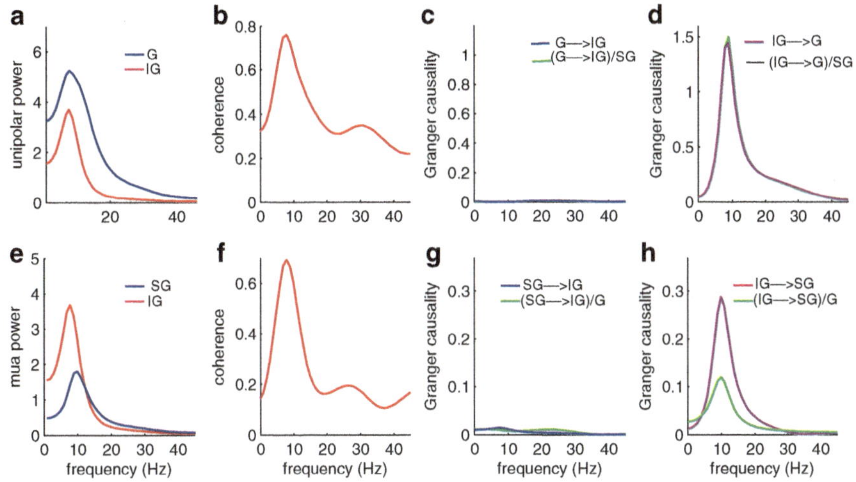

Fig. 8 Spectral analysis based on bipolar LFP data. (**a**) Power spectra of bipolar LFP signals from granular (G) and infragranular (IG) layers. (**b**) Coherence spectrum between the two bipolar signals in (**a**). (**c**) and (**d**) Granger causality spectra between G and IG layers. Here $x \rightarrow y$ denotes x driving y and $(x \rightarrow y)/z$ denotes x driving y after conditioning out z. (**e**) Power spectra of the bipolar LFP signals from supragranular (SG) and IG layers. (**f**) Coherence spectrum between the two bipolar signals in (**e**). (**g**) and (**h**) Granger causality spectra between SG and IG.

any prominent features, suggesting that the process is white. In addition the Durbin–Watson test was used to check the goodness of fit. The whiteness of the residuals was confirmed at the $p = 0.05$ significance level.

After verifying the adequacy of the AR model representation of the bipolar LFP data, power, coherence, and Granger causality analysis was carried out for the three bipolar signals in V4. The results are contained in Fig. 8. For IG and G layers, the bipolar LFP power spectra exhibit clear peaks around 10 Hz (Fig. 8a). The coherence between the two layers has a pronounced peak at 9 Hz, where the peak value is 0.76 ($p < 0.001$), as shown in Fig. 8b. This suggests that the alpha currents in these layers are highly synchronized. The Granger causality spectrum of $IG \rightarrow G$ shows (Fig. 8d) a strong peak at 10 Hz with a peak value 1.48 ($p < 0.001$), whereas the causality in the opposite direction ($G \rightarrow IG$) is not significant (Fig. 8c), indicating that neural activity in the G layer is strongly driven by that in the IG layers. To examine the influence of the SG layers on the interaction between the G and IG layers, we included the bipolar signal from the SG layer and performed conditional Granger causality analysis. The Granger causality from IG to G layer after conditioning out SG layer activity is nearly identical to the bivariate case (Fig. 8d), suggesting that the SG layers has no influence on the interaction between the IG and G layers. This is an expected result as the CSD–MUA coherence analysis has already demonstrated that the SG alpha current generator is not accompanied by rhythmic firing and thus not capable of pacemaking.

The interaction between the IG and SG layers was studied by first performing a bivariate analysis. Figure 8e and f show the power and coherence spectra, respectively. The power of the bipolar LFP signal for the SG layer has a clear peak at 10 Hz. The coherence spectrum peaked at 10 Hz with a peak value of 0.67 ($p < 0.001$), indicating a significant synchrony between the local alpha currents in these two layers. Granger causality again reveals IG as the driver of the SG current with the peak value of 0.28 ($p < 0.001$) at 10 Hz (Fig. 8h). The causal influence in the opposite direction ($SG \rightarrow IG$) is not significant (Fig. 8g). Finally, the role of the G layer on the interaction between IG and SG alpha activities was studied by performing conditional causality analysis. After conditioning out the influence of the G layer, the peak (10 Hz) Granger causality of the IG driving the SG layer is significantly reduced from 0.28 to 0.12 ($p < 0.001$) (Fig. 8h), suggesting that part of IG influence on SG layers could be mediated by the G layer. The significance testing here was performed using the bootstrap resampled method [2]. These results, together with laminar pattern of CSD (Fig. 5a) and CSD–MUA coherence analysis (Fig. 5c), support the hypothesis that alpha rhythm is of cortical origin with layer 5 pyramidal neurons acting as pacemakers [14, 21]. Moreover, the laminar organization revealed by Granger causality analysis is consistent with the anatomical connectivity within the cortical column [15].

The Choice of Neural Signals for Neuronal Interaction Analysis

In the previous section, bipolar LFP signals were used for coherence and Granger causality analysis. Three other choices of signals are possible for the present experiment: original unipolar LFP data, single-trial CSDs, and MUAs. Here we consider the appropriateness of these three types of signals for analyzing the interaction between different alpha current generators in V4.

Single-trial CSDs were derived at electrode contacts 5, 9, and 12 where strong alpha current generators have been identified (Fig. 5a). As shown in Fig. 9, Granger causality analysis results based on this type of signal are nearly identical to those using bipolar LFP data. CSD power spectra at IG, G, and SG layer contacts have a clear peak at 10 Hz (Fig. 9a, e). Coherence spectrum shows (Fig. 9b, f) that the transmembrane currents in G and SG layers are coherent with that at IG layer. Granger causality analysis revealed that IG layer drives both G and SG layers (Fig. 9d, h), whereas the Granger causality in the opposite directions ($G \rightarrow IG$, $SG \rightarrow IG$) are not significant at $p = 0.05$ level (Fig. 9c, g). Conditional Granger causality analysis further revealed that SG layer activity has no influence on the interaction between IG and G layer generators (Fig. 9d), whereas $IG \rightarrow SG$ is partly mediated by the G layer. Thus, Granger causality analysis based on either single-trial bipolar LFPs or single-trial CSDs yielded identical laminar organization for the alpha rhythm in the cortical area V4.

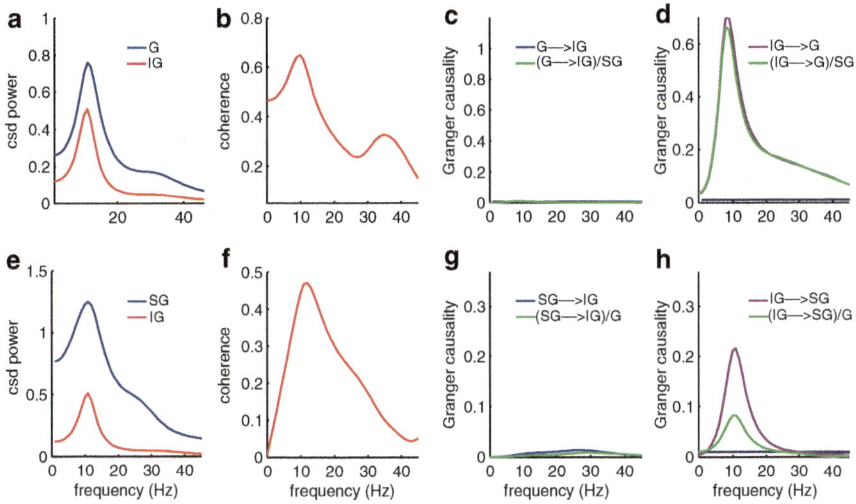

Fig. 9 Spectral analysis based on single-trial CSD data. (**a**) Power spectra of CSD signals from granular (G) and infragranular (IG) layers. (**b**) Coherence spectrum between the two CSD signals in (**a**). (**c**) and (**d**) Granger causality spectra between G and IG layers. (**e**) Power spectra of the CSD signals from supragranular (SG) and IG layers. (**f**) Coherence spectrum between the CSD signals in (**e**). (**g**) and (**h**) Granger causality spectra between SG and IG.

Unipolar LFPs are vulnerable to volume-conducted far-field effects, and they also contain the common reference, which is the electrode against which all differences in electrical potentials are measured. It is thus expected that interdependence analysis based on this type of signal will be adversely affected. The spectral analysis using unipolar LFPs (at electrode contacts 5, 9, and 12; see Fig. 10a) shows very high coherence over a broad frequency range (Fig. 10b). In addition, Granger causality analysis shows bidirectional causal influence between IG and G layers (Fig. 10c, d). This is not consistent with the unidirectional driving from IG to G layer revealed by bipolar LFP and single-trial CSD based analysis.

The MUA signal contains action potentials fired by both neurons participating in alpha activity and neurons not related to it. Figure 10e shows the power spectra of the mean centered MUA activity at the current generators in G and IG layers. No peak in the alpha frequency range is observed, indicating that much of MUA signals is not related to alpha frequency firing. The same type of spectral form is also seen for coherence (Fig. 10f) and Granger causality. In particular, the latter is found to be bidirectional (Fig. 10g, h).

Contrasting Figs. 8 and 9 with Fig. 10, and taking into account of the appropriate physiological interpretation, it is clear that bipolar LFPs or single-trial CSDs are good indices of local synchronous neuronal activity. They are preferred variables compared to unipolar LFPs or MUAs in the study of neuronal interactions between different generators of alpha oscillation in the cortical column.

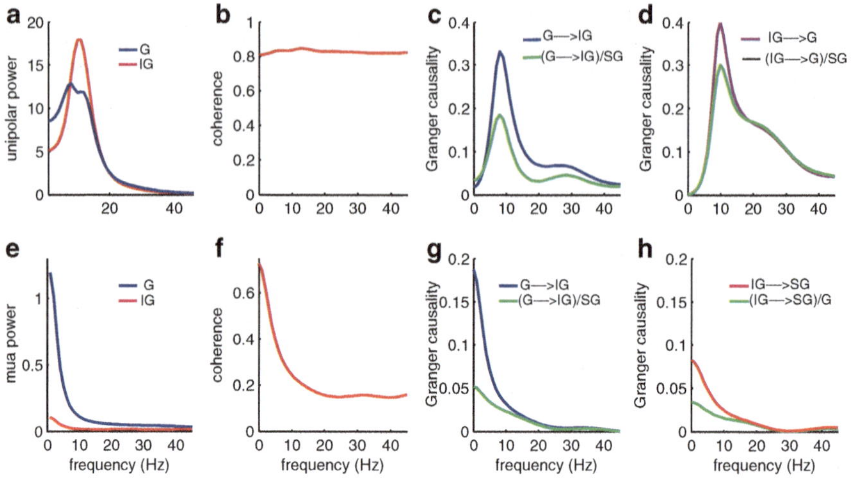

Fig. 10 Spectral analysis based on unipolar LFP and MUA data. (**a**) Power spectra of unipolar LFP signals from granular (G) and infragranular (IG) layers. (**b**) Coherence spectrum between the two unipolar signals in (**a**). (**c**) and (**d**) Granger causality spectra between G and IG. (**e**) Power spectra of the MUA signals at supragranular (SG) and IG layers. (**f**) Coherence spectrum between the MUA signals in (**e**). (**g**) and (**h**) Granger causality spectra between SG and IG.

Summary

In this chapter a framework for the analysis of multivariate neuronal time series centered on Granger causality is outlined. The mathematical essentials of Granger causality analysis is given. Three simulation examples are used to illustrate the method. The technique is then applied to study the laminar organization of the cortical alpha rhythm. It is shown that, in area V4, alpha rhythm is of a cortical origin with layer 5 pyramidal neurons acting as pacemakers. Our results suggest that Granger causality analysis, when combined with traditional techniques like current source density analysis, can improve our ability to understand the dynamical organization of synchronous oscillatory cortical networks.

Acknowledgments This work was supported by NIH grants MH070498, MH079388, and MH060358.

References

1. Andersen P, Andersson SA (1968) Physiological basis of the Alpha Rhythm. New York: Appleton-Century-Crofts
2. Bollimunta A, Chen Y, Schroeder CE, Ding M (2008) Neuronal mechanisms of cortical alpha oscillations in awake-behaving macaques. J Neurosci 28(40):9976–88

3. Brovelli A, Ding M, Ledberg A, Chen Y, Nakamura R, Bressler SL (2004) Beta oscillations in a large-scale sensorimotor cortical network: directional influences revealed by Granger causality. Proc Natl Acad Sci USA 101(26):9849–54
4. Chen Y, Bressler SL, Ding M (2006) Frequency decomposition of conditional Granger causality and application to multivariate neural field potential data. J Neurosci Methods 150(2):228–37
5. Connors BW, Amitai Y (1997) Making waves in the neocortex. Neuron 18(3):347–9
6. Ding M, Bressler SL, Yang W, Liang H (2000) Short-window spectral analysis of cortical event-related potentials by adaptive multivariate autoregressive modeling: data preprocessing, model validation, and variability assessment. Biol Cybern 83(1):35–45
7. Ding M, Chen Y, Bressler SL (2006) Granger causality: Basic theory and application to neuroscience. In: Winterhalder M, Schelter B, Timmer J (eds) Handbook of Time Series Analysis. Berlin: Wiley-VCH Verlag, pp 437–460
8. Geweke J (1982) Measurement of linear-dependence and feedback between multiple time-series. J Am Stat. Assoc 77:304–313
9. Geweke J (1984) Measures of conditional linear-dependence and feedback between time-series. J Am Statist Assoc 79:907–915
10. Granger CWJ (1969) Investigating causal relations by econometric models and cross-spectral methods. Econometrics 37:424–38
11. Gray CM, McCormick DA (1996) Chattering cells: superficial pyramidal neurons contributing to the generation of synchronous oscillations in the visual cortex. Science 274(5284):109–13
12. Kaminski M, Ding M, Truccolo WA, Bressler SL (2001) Evaluating causal relations in neural systems: granger causality, directed transfer function and statistical assessment of significance. Biol Cybern 85(2):145–57
13. Le Van Quyen M, Bragin A (2007) Analysis of dynamic brain oscillations: methodological advances. Trends Neurosci 30(7):365–73
14. Lopes da Silva FH (1991) Neural mechanisms underlying brain waves: from neural membranes to networks. Electroencephalogr Clin Neurophysiol 79(2):81–93
15. Lund JS (2002) Specificity and non-specificity of synaptic connections within mammalian visual cortex. J Neurocytol 31(3-5):203–9
16. Mehta AD, Ulbert I, Schroeder CE (2000) Intermodal selective attention in monkeys. I: distribution and timing of effects across visual areas. Cereb Cortex 10(4):343–58
17. Mitzdorf U (1985) Current source-density method and application in cat cerebral cortex: investigation of evoked potentials and EEG phenomena. Physiol Rev 65(1):37–100
18. Ritz R, Sejnowski TJ (1997) Synchronous oscillatory activity in sensory systems: new vistas on mechanisms. Curr Opin Neurobiol 7(4):536–46
19. Schroeder CE, Steinschneider M, Javitt DC, Tenke CE, Givre SJ, Mehta AD, Simpson GV, Arezzo JC, Vaughan HGJ (1995) Localization of ERP generators and identification of underlying neural processes. Electroencephalogr Clin Neurophysiol Suppl 44(NIL):55–75
20. Shaw JC (2003) Brain's Alpha Rhythm and the mind. Amsterdam: Elsevier
21. Silva LR, Amitai Y, Connors BW (1991) Intrinsic oscillations of neocortex generated by layer 5 pyramidal neurons. Science 251(4992):432–5
22. Steriade M (2004) Neocortical cell classes are flexible entities. Nat Rev Neurosci 5(2):121–34
23. Steriade M, Gloor P, Llinas RR, Lopes da Silva FH, Mesulam MM (1990) Basic mechanisms of cerebral rhythmic activities. Electroencephalogr Clin Neurophysiol 76(6):481–508
24. Weiner N (1956) The theory of prediction. In: Beckenbach EF (ed) Modern mathematics for the engineer. New York: McGraw-Hill

Neurophysiology of Interceptive Behavior in the Primate: Encoding and Decoding Target Parameters in the Parietofrontal System

Hugo Merchant and Oswaldo Pérez

Abstract This chapter describes the encoding and decoding properties of target parameters in the primate parietofrontal system during an interception task of stimuli in real and apparent motion. The stimulus moved along a circular path with one of 5 speeds (180–540 degrees/s), and was intercepted at 6 o'clock by exerting a force pulse on a joystick that controlled a cursor on the screen. The real stimuli moved smoothly along the circular path, whereas in the apparent motion situation five stimuli were flashed successively at the vertices of a regular pentagon. First, we include a description of the neural responses associated with temporal and spatial aspects of the targets with real and apparent motion. Then, using a selected population of cells that encoded the target's angular position or time-to-contact, we tested the decoding power of the motor cortex and area 7a to reconstruct these variables in the real and apparent motion conditions. On the basis of these results, we suggest a possible neurophysiological mechanism involved in the integration of target information to trigger an interception movement.

Introduction

People and animals usually interact with objects in relative motion, that is, organisms are moving in the environment and/or objects are moving within the visual field toward or away from organisms. Thus, there are two main types of interactions between subjects and objects in relative motion: collision avoidance and the opposite, an interception. Successful control of these interactions is essential for survival. Fatal encounters can happen if the organism is not able to avoid collision or a predator, and a predator will eventually die if unable to catch its prey. This huge adaptative pressure suggests that the neural mechanisms underlying collision avoidance and interception have been sculpted by evolution throughout millions of years in different vertebrate and invertebrate species.

H. Merchant (✉)
Instituto de Neurobiología, UNAM, Campus Juriquilla, Querétaro Qro. 76230, México,
e-mail: merchant@inb.unam.mx

K. Josić et al. (eds.), *Coherent Behavior in Neuronal Networks*, Springer Series in Computational Neuroscience 3, DOI 10.1007/978-1-4419-0389-1_10,

Although numerous studies have characterized different aspects of the interceptive behavior using an experimental psychology approach [26], few neurophysiological studies have determined the neural underpinnings of such an important action. However, recent findings from our group have provided the first clues regarding the neural mechanism of target interception [21–23, 30]. The present chapter focuses on the neurophysiological properties of the parietofrontal system in monkeys trained to intercept circularly moving targets in real or apparent motion. We describe the neural encoding and decoding abilities of the motor cortex and area 7a to represent different target attributes during this interception task.

Behavioral Aspects of an Interceptive Action

In manual hitting interceptions, the control of movement is done under explicit representations of where to go (the interception location or zone IZ) and how long it will take to get there (time-to-contact or TTC). Thus, in this type of behavior the time and position information are clearly distinguishable. Predictive and reactive models have formalized the integration of the temporal and spatial variables involved in the perceptual and motor components of manual hitting interceptions. In the predictive model, the interception movement is predetermined and is not influenced by visual information after the motor command is triggered. This model accounts for manual hitting interceptions with fast and ballistic movements and assumes that the programmed movement time is triggered after a key target parameter reaches a particular threshold. In some circumstances, it has been observed that the distance remaining to reach the interception zone (DTC_{tar}) can be a key parameter. In many others, the TTC_{tar} is the key parameter of preference [14, 34]. In contrast, the reactive strategy assumes that the interception movement starts at a target traveling time or distance, and then is further modulated in an ongoing fashion [8, 15]. Target pursuit is a behavior well explained by this model.

Summarizing the psychophysics of manual interception, a set of requirements must be satisfied to intercept a moving target. First, it is necessary to process the visual motion information of the target, including its actual position, TTC_{tar}, DTC_{tar}, and velocity. Second, the subject uses a predictive or reactive strategy to control the initiation of the interception movement, so that at the end of the movement the target is intercepted. Third, an interception movement should be implemented. This can be a ballistic movement with a predetermined direction and kinetics, or it can be a complex movement divided into submovements that can be regulated to optimize the precision of the interception. Finally, it is necessary to evaluate the end result of the interception, i.e., how precise it was. This information can be used to correct the strategy and the interception movement properties.

The neurophysiology of several of these behavioral components has been studied separately. It is well known that different cortical and subcortical areas, such as the middle temporal area (MT), process visual motion information. It has also been demonstrated that the different premotor areas and the primary motor cortex are involved in the preparation and execution of voluntary movements [13, 36]. Finally, it has been suggested that different areas of the parietal and frontal lobes are engaged in visuomotor transformations [2]. Indeed, here we show how the visuomotor information is integrated in two areas of the parietofrontal system during an interception task [25]. Initially, however, we describe the cortical network engaged in visual motion processing.

Visual Motion Processing

Visual motion is a powerful stimulus for activating a large portion of the cerebral cortex. Neurophysiological studies in monkeys [1,28] and functional neuroimaging studies in human subjects [5, 38] have documented the involvement of several areas in stimulus motion processing, including the MT [37], medial superior temporal area (MST) [35], superior temporal polysensory area [4], area 7a [18, 27, 33], and the ventral intraparietal area [6]. More detailed analyses of the neural mechanisms underlying visual motion processing have been performed in monkey experiments, the results of which indicate that different areas relate to different aspects of this processing. The direction of rectilinear motion is explicitly represented in the neural activity of the MT, a structure that projects to the MST, areas 7a and 7m, and VIP. These target areas are part of the posterior parietal cortex (PPC). Cells in the MST and area 7a not only respond to rectilinear motion, but also to optic flow stimuli, including stimulus motion in depth [9, 18, 33]. Neurons in the MST are tuned to the focus of expansion and can code for the direction of heading [3, 10]. The responses of area 7a neurons to optic flow stimuli appear to be more complex than those in the MST, since individual neurons respond similarly to opposed directions of motion, like clockwise (CW) and counterclockwise (CCW) rotations, upward and downward motions, or rightward and leftward translations [19]. Interestingly, optical expansion is the most prominent stimulus driving the activity of neurons in this area. Thus, the PPC can process optic flow information in a very complex fashion. It is reasonable to expect, then, that the PPC is a good candidate for the neural representation of TTC in primates. In fact, our group was the first to characterize the neural correlates of TTC_{tar} in area 7a and the motor cortex in the monkey [22]. Furthermore, in a recent fMRI study, it was demonstrated that the parietofrontal system in humans is specifically activated during perception of TTC judgments [12]. Besides the representation of TTC and direction of motion, areas such as the MT, MST, and area 7a also code for the speed of visual motion [11, 17, 29].

Overall, the current knowledge of visual motion processing indicates that the motor system has access to TTC_{tar}, DTC_{tar}, and target velocity to drive the interceptive response. This visual information travels to premotor areas and then to the primary motor cortex from different areas of the PPC. Therefore, the anatomic evidence indicates that the neural substrate of interceptive actions may be a distributed network engaging the parietofrontal system. In the following sections, we review some neural correlates of target interception in two important nodes of the parietofrontal system: area 7a and the motor cortex. We begin by describing the interception task used in these studies.

The Interception Task

The task required the interception of a moving target at 6 o'clock in its circular trajectory by applying a downward force pulse on a pseudoisometric joystick that controlled a cursor on the computer monitor (Fig. 1a) [20]. The target moved CCW with one of five speeds, ranging from 180 to 540 degrees/s. In addition to the real motion condition where the targets moved smoothly along a low contrast circular path, we also used an apparent motion situation where the target was flashed successively at the vertices of a regular pentagon [32]. In the latter condition, an illusion of a stimulus continuously moving along the circular path was obtained at target speeds above ~315 degrees/s in human subjects [24]. We included path-guided apparent motion because we were interested in comparing the behavioral strategy and the overall neural mechanisms during the interception of stimuli with real and apparent motion. The hypothesis here was that the neural underpinnings of target interception is different during real and apparent motion conditions.

In this task the monkeys used a predictive strategy for interception, producing predetermined ballistic movements. We, therefore, could investigate the possible key parameter used to control the initiation of the interception movement. For that purpose, we calculated TTC_{tar} and DTC_{tar} at the beginning of the effector movement. We found that DTC_{tar} increased asymptotically as a function of the stimulus speed in both motion conditions (Fig. 1b, *top*). In addition, the movement time (which corresponded to TTC_{tar} in these conditions) decreased slightly as a function of the stimulus speed, and it was larger in the real than in the apparent motion condition (Fig. 1b, *bottom*). Despite these results it was difficult to unambiguously identify the key parameter used for interception. Nevertheless, as we will show later, the neurophysiological data collected in the parietofrontal system suggest that TTC_{tar} is used to trigger the interception movement in both the real and apparent motion conditions [22].

Sensorimotor Processing During the Interception of Circularly Moving Targets

In a previous study, we determined quantitatively the relation between the temporal pattern of neural activation and different aspects of the target and the motor execution during the interception task [22]. We designed a general multiple linear regression model to test the effects of different parameters on the time-varying neural activity. These parameters were the direction cosines of the stimulus angle, TTC_{tar}, the vertical hand force, and the vertical hand force velocity. This analysis revealed that the time-varying neuronal activity in area 7a and in the motor cortex was related to various aspects of stimulus motion and hand force in conditions of both the real and apparent motion, with stimulus-related activity prevailing in area 7a and hand-related activity prevailing in the motor cortex (Fig. 2). The most important

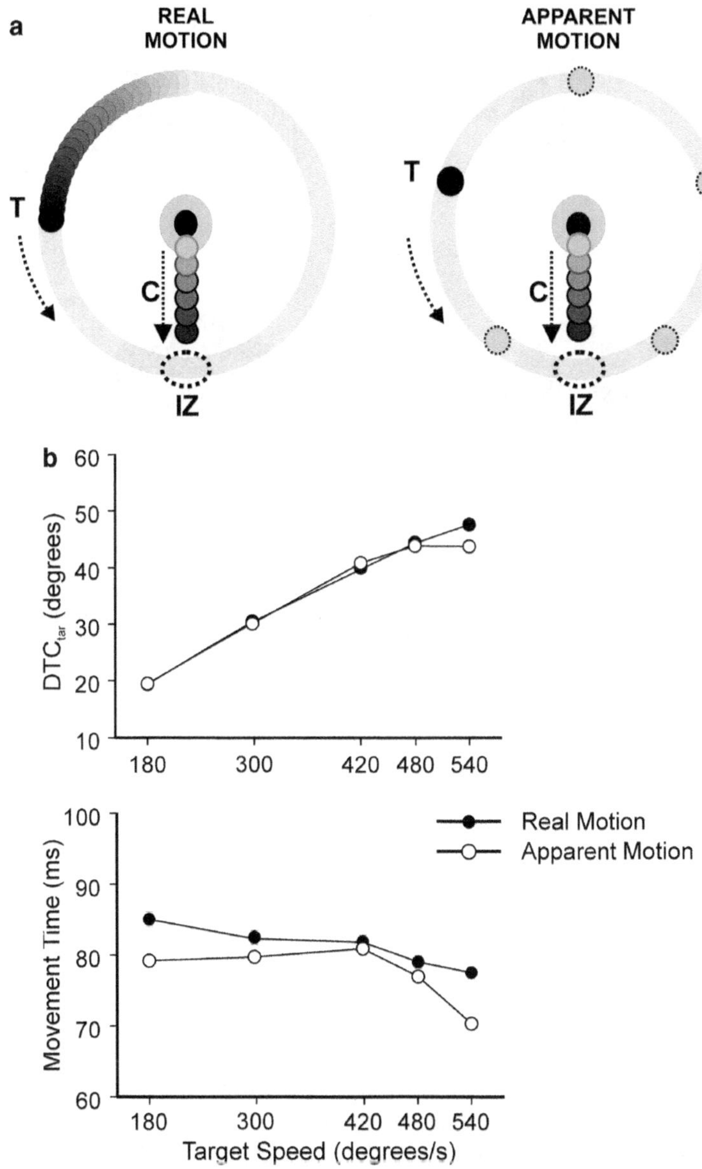

Fig. 1 (**a**) Interception task of circularly moving targets. T represents the smoothly moving target in the real motion condition, or the flashing stimulus at the vertices of a regular pentagon in the apparent motion condition; *C* cursor, *IZ* interception zone. (**b**) Behavioral performance during the interception task. *Top*, target distance to contact (DTC_{tar}) at the beginning of the interception movement; *bottom*, movement time is plotted as a function of the stimulus speed. *Filled circles* correspond to the real motion and *open circles* to the apparent motion condition. Modified from [20]

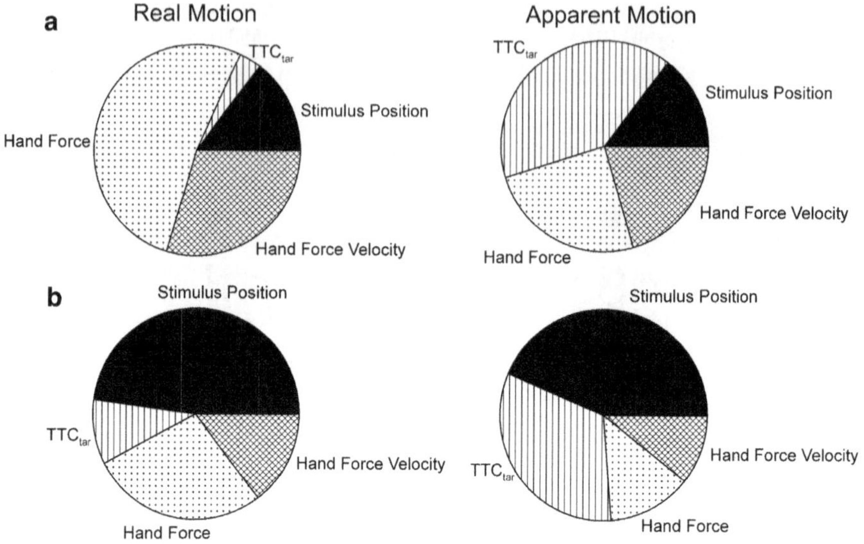

Fig. 2 Percentages of neurons in the real and apparent motion conditions, for which the noted parameter was ranked first using the standardized coefficients obtained from the multiple regression analysis. (**a**) Motor cortex. (**b**) Area 7a. Modified from [22]

finding was that the neural activity was selectively associated with the stimulus angle during real motion, whereas it was tightly correlated with TTC_{tar} in the apparent motion condition, particularly in the motor cortex (Fig. 2).

Encoding of Angular Position and Time-to-Contact During the Interception Task

As a following step, we compared how the time-varying neural activity was specifically related to the stimulus angle or the TTC_{tar}, in both motor cortex and area 7a. Independent linear regression models were carried out to test which target parameter was better explained by the temporal profile of activation for each target speed, in both the real and apparent motion conditions. The first model was defined as:

$$f_{t+\Delta} = b_0 + b_1 \cos\theta_t + b_2 \sin\theta_t + \varepsilon_t, \tag{1}$$

where f_t is the mean spike density function at time t (20 ms window), Δ was the time lag between the neural activity and the independent variables and varied from −160 to +160 ms, b_0 is a constant, b_1 and b_2 are the regression coefficients for the stimulus angle (also referred to as θ or theta), and ε_t is the error. The second model was defined as:

$$f_{t+\Delta} = b_0 + b_1\tau_t + \varepsilon_t \tag{2}$$

with the same parameter definitions as (1), except that here b_1 is the regression coefficient for TTC_{tar} (also referred to as τ).

The adjusted R^2 was used to compare the goodness of fit, and the model with the highest value (1) vs. (2) was used for further analysis if the winning regression ANOVA was significant ($p < 0.05$). A total of 587 neurons in the motor cortex and 458 neurons in area 7a were analyzed using the two models, since these cells showed significant effects in the multiple linear regression model described earlier [22]. These analyses revealed again that the time-varying neuronal activity in area 7a and the motor cortex was related to different aspects of the target motion in both the real and apparent motion conditions. Neurons in area 7a showed that the target angle was the best parameter to explain the time-varying neural activity in both motion conditions (Table 1). In contrast, in the motor cortex the neural activity was selectively associated with the TTC_{tar}, in both the real and the apparent motion (Table 2).

Table 1 Percent and total number of neurons in area 7a that showed significant regression models from (1) or (2) and where the target angular position (θ) or the time-to-contact (TTC_{tar}) was the best parameter to explain the temporal profile of activation

Motion condition	Target velocity	% theta	% TTC_{tar}	Total neurons
Real Motion	180	70.85	29.15	247
	300	73.21	26.79	265
	420	74.45	25.55	227
	480	73.68	26.32	209
	540	73.14	26.86	175
Apparent motion	180	83.33	16.67	396
	300	79.40	20.60	369
	420	69.00	31.00	400
	480	69.04	30.96	394
	540	75.34	24.66	373

Table 2 Percent and total number of motor cortical cells that showed significant regression models from (1) or (2) and where the target angular position (θ) or the time-to-contact (TTC_{tar}) was the best parameter to explain the temporal profile of activation

Motion condition	Target velocity	% Theta	% TTC_{tar}	Total neurons
Real motion	180	27.00	73.00	337
	300	23.12	76.88	346
	420	24.41	75.59	299
	480	26.92	73.08	312
	540	22.97	77.03	296
Apparent motion	180	16.37	83.63	452
	300	21.70	78.30	447
	420	24.46	75.54	462
	480	27.20	72.80	478
	540	18.92	81.08	465

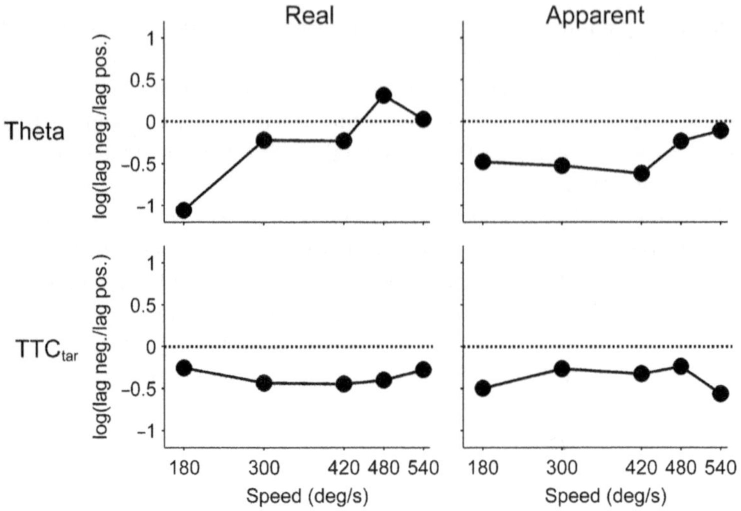

Fig. 3 The logarithm of the quotient between positive and negative lags plotted against the target speed for significant regressions in (1) (*Theta, top*) or (2) (TTC_{tar}, *bottom*) for the real (*right*) and apparent (*left*) motion conditions in the motor cortex

A different question concerns the time shifts of the stimulus angle and TTC for which the highest adjusted R^2 values were obtained across cells, target speeds, and motion conditions. Since the neural activity was shifted with respect to the independent variables: a negative shift indicated that the neural activity was leading the variable (predictive response), whereas a positive shift indicated that the variable was leading the neural activity (sensory response). In the motor cortex, the neural time shift distributions were skewed toward the predictive side. The overall median of the distribution of lags for all target speeds was −20 ms for both real and apparent motion conditions. To further analyze the time shifts for the best regression models in (1) and (2), we plotted the logarithm of the quotient (log-ratio) between all positive and all negative lags against the target speed (Fig. 3). In the apparent motion condition, both the target angle (θ) and TTC_{tar} showed negative log-ratio values, indicating that the best time shifts were predictive across the target speeds. The same was observed for the TTC_{tar} in the real motion condition; however, the target angle showed positive log-ratio values at the highest target speeds in this motion condition. Nevertheless, no significant differences between the lag distributions in the real and apparent motion conditions were observed for target angle or TTC_{tar} (Kolmogorov–Smirnov test, $p > 0.05$). Therefore, these findings suggest that the time-varying activity of the motor cortex can encode the TTC_{tar} and the target angle in a predictive fashion in both motion conditions.

In area 7a the neural time shift distributions for the highest adjusted R^2 models were skewed toward positive values (medians: 40 ms apparent, 20 ms real motion condition), indicating that area 7a neurons were responding to the change in the

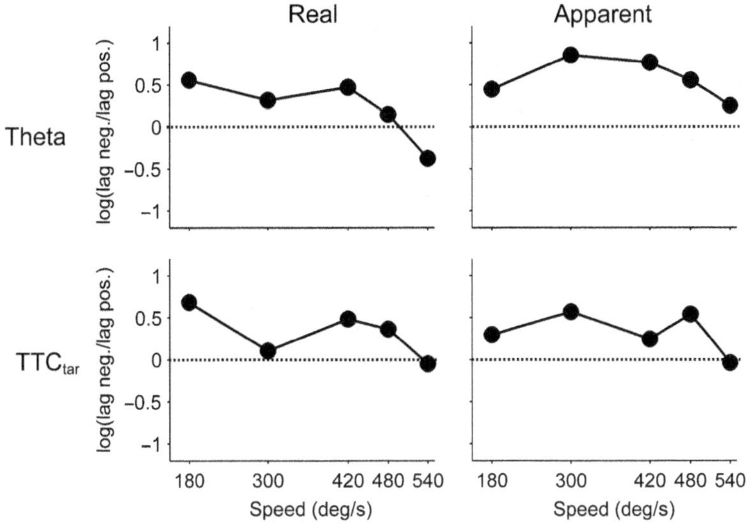

Fig. 4 The logarithm of the quotient between positive and negative lags plotted against the target speed for significant regressions in (1) (Theta, *top*) or (2) (TTC_{tar}, *bottom*) for the conditions of real (*right*) and apparent (*left*) motion in area 7a

target angle and TTC_{tar}. Actually, the log-ratio values were positive for most motion conditions and target speeds, with the exception of the target angle at the highest target speed, which showed a negative value, and hence predictive responses, in the real motion condition (Fig. 4). However, again, no significant differences were found between the target angle or the TTC_{tar} lag distributions in the real and apparent motion conditions in this parietal area (Kolmogorov–Smirnov test, $p > 0.05$). Overall, these results emphasize the sensory role of area 7a in visual motion processing, with an initial reconstruction of the target TTC for real and apparent moving targets that could be transferred to the frontal areas for further processing.

As a final point, it is important to mention that the results of regressions from (1) and (2) were not totally consistent with the multiple regression analysis of the previous section. Specifically, during the real motion condition more neurons showed better fittings for the target angle in the previous analysis, whereas for (1) and (2), TTC_{tar} was the best explanatory parameter in both areas. The most probable cause for this discrepancy is the fact that in the previous multiple regression model, we included the hand force and hand force velocity, which have some degree of collinearity with the TTC_{tar}. Therefore, the discrepancy probably reflects a competition between TTC_{tar} and the arm movement parameters in the regression model of (2), competition that is quite relevant in the motor cortical cell activity. Then, to explore whether the activity of both areas carried enough information regarding the target angle and the TTC_{tar}, in the following section, we performed a detailed decoding analysis on these parameters.

Decoding of Angular Position and Tau During Interception of Circularly Moving Targets

Once we had determined the dependence of the neural responses on the target angle or TTC_{tar}, we used a Bayesian analysis approach to directly address the inverse problem: given the firing rates of these cells, how can we infer the spatial and temporal parameters of the target. The basic method assumes that we know the encoding functions $f_1(x), f_2(x), \ldots, f_N(x)$ associated with the time series for the target parameter (angle or TTC_{tar}) of a population of N cells from (1) or (2). Given the number of spikes fired by the cells within a time interval from $T - \Delta/2$ to $T + \Delta/2$, where Δ is the length of the time window (20 ms), the goal is to compute the probability distribution of the target angle or TTC_{tar} at time T. Notice that what is to be computed here is a distribution of the target parameter, not a single value. Thus, we always can take the most probable value, which corresponds to the peak of the probability distribution, as the most likely reconstructed target angle or TTC_{tar}.

Let the vector $\mathbf{x}$ be the target parameter, and the vector $n = (n_1, n_2, \ldots, n_N)$ be the numbers of spikes fired by our recorded cells within the time window t, where n_i is the number of spikes of cell i. The reconstruction is based on the standard Bayes formula of conditional probability:

$$P(x|n) = \frac{P(n|x)p(n)}{P(r)}. \tag{3}$$

The goal is to compute $P(\mathbf{x}|\mathbf{n})$, that is the probability for the target parameter to be at the value $\mathbf{x}$, given the number of spikes $\mathbf{n}$. $P(\mathbf{x})$ is the probability for the target to be at a particular value $\mathbf{x}$, which was fixed during the experiment. The probability $P(\mathbf{n})$ for the occurrence of the number of spikes $\mathbf{n}$ is equal to the mean of the conditional probability $P(\mathbf{n}|\,\mathbf{x})$ since $\mathbf{x}$ is deterministic in this experiment. Therefore, $P(\mathbf{n})$ is fixed and does not have to be estimated directly. Consequently, given that $P(\mathbf{n})$ and $P(\mathbf{x})$ are constant in this experiment then $P(\mathbf{x}|\mathbf{n})$ is a constant multiple of $P(\mathbf{n}|\mathbf{x})$.

Thus, the key step is to evaluate $P(\mathbf{n}|\mathbf{x})$, which is the probability for the numbers of spikes $\mathbf{n}$ to occur, given that we know the target parameter $\mathbf{x}$. It is intuitively clear that this probability is determined by the estimated firing rates from (1) or (2). More precisely, if we assume that the spikes have a Poisson distribution and that different cells are statistically independent of one another, then we can obtain the explicit expression:

$$P(n|x) = \prod_{i=1}^{N} \frac{(f_i(x)T)^{n_i}}{n_i!} e^{-f_i(x)T}, \tag{4}$$

where $f_i(x)$ is the average predicted firing rate of cell i of a population of N cells, $\mathbf{x}$ is the target parameter, and T is the length of the time window.

The Bayesian reconstruction method uses (4) to compute the probability $P(\mathbf{n}|\mathbf{x})$ for the target parameter to be at the value $\mathbf{x}$, given the numbers of spikes $\mathbf{n}$ of all the cells within the time window. In this probability distribution, the peak value is taken as the magnitude of the reconstructed target parameter. In other words:

$$\hat{x}\text{Bayes} = \arg\max_{x} P\,\langle n|x\rangle\,.$$

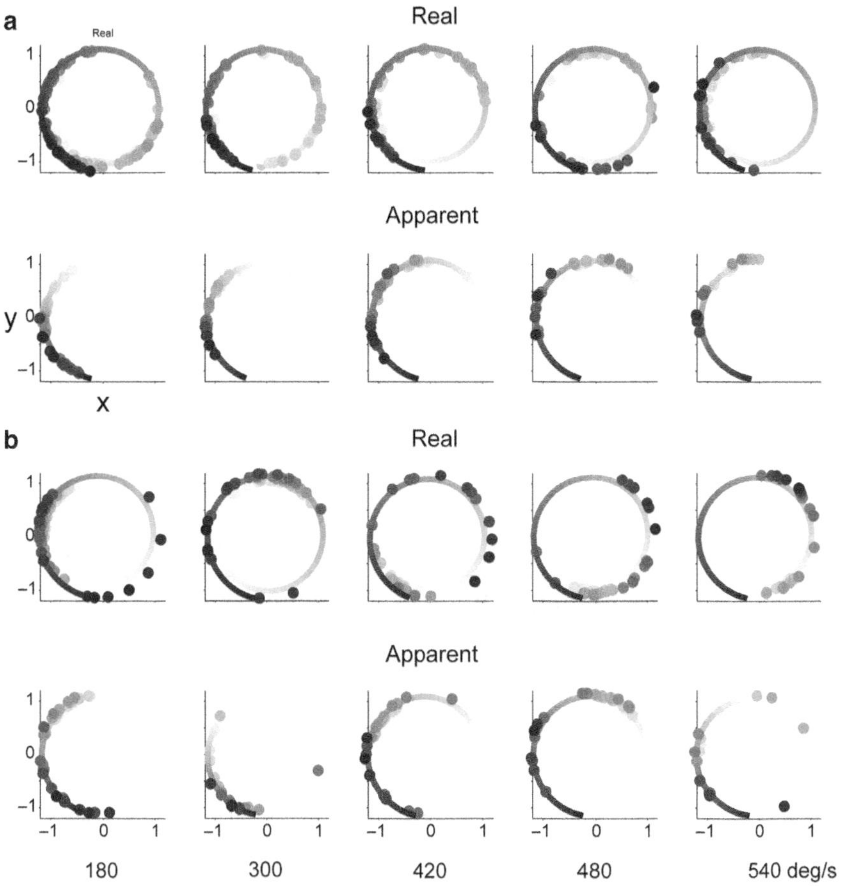

Fig. 5 Mean predicted angular position over time using ensembles of 50 neurons during the real and apparent motion conditions and for the five different stimulus speeds. (**a**) Area 7a. (**b**) Motor cortex. The real (*line*) and predicted (*circles*) positions are color coded in a gray scale, starting at time zero (target onset) in light gray, and ending in black at the last time bin (interception time)

By sliding the time window forward, the entire trajectory of the target parameter can be reconstructed from the time-varying activity in the neural population.

To systematically decode both target parameters, we used the cells with significant regressions from (1) or (2). However, since the number of significant cells varied across motion conditions, target speeds, and cortical areas, we used a constant population of 50 cells to decode both target parameters across all these conditions, to avoid a population-size effect in the reconstructed angular position or TTC_{tar}. In fact, we carried out 100 decodifications for each condition using permuted populations of 50 cells (from the total number of neurons) and cross-validation (across trials) with the purpose of sampling the reconstruction accuracy (variance and bias, see eqs. 3.38 and 3.39 of [7]) within the overall cell population.

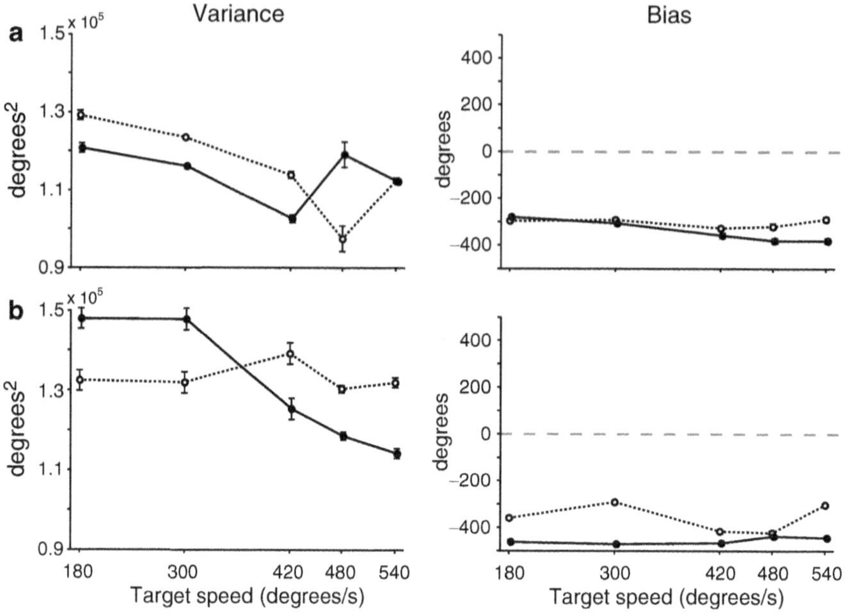

Fig. 6 Mean (±SEM) decodification variance (*left*) and bias (*right*) of the angular position as a function of the target speed. (**a**) Area 7a. (**b**) Motor cortex. *Filled circles* and continuous line correspond to real motion; *open circles* and *dashed line* correspond to apparent motion

The mean decoded angular position over time across target speeds and motion conditions are depicted in Fig. 5 for area 7a and the motor cortex. It is evident that the resulting reconstruction was quite accurate across target speeds and motion conditions in area 7a (Fig. 5a), but deficient in the motor cortex (Fig. 5b). In fact, the mean decoding variability and the mean bias for angular position in both motion conditions were large in the motor cortex (Fig. 6b), but closer to zero in area 7a (Fig. 6a). These results confirm that area 7a is an important node for visual-motion processing [27, 33], and that the neurons in this area can properly represent the change in angular position of the target over time, not only in the real but also in the apparent motion condition [24]. In addition, the results suggest that the motor cortex has limited access to the spatial position of the target during the interception task in both motion conditions. Finally, in accord with the encoding results from the previous section, the decoding from motor cortical activity suggests that the target angle (DTC_{tar}) is probably not the variable used to trigger the interception movement under these conditions.

Figure 7 show the reconstructed TTC_{tar} across target speeds and motion conditions for the motor cortex and area 7a. Again, area 7a (Fig. 7a) shows a decoded TTC that is close to the actual TTC_{tar} for every target speed of the real and apparent

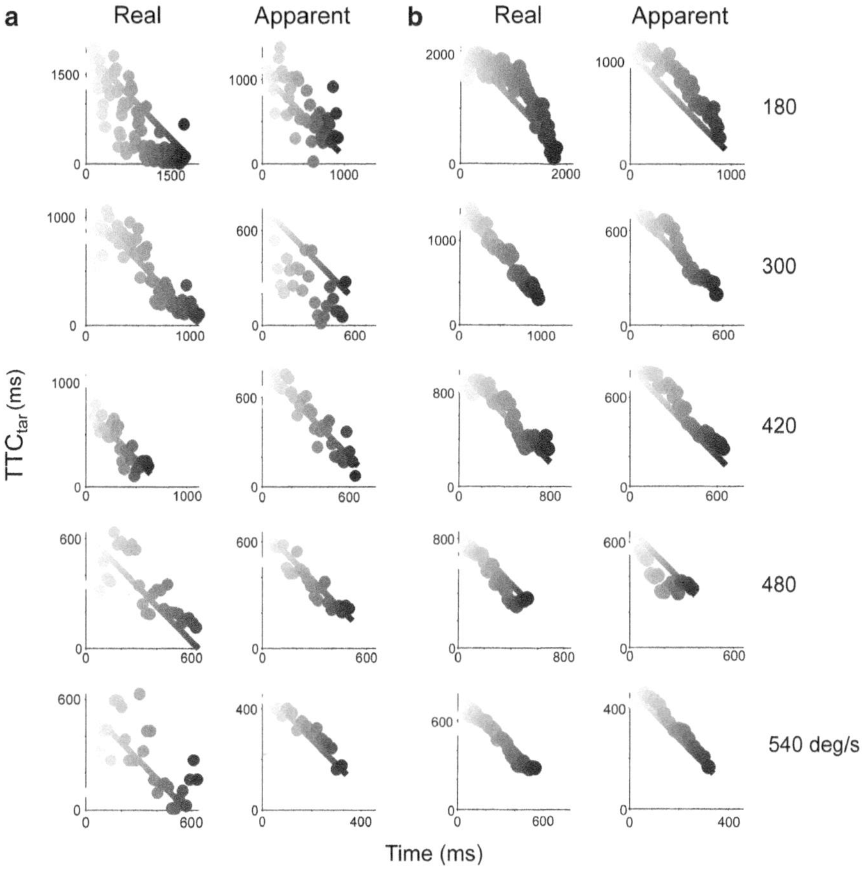

Fig. 7 Mean predicted TTC$_{tar}$ using ensembles of 50 neurons during the real and apparent motion conditions and for the five different stimulus speeds. (**a**) Area 7a. (**b**) Motor cortex. Same notation as in Fig. 5

motion. In addition, the motor cortex (Fig. 7b) shows also an accurate TTC$_{tar}$ decodification during both motion conditions. Actually, the mean decoding variability and mean bias for the target TTC was close to zero in the real and apparent motion conditions using populations of motor cortical (Fig. 8b) or area 7a (Fig. 8a) cells, particularly for the highest speeds. These results indicate, first, that the motor cortex had access to an accurate representation of TTC$_{tar}$ information. This temporal information is probably coming from premotor and posterior parietal areas. Second, these results strengthen the evidence for the hypothesis that TTC$_{tar}$ is the critical target parameter used to trigger the interception movement in this particular task.

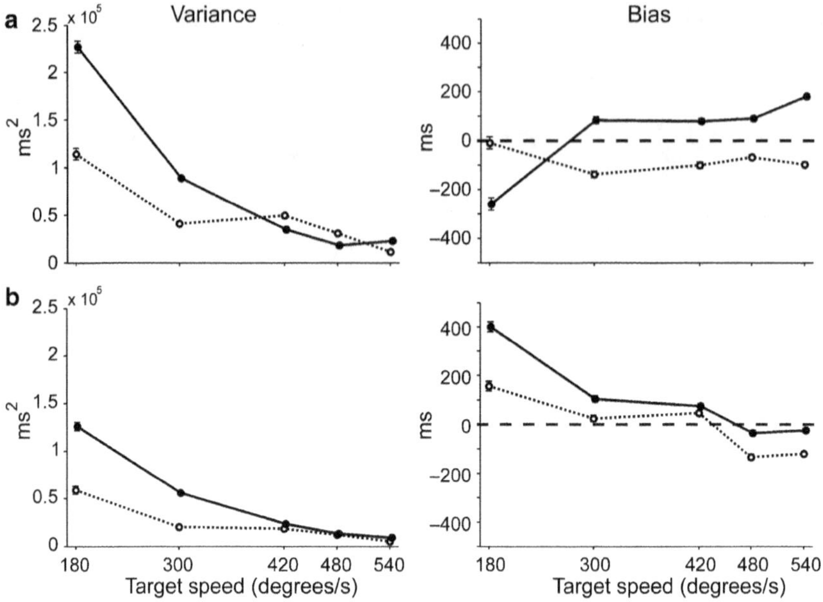

Fig. 8 Mean (±SEM) decodification variance (*left*) and bias (*right*) of the TTC_{tar} as a function of the target speed. (**a**) Area 7a. (**b**) Motor cortex. Filled circles and continuous line correspond to real motion; open circles and dashed line correspond to apparent motion

Concluding Remarks

The neurophysiological experiments using our target interception task revealed that the parietofrontal system of primates is engaged in the representation of spatial and temporal parameters of the target motion. Area 7a processed the target angle and TTC as a sensory area, with a clear preference for the spatial parameter. These findings not only emphasize the role of area 7a in visual motion processing, but also suggest that the representation of TTC begins in the parietal lobe. Actually, imaging and neurophysiological studies have demonstrated that the PPC is involved in temporal information processing [16, 31].

A larger population of motor cortical cells encoded TTC_{tar} than target angle. This information was represented in a predictive rather than a sensory fashion. In addition, the estimated TTC using the activity of motor cortical cells was more accurate than the angular trajectory of the target. Therefore, it is feasible that the motor system uses this type of temporal information to trigger the interception movement in both motion conditions. In fact, we suggest that the motor cortex has the ability not only to represent in a predictive way the TTC_{tar}, but also to detect when it reaches a specific magnitude in order to trigger the interception movement.

Our initial observations using a multiple linear regression model suggested that in the real motion condition the angular position of the target was the critical interception variable, whereas in the apparent motion condition it was the TTC. The current encoding and decoding results indicate that the nervous system represents spatial and temporal parameters of the moving target in the parietofrontal circuit. However, the present findings also suggest that in both the real and apparent motion conditions, the motor system may use the TTC to control the initiation of the interception movement. Since the present encoding models are more specific and were supported by the decoding results, it is more likely that the key interception parameter was temporal rather than spatial in both motion conditions.

Taken together, these results indicate that neurons in the motor cortex and area 7a are processing different target parameters during the interception task. However, the predictive representation of the target TTC is the most probable variable used in the motor cortex to trigger the interception movement.

Acknowledgments We thank Dr. A. P. Georgopoulos for his continuous support throughout the experimental part of the studies and during the writing of this chapter. We also thank Luis Prado and Raúl Paulín for their technical assistance, and Dorothy Pless for proofreading the manuscript. The writing of this manuscript was supported by PAPIIT grant IN206508–19, FIRCA: TW007224–01A1, and CONACYT grant 47170.

References

1. Andersen RA. Neural mechanisms of visual motion perception in primates. *Neuron* 18: 865–872, 1997.
2. Battaglia-Mayer A, Ferraina S, Genovesio A, Marconi B, Squatrito S, Molinari M, Lacquaniti F, Caminiti R. Eye-hand coordination during reaching. II. An analysis of the relationships between visuomanual signals in parietal cortex and parieto-frontal association projections. *Cereb Cortex* 11: 528–544, 2001.
3. Bradley DC, Maxwell M, Andersen RA, Banks MS, Shenoy KV. Neural mechanisms of heading perception in primate visual cortex. *Science* 273: 1544–1547, 1996.
4. Bruce C, Desimone R, Gross CG. Visual properties of neurons in a polysensory area in superior temporal sulcus of the macaque. *J Neurophysiol* 46: 369–384, 1981.
5. Cheng K, Fujita H, Kanno I, Miura S, Tanaka K. Human cortical regions activated by wide-field visual motion: an H2(15)O PET study. *J Neurophysiol* 74: 413–427, 1995.
6. Colby CL, Duhamel JR, Goldberg ME. Ventral intraparietal area of the macaque: anatomic location and visual response properties. *J Neurophysiol* 69: 902–914, 1993.
7. Dayan P, Abbott LF. Theoretical Neuroscience, computational and mathematical modeling of neural systems. MIT press. London England, 2001.
8. Donkelaar P van, Lee RG, Gellman RS. Control strategies in directing the hand to moving targets. *Exp Brain Res* 91: 151–161, 1992.
9. Duffy CJ, Wurtz RH. Sensitivity of MST neurons to optic flow stimuli. I. A continuum of response selectivity to large-field stimuli. *J Neurophysiol* 65: 1329–1345, 1991.
10. Duffy CJ, Wurtz RH. Response of monkey MST neurons to optic flow stimuli with shifted centers of motion. *J Neurosci* 15: 5192–5208, 1995.
11. Duffy CJ, Wurtz RH. Medial superior temporal area neurons respond to speed patterns of optic flow. *J Neurosci* 17: 2839–2851, 1997.
12. Field DT, Wann JP. Perceiving time to collision activates the sensorimotor cortex. *Curr Biol* 15: 453–458, 2005.
13. Georgopoulos AP. Neural aspects of cognitive motor control. *Curr Opin Neurobiol* 10: 238–241, 2000.
14. Lee DN, Reddish PE. Plummeting gannets: a paradigm of ecological optics. *Nature* 293: 293–294, 1981.

15. Lee DN. Guiding movement by coupling taus. *Ecological Psychol* 10: 221–250, 1998.
16. Leon MI, Shadlen MN. Representation of time by neurons in the posterior parietal cortex of the macaque. *Neuron* 38: 317–327, 2003.
17. Maunsell JH, Van Essen DC. Functional properties of neurons in middle temporal visual area of the macaque monkey. I. Selectivity for stimulus direction, speed, and orientation. *J Neurophysiol*. 49:1127–1147, 1983.
18. Merchant H, Battaglia-Mayer A, Georgopoulos AP. Effects of optic flow in motor cortex area 7a. *J Neurophysiol* 86: 1937–1954, 2001.
19. Merchant H, Battaglia-Mayer A, Georgopoulos, AP. Functional organization of parietal neuronal responses to optic flow stimuli. *J Neurophysiol* 90: 675–682, 2003a.
20. Merchant H, Battaglia-Mayer A, Georgopoulos, AP. Interception of real and apparent circularly moving targets: Psychophysics in Human Subjects and Monkeys. *Exp Brain Res* 152: 106–112, 2003b.
21. Merchant H, Battaglia-Mayer A, Georgopoulos, AP. Neural responses in motor cortex and area 7a to real and apparent motion. *Exp Brain Res* 154: 291–307, 2004a.
22. Merchant H, Battaglia-Mayer A, Georgopoulos, AP. Neural responses during interception of real and apparent circularly moving targets in motor cortex and area 7a. *Cereb Cortex* 14: 314–331, 2004b.
23. Merchant H, Battaglia-Mayer A, Georgopoulos, AP. Neurophysiology of the parieto-frontal system during target interception. *Neurol. Clin Neurophysiol* 1: 1–5, 2004c.
24. Merchant H, Battaglia-Mayer A, Georgopoulos, AP. Decoding of path-guided apparent motion from neural ensembles in posterior parietal cortex. *Exp Brain Res* 161: 532–540, 2005.
25. Merchant H, Georgopoulos AP. Neurophysiology of perceptual and motor aspects of interception. *J Neurophysiol* 95: 1–13, 2006.
26. Merchant H, Zarco W, Prado L, Perez O. Behavioral and neurophysiological aspects of target interception. *Adv Exp Med Biol* 629: 199–218, 2008.
27. Motter BC, Mountcastle VB. The functional properties of the light-sensitive neurons of the posterior parietal cortex studied in waking monkeys: Foveal sparing and opponent vector organization. *J Neurosci* 1: 3–26, 1981.
28. Newsome WT, Britten KH, Salzman CD, Movshon JA. Neuronal mechanisms of motion perception. *Cold Spring Harb Symp Quant Biol* 55: 697–705, 1990.
29. Phinney RE, Siegel RM. Speed selectivity for optic flow in area 7a of the behaving monkey. *Cereb Cortex* 10: 413–421, 2000.
30. Port NL, Kruse W, Lee D, Georgopoulos AP. Motor cortical activity during interception of moving targets. *J Cogn Neurosci* 13: 306–318, 2001.
31. Rao SM, Mayer AR, Harrington DL. The evolution of brain activation during temporal processing. *Nat Neurosci* 4(3): 317–323, 2001.
32. Shepard RN, Zare SL. Path-guided apparent motion. *Science* 220: 632–634, 1983.
33. Siegel RM, Read HL. Analysis of optic flow in the monkey parietal area 7a. *Cereb Cortex* 7: 327–346, 1997.
34. Tresilian JR. Hitting a moving target: Perception and action in the timing of rapid interceptions. *Percept Psychophys* 67: 129–149, 2005.
35. Van Essen DC, Maunsell JH, Bixby JL. The middle temporal visual area in the macaque: Myeloarchitecture, connections, functional properties and topographic organization. *J Comp Neurol* 199: 293–326, 1981.
36. Wise SP, Boussaoud D, Johnson PB, Caminiti R. Premotor and parietal cortex: corticocortical connectivity and combinatorial computations. *Annu Rev Neurosci* 20: 25–42, 1997.
37. Zeki, SM. Functional organization of a visual area in the posterior bank of the superior temporal sulcus of the rhesus monkey. *J Physiol* 236: 549–573, 1974.
38. Zeki S, Watson JD, Lueck CJ, Friston KJ, Kennard C, Frackowiak RS. A direct demonstration of functional specialization in human visual cortex. *J Neurosci* 11: 641–649, 1991.

Noise Correlations and Information Encoding and Decoding

Bruno B. Averbeck

Abstract Neuronal noise is correlated in the brain, and these correlations can affect both information encoding and decoding. In this chapter we discuss the recent progress that has been made, both theoretical and empirical, on how noise correlations affect information encoding and decoding. Specifically, we discuss theoretical results which show the conditions under which correlations either do or do not cause the amount of encoded information to saturate in modestly large populations of neurons. Correspondingly, we also describe the conditions under which information decoding can be affected by the presence of correlations. Complementing the theory, empirical studies have generally shown that the effects of correlations on both encoding and decoding are small in pairs of neurons. However, theory shows that small effects at the level of pairs of neurons can lead to large effects in populations. Thus, it is difficult to draw conclusions about the effects of correlations at the population level by studying pairs of neurons. Therefore, we conclude the chapter by briefly considering the issues around estimating information in larger populations.

Introduction

Information is coded in the brain by populations of neurons using distributed representations [15, 17, 31]. Interestingly, this conceptual advance was brought about in the late 1980s by studying one neuron at a time, although the theory was developed by thinking about populations of neurons. This leads to the question of whether or not we can learn anything about neural representations by studying more than one neuron at a time. Is there any information in the population code that exists only at the population level? This question has been considered from many perspectives and for many years [26]. However, in the last 15 years considerable progress has been made. Much of this progress has been brought about by theoretical work that

B.B. Averbeck (✉)
Sobell Department of Motor Neuroscience and Movement Disorders, Institute of Neurology, UCL, London WC1N 3BG, UK
e-mail: b.averbeck@ion.ucl.ac.uk

K. Josić et al. (eds.), *Coherent Behavior in Neuronal Networks*, Springer Series in Computational Neuroscience 3, DOI 10.1007/978-1-4419-0389-1_11,

has generated well-defined questions and analytical tools for considering the possibility of coding effects that exist only at the population level. The application of these tools to experimental data has also generated a number of consistent findings.

In this chapter, we first review much of what has been learned about the role of correlated neural activity in information coding from both empirical and theoretical perspectives. Then we consider approaches to analyzing data using these tools and examine the findings that have been put forth in empirical data and compare this to the predictions of theoretical models. Finally, we consider the outstanding questions and the potential experimental and theoretical problems associated with answering these questions.

Defining Noise Correlations

When examining information that can only exist at the population level, one normally looks for *patterns* of activity across neurons. However, single neurons in the brain are noisy. If the same stimulus is shown repeatedly to an animal, neurons in visual cortex respond differently in different trials. Similarly in the motor system, when the same movement is repeated, the response of single neurons differs across trials. Thus, patterns of activity in the brain manifest as stochastic correlations between neurons. In this chapter, we focus on the role of noise correlations. Noise correlations are trial-by-trial correlations in the variability of neural responses for pairs of neurons (Fig. 1). When they are measured using large bins of neural activity (lets say >25 ms), they are called spike count correlations, and can be thought of as correlations in rate coded information. However, one can always divide the neural activity into smaller time bins and measure the correlation in responses at a finer grain [4]. In this case, the noise correlations are formally equivalent to shift-predictor corrected cross correlations [5], and as such, the effects of synchrony or oscillations on information coding can be assessed. Thus, the tools that we use to measure the impact of noise correlations on information coding can be used at any time-scale, and whether one is looking at spike-count correlations or synchrony can be more precisely understood as a question of the bin-size used for data analysis.

Throughout much of this chapter, we examine the simple case of the neural representation and information coding of two stimuli or movements. This case, although simplified, is often sufficiently complex to illustrate the necessary points. Given that we have two targets, we will also make a distinction between signal independent (Fig. 1c) and signal dependent (Fig. 1d) noise. Signal-independent noise describes the case where the noise correlations are the same for both targets. Signal-dependent noise, as the name suggests, describes the case where the noise is different for different targets.

The effect of noise correlations on information coding can be studied from both the encoding and decoding perspectives, and one obtains different answers depending upon the perspective. When the effect of noise correlation on information encoding is studied, one is considering the total information encoded by the population, without consideration of how downstream brain networks would

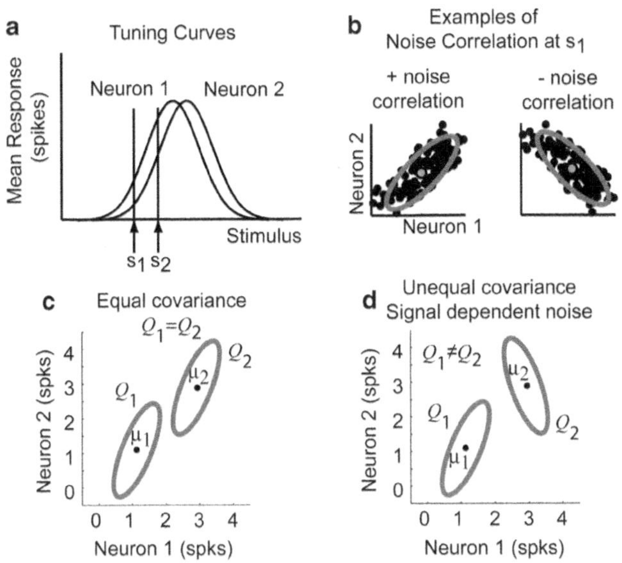

Fig. 1 Definition of noise correlations: (**a**) Tuning curves for a pair of neurons. The tuning curves represent the average response of the neurons to a particular target direction across trials. *Vertical lines* indicate two stimuli used to drive neural responses. (**b**) Noise correlations. Correlation in neuronal variability at a single stimulus value. (**c**) Example noise correlations for only signal-independent noise. In this case, the covariance (measured by the covariance matrix Q) is the same for both targets. (**d**) Example of noise correlation for signal-dependent noise. In this case, the covariance differs between targets

actually extract that information. One normally assumes that all of the information in the population can be extracted. When information decoding is studied, one is focusing on the effect of noise correlations on strategies for extracting or decoding information. Specifically, in this case one asks whether or not ignoring noise correlations with a decoding algorithm, which is a simplifying assumption, leads to a loss of information.

Some confusion can arise from the fact that one often studies encoding by decoding the neural responses. However, when one studies encoding by decoding, one always uses a technique which should, at least in principle, extract all of the information. Thus, one can decode the neural response and make statements about how much information was encoded. Techniques exist, which are guaranteed to extract all of the information that is encoded in neural responses, under particular assumptions. This issue also relates to two different ways of studying information coding in neurons. In the first approach, one carries out information theory calculations (not necessarily Shannon information). This is normally done in theoretical studies. The other approach is to actually decode neural responses on a trial-by-trial basis. This is normally done on empirical data. These two approaches are linked, however, by the fact that information theoretic calculations make predictions about how well one should be able to extract information from neural responses. Thus, one can often make an information theoretic calculation, use that to predict how well one can extract information from neural responses, and then try to extract that information. In this way, theory and experiment can be brought together.

Theoretical Studies: Noise Correlations and Information Encoding

There has been considerable theoretical work on the effect of noise correlations on information encoding. This work often proceeds by using empirical data on how noise correlations between neurons are related to the tuning curves of the neurons, to constrain a model of a neuronal population. The model is then used to predict how information coding scales with population size. In one of the early examples of this approach, Zohary and colleagues noted that the large noise correlation (about 0.2 for neurons with a similar preferred direction) between pairs of MT neurons would strongly constrain the relevant population size for representing movement direction to about 100 neurons [40]. Specifically, if one looks at how accurately movement direction could be estimated from population neural responses, their model suggested that one could not improve accuracy estimation if the population size was increased beyond about 100 neurons. This defined an important question that in some respects it is still being considered: does the correlation structure in a population of neurons cause the information to saturate as the size of the population is increased?

For at least four reasons, however, the coding estimates in their study were too conservative. These four reasons relate to assumptions of their model. This is not meant to be a criticism of their approach. Rather it allows us to illustrate how the theoretical models have evolved since this early study. First, the stimuli used in the experiments in which the noise correlations were measured were themselves noisy [8]. Noise correlations, as we have defined them, are correlations in the variability of neural responses across trials in which the stimuli or movements are identical. In their experiments, monkeys were being shown random patterns of moving dots, and on each trial a different random dot pattern was shown to the monkey. However, all random dot patterns with a particular underlying average direction were treated as equivalent. Thus, much of the variability in the neural response that was treated as noise likely came from variability in the stimuli, because MT neurons are very sensitive to small variations in stimuli [9] and therefore the neurons were likely responding to the stochastic variability in the stimuli in a deterministic way. Thus, some of the noise was likely signal and this signal would tend to be correlated between neurons with similar preferred directions, because such neurons would tend to either increase or decrease their response in a similar manner, creating artificially large noise correlations. For example, if we examine the tuning curves in Fig. 1, if the stimulus is jittered around s_1 from trial to trial, the two neurons will tend to increase or decrease their firing rates similarly. If one assumed that the identical stimulus was being shown, one would treat this variability as noise, and it would be correlated between these neurons. This is also a possible explanation for the fact that the mean noise correlations are very near zero in our data (0.01 at 40 ms; [4]), which is much smaller than the average correlations seen in the data analyzed by Zohary et al.

The second reason the estimates in Zohary et al. were likely too conservative is that they used a simplified tuning curve model. Specifically, instead of assuming tuning curves across a continuum of directions, they assumed that neurons only coded 2 opposing directions of stimuli, as this is what was used in their experiments. Although this assumption is not always a problem, and we make it in many places in this chapter, it can be problematic if one assumes the pattern of noise correlations used by Zohary et al. When Abbott and Dayan examined the effect of noise correlations in population codes using neurons with full tuning curves, subsequent to the Zohary et al. study, they found that in general information did not saturate, although it still could in specific cases [1, 35]. The study of Abbott and Dayan was also important in using Fisher Information to assess the effects of noise correlations in population of neurons, an approach that has been used in most subsequent theoretical studies.

The third reason that the estimates were too conservative in Zohary et al. is that they did not consider signal dependent noise in their model. Specifically, when spikes are counted in a reasonably large bin, the variability in the neural response scales with the spike rate [4, 37], and the covariability may scale as well. This variance scaling generates a situation in which additional information can be encoded by the variability of the neural response in the population [33]. It is, however, very difficult to extract this information, and it only becomes relevant in large populations making it difficult to demonstrate in empirical data. Nevertheless, it may be an important component of the neural code.

The final reason for the underestimate is that the study of Zohary et al. and the other studies we have discussed assumed that the tuning curves in all of the neurons in the population were identical in shape, differing only in their preferred direction or the stimulus to which they would respond maximally. When one relaxes this assumption, information coding does not saturate [34, 38].

Thus, while the study by Zohary et al. was important for generating an interesting question, subsequent studies have extended their model in several relevant directions and shown that the original results do not necessarily hold. In later sections, we consider in more detail what these models do tell us, somewhat independent of their details, and how subsequent modeling efforts might proceed. Finally, while all of these studies are interesting, there is one theoretical point that has not been addressed in detail (but see [32]). Specifically, there is a theorem from communications engineering known as the data processing inequality [11], which states that information cannot be created, it can only be destroyed by processing [3]. The data processing inequality implies that one cannot increase information indefinitely by increasing the size of the population, for any population of neurons downstream from peripheral sensory receptors. For example, one cannot generate more information in the lateral-geniculate nucleus (LGN), than that exists in the retina, by increasing the population size in the LGN. If one continues to increase the population size, information would eventually saturate to the quantity of information in the retina. As an aside, this fact is intriguing given that there is often a large increase in the size of the population representation as one moves from the thalamus to the cortex in sensory systems. There are two possible reasons for this.

First, the thalamus is not the only input to any given cortical area, and as such the cortical area is likely processing information from more sources than the feed-forward sensory input. Second, the representation generated in the cortex is likely useful for subsequent computations, similar to the responses used for classification in a support vector machine. At this point, however, this is just speculation, and little is known about why sensory representations increase in dimensionality as one goes from the thalamus to cortex.

Theoretical Studies: Noise Correlations and Information Decoding

As mentioned earlier, when one studies the effects of noise correlations on information decoding, one is considering whether or not ignoring correlations with a decoding algorithm leads to a loss of information. This question also breaks down into a question of whether or not one ignores signal-dependent noise correlations [33] or signal-independent correlations [39]. Although it is difficult to make general statements, the study of Shamir and Sompolinsky has shown that, when one is dealing with a correlation structure that causes linear information to saturate, there can be considerable information in signal-dependent correlations. Thus, on the one hand, if one employs a decoding strategy, which ignores or cannot extract this information, one can lose a lot of the information contained in the population. On the other hand, the study of Wu et al. has shown that information loss can be minimal, when signal-dependent correlations are not present, if one ignores noise correlations. As with many of the issues that have been raised by these theoretical studies, we do not understand the noise structure in various neural systems in sufficient detail to know whether or not either of these models is relevant. However, these models define interesting questions, and provide us with tools for examining those questions in empirical data, at least in principle.

Empirical Studies: Noise Correlations and Information Encoding and Decoding

In parallel with the theoretical studies, there has been considerable empirical work on the effect of noise correlations on information coding. In contrast to the theoretical work, which has focused on populations, the empirical work has focused on whether or not noise correlations increase or decrease information encoding or lead to a loss of information when ignored by decoding algorithms, in *pairs* of neurons. This is largely due to the limitation of recording from more than a few neurons simultaneously. The earliest work on this problem was by Richmond and colleagues, and it introduced the concept of spike count correlations and studied their effect on information encoding in visual cortex [13, 14]. Following this, Panzeri, Shultz, Treves and their colleagues developed and applied an expansion of

Shannon information into separate terms, which allowed one to separately examine the effects of signal-dependent and signal-independent noise on the total Shannon information encoded [16, 20, 23–25, 27–29]. This information breakdown motivated much of our approach detailed later. Other groups examined whether correlations could be ignored during decoding without a loss of information [2, 12, 19, 21, 22]. These studies have generally found that noise correlations have little impact ($<10\%$) on information coding, whether considered from the encoding or the decoding perspective [5]. However, they have only analyzed interactions at the level of pairs of neurons in most cases but see [2, 6].

Theoretical Analysis of the Effects of Correlations on Encoding and Decoding in Pairs

In our own work, we have tried to clarify the relationship between encoding and decoding and also bring together as much as possible the empirical and theoretical work, by using both information metrics and decoding analyses [6]. To do this, we have often used d', which is a measure of the signal to noise ratio in the system, as our information measure. We have several motivations for using d'. First, most of the theory that has been done has used Fisher Information to analyze information in neural populations. Fisher Information applies to information coding of continuous variables and the corresponding problem of estimating their continuous value from neural activity, while d' applies analogously to discrete variables and the corresponding problem of classifying which of a discrete set of stimuli or movements occurred on a single trial. Thus, d' is the analog of Fisher information, if one wants to work with classification. Second, d' under the Gaussian assumption is relatively simple to compute and understand, and therefore it allows one to build an intuition for how correlations are affecting information encoding and decoding, and how these are related.

The main drawback to using d' is that one has to make several assumptions about neural responses that are not strictly correct. First, we have to assume that the variance and covariance is the same for all targets considered. That is to say, d' only applies to signal-independent noise. As we discussed earlier, the variance of neural responses scales with the mean response, so the assumption of no signal-independent noise is violated in neural data. Second, we have to assume that the neural responses follow a Gaussian distribution. Although this is not necessarily a bad assumption, and it is often better than assuming that neural responses follow a Poisson distribution [4], it also does not strictly hold, as we show later. This is, however, also the assumption made by every theoretical study cited above, mainly because the Gaussian distribution is one of the few analytically tractable distributions, which allows one to model covariances. We ultimately validate the use of d' by relaxing each of these assumptions and showing that we get the same information estimates that we got when we made the assumptions, at least in pairs.

The information measure d', or its square, d^2, is perhaps the simplest information measure. It is given by the signal to noise ratio:

$$d^2 = \frac{(\mu_2 - \mu_1)^2}{\sigma^2} = \frac{\Delta\mu^2}{\sigma^2}, \tag{1}$$

where μ_i indicates the mean spike count of a neuron to target i and σ^2 is the variance of the spike count around the mean response. As we are mostly interested in the responses of multiple neurons recorded simultaneously, we will use the multivariate generalization given by [30]:

$$d^2 = \Delta\mu^{\mathrm{T}} Q^{-1} \Delta\mu, \tag{2}$$

where Q is the covariance matrix that describes the variance and covariance of the neural responses. The off-diagonal elements of Q are the noise covariance (unnormalized correlations) between neurons and with d^2 we assume they are the same for all targets, as mentioned earlier.

Our interest is in estimating the effects of noise correlations on encoding and decoding. To estimate the effects of noise correlations on information encoding, we can compare the information in the correlated neural responses, give by d^2 in (2), to the information that would be in the same population if it were uncorrelated. This is given by an analogous quantity:

$$d^2_{\mathrm{shuffled}} = \Delta\mu^{\mathrm{T}} Q_{\mathrm{d}}^{-1} \Delta\mu, \tag{3}$$

where, Q_{d} is the matrix obtained by setting the off-diagonal terms of Q to zero, which is the same as setting the noise correlations between neurons to zero. This is called d^2_{shuffled}, because experimentalists often shuffle trials between simultaneously recorded neurons to destroy noise correlations. We can then define a quantity which measures the effect of noise correlations on information encoding:

$$\Delta d^2_{\mathrm{shuffled}} = d^2 - d^2_{\mathrm{shuffled}}. \tag{4}$$

This quantity can be positive or negative depending upon whether noise correlations increase or decrease the information encoded with respect to an uncorrelated population, which we will see in more detail later.

To determine the effect of noise correlations on information decoding, we need a quantity analogous to (3), which describes the amount of information that would be extracted by a decoding algorithm that ignored correlations. This is given by:

$$d^2_{\mathrm{diag}} = \frac{\left(\Delta\mu^{\mathrm{T}} Q_{\mathrm{d}}^{-1} \Delta\mu\right)^2}{\Delta\mu^{\mathrm{T}} Q_{\mathrm{d}}^{-1} Q Q_{\mathrm{d}}^{-1} \Delta\mu}. \tag{5}$$

This measures the amount of information that would be extracted by using a decoding algorithm that ignored correlations on the original *unshuffled* or correlated dataset. We refer to this as d^2_{diag} since it is equivalent to assuming a diagonal covariance matrix for the neural responses when deriving the decoding algorithm. In this case, the decoding algorithm is suboptimal. This quantity was derived for Fisher information by [39] as a local linear approximation. In our case, the formula is exact, since the difference in the mean responses is necessarily linear. We can then define a quantity which estimates the amount of information lost by a decoding algorithm that ignored correlations:

$$\Delta d^2_{\mathrm{diag}} = d^2 - d^2_{\mathrm{diag}}. \tag{6}$$

We can use these quantities to examine the theoretical effects of noise correlations on information encoding and decoding in pairs of neurons (Fig. 2). The effects of

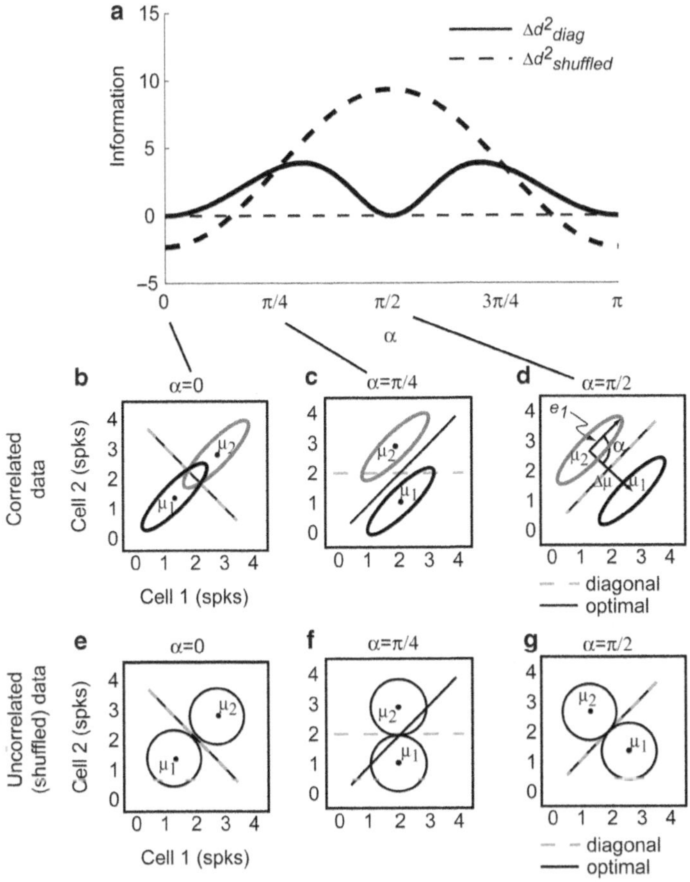

Fig. 2 Effects of noise correlations on information encoding and decoding. (**a**) Plot shows effects of noise correlations as a function of α. (**b–d**) Examples of uncorrelated (shuffled) response distributions at different values of α, which is defined in **d**. Magnitude of noise and signal are constant across plots. Classification boundaries are shown as indicated. (**e–g**) Examples of response distributions at different values of α. The variance of these response distributions are the same as those in **b–d**, but these example neurons are uncorrelated

noise correlations on both factors are controlled by the relationship between the signal and the noise. Specifically, if the signal is represented by $\Delta\mu$, and the covariance matrix is represented by its eigenvectors,e_i and eigenvalues, the effect of the noise correlations can be characterized by the angle, α between the eigenvector associated with the largest eigenvalue of the covariance matrix, e_1 and $\Delta\mu$. Of course, the length of $\Delta\mu$ as well as the size of the eigenvalues associated with the noise modifies the effects as well, but they only scale the effects.

We first focus on the effects of noise correlations on information encoding measured by $\Delta d^2_{\mathrm{shuffled}}$. The information encoded is maximal when response distributions have minimal overlap or are far apart (compare Fig. 2b and d), and this varies continuously as a function of α (Fig. 2a, $\Delta d^2_{\mathrm{shuffled}}$). In the case being depicted here, the fictional uncorrelated responses are circles instead of ellipses (Fig. 2e–g). Thus, if we compare the overlap in the response distributions between the correlated and uncorrelated data, it can be seen that there is less overlap in the distributions for uncorrelated data when α is near zero (i.e., more information in the uncorrelated population; compare Fig. 2b and e) and more overlap when α is near $\pi/2$ (i.e., less information in the uncorrelated population; compare Fig. 2d and g). From another perspective, when the signal in the system, measured by $\Delta\mu$ lies in the same direction as the larger component of the noise (i.e., when $\Delta\mu$ and e_1 have a similar direction) information is decreased, and when they lie in different directions information is increased.

We next turn to the effects of noise correlations on information decoding, measured by $\Delta d^2_{\mathrm{diag}}$. This metric is large when very different classification boundaries are derived under the correlated or the uncorrelated assumptions and it is small when similar classification boundaries are derived. The boundaries derived under the diagonal, uncorrelated model are the same as the boundaries derived under the correlated model when α is near zero or $\pi/2$ and different when α is near $\pi/4$ (Fig. 2b–d compare dashed grey and solid black lines). Thus, if the boundary derived under the diagonal assumption at $\pi/4$ is applied to the correlated data, it is suboptimal, which is to say that less information is extracted from the responses. This is because more of the response distributions lie on the wrong side of the suboptimal classification boundary and the response distributions are better separated by the optimal classification boundary. In Figs. 2e–g, it can be seen that the diagonal boundary would indeed be optimal if it were being applied to uncorrelated data. However, $\Delta d^2_{\mathrm{diag}}$ measures the effect of applying this suboptimal boundary to correlated data.

Empirical Validation

We have used d^2 to examine the effects of noise correlations on information encoding and decoding and to see how these quantities are related. We had to make several assumptions, however, and these assumptions are not necessarily valid. This raises the question of whether or not d^2 is giving us an accurate estimate

Table 1 Relationship between assumptions, information measures, and classifiers

Assumptions	Information measure	Classifier
Gaussian, equal covariances	d^2	Linear Gaussian $(Q_i = Q_j)$
Gaussian, unequal covariances	Battacharyya distance	Quadratic Gaussian $(Q_i \neq Q_j)$
None		Multinomial

of the information coded in the neural responses. To examine this, we proceeded in several steps (Table 1). In all examples shown here, where classifiers are being compared, analyses were done using twofold cross validation. First, we compared the classification performance that was predicted by d^2 to the actual classification performance that we were able to achieve and found that there was a close correspondence between d^2 values and how well we were able to decode data. Thus, the predictions of the information metric closely paralleled our actual performance. Importantly, however, in the first analysis we used a linear classifier which is optimal under the same assumptions as d^2, and therefore both d^2 and the classifier may have missed important information, because neither assessed information in signal dependent correlations. Thus, in the second step we relaxed the assumption of only signal-independent noise, and examined the amount of information in signal-dependent noise in our neural data using an information measure that can capture this information. We then used a quadratic classifier, which is able to extract all the information when there is signal-dependent noise and compared its performance to the linear classifier. Finally, we relaxed both the assumption that the covariances were the same and that the data followed a Gaussian distribution. By using a classifier that relaxed both of these assumptions and comparing its performance to the linear classifier, we could see if relaxing all of the assumptions made by d^2 led to different information estimates.

We began by comparing the classification performance predicted by d^2 to the actual classification performance achieved by a classifier that made the same assumptions. (For details of the experimental task and other procedures see [4,6].) To predict the classification accuracy, we measured d^2 for each pair of neurons, and used it to predict the classification performance using the following equation [6]:

$$p\left(\hat{t} = 2 | t = 1\right) = (2\pi)^{-1/2} \int_{d'/2}^{\infty} \exp\left(\frac{-x^2}{2}\right) \mathrm{d}x. \tag{7}$$

The predicted target is $\hat{t}$ and the actual target is t. The metric d' enters as the lower boundary of the integral. Therefore, this equation gives us the probability that we misclassify each response for a particular value of d'. We then fit a classifier to the data to see if our actual decoding performance matched the performance predicted by (7). Under the assumptions of d^2, which are that neural responses follow a Gaussian distribution and that the variance is the same for both targets to be decoded, one can show that a Gaussian linear classifier can extract all of the information [18]. Specifically, we assumed the following likelihood function:

$$p\left(r|t=i\right)=\left|2\pi Q\right|^{-\frac{1}{2}}\exp\left(-\frac{1}{2}\left(r-\mu_i\right)^{\mathrm{T}}Q^{-1}\left(r-\mu_i\right)\right). \tag{8}$$

In this equation, r is the response on an individual trial, and μ_i and Q are the same as those defined above for d^2. Individual neural responses can be entered into the equation for each target (i.e., $t = 1$ or $t = 2$, where only μ_i is different in each equation), and the probability that the response comes from the corresponding distribution can be calculated. The response is then classified to the target for which the probability of the response is the highest. Because (8) is the likelihood, this means we are carrying out maximum likelihood estimation. Because we have a flat prior, identical decoding performance is obtained if one does maximum-a-posteriori classification. By comparing the actual classification performance to the performance predicted using (7), we were able to test whether or not d^2 was accurately representing the information in the neural responses. We found that the correlation between measured and predicted classification performance was very strong under d^2, d^2_{shuffled} and d^2_{diag} (Fig. 3, top row) for pairs of neurons. Similar analyses carried out on

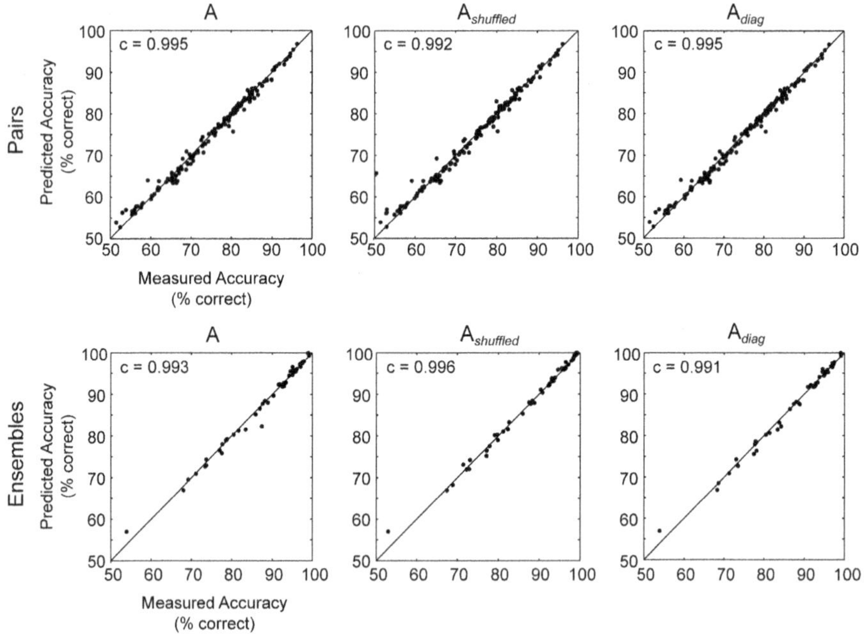

Fig. 3 Comparison of measured and predicted classification accuracy for pairs and ensembles. *Top row* is measured and predicted for correlated data (A), shuffled (A_{shuffled} uncorrelated data) and classifier that ignores correlations (A_{diag}), applied to the correlated data. *Bottom row* is same data for ensembles of 3–8 simultaneously recorded neurons

ensembles of 3–8 simultaneously recorded neurons also showed that d^2 predicted actual classification performance accurately (Fig. 3, bottom row). Thus, classification performance estimated using (7) closely matched the classification performance obtained by actually carrying out the classification analysis trial-by-trial.

In our next step, we extended our analyses by relaxing the assumption of only signal-independent noise. In practice, this is done by estimating a separate covariance matrix for each target. First, we used an information measure to estimate whether or not there was extra information to be obtained from the neural responses if we allowed for differences in the covariance matrices. To measure this, we used the Battacharyya distance (BD) [7] which, for Gaussian distributions is given by:

$$\mathrm{BD_G} = \frac{1}{4}\Delta\mu^{\mathrm{T}}\left(Q_1 + Q_2\right)^{-1}\Delta\mu + \frac{1}{2}\log\frac{|Q_1 + Q_2|}{2\sqrt{|Q_1 Q_2|}}. \tag{9}$$

The first term on the RHS is equal to $1/8d^2$. The second term captures information present in signal-dependent correlations. As such, if we plot d^2 vs. 8*BD all points lying above the line are cases for which BD predicts that there should be additional information in the signal-dependent noise. We found that there were many cases for which this was true (Fig. 4a). In theory, signal-dependent noise can only add information, it cannot decrease information. The points below the line in Fig. 4a are due to numerical errors in estimating (9). BD suggested that there was additional information in the signal-dependent noise. Unlike d^2, however, the BD cannot be used to predict classification performance. It can be used to put bounds on the classification performance [7], but the bounds are rather loose. However, under the assumption that the response distributions are Gaussian and the covariances are not identical across targets, we can derive a maximum likelihood estimator [18]. In this case, it is a quadratic estimator, because the classification boundaries are quadratic and not linear. Effectively, (8) is used, just as in the case of the linear classifier. However, for linear classification, a single covariance matrix, Q, is estimated by pooling across both targets. For quadratic classification, a separate Q_i is estimated for each target.

Thus, we can define an estimator that can, in principle, extract all of the information from the neural responses under the Gaussian assumption with unequal covariance matrices. We did this, and compared its performance to the linear classifier. If there was additional information in the signal-dependent noise, we should do better using the quadratic classifier. In fact, we found that our classification performance was essentially identical (Fig. 4b). This is an interesting disconnect between the BD and classification performance. As mentioned earlier, a predicted increase in the classification performance by BD will not necessarily translate into increased information, as the BD can only be used to place bounds on the classification performance. However, we also examined some specific examples, to see if we could gain insight into why we were not extracting additional information with the quadratic classifier. We examined examples in which the BD predicted a large improvement in classification performance, but little improvement was actually realized by the quadratic classifier. When we looked at specific examples, we found that it was often the case that, despite the very different shapes of the classification boundaries

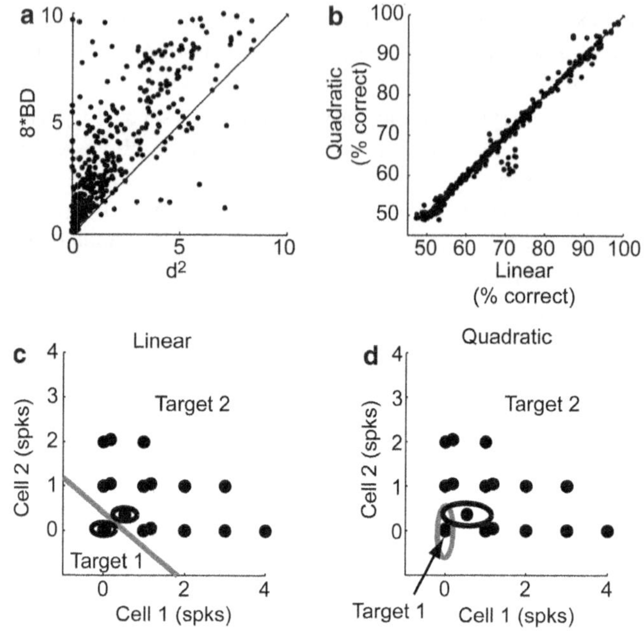

Fig. 4 (**a**) Bhattacharyya distance (*BD*) vs. d^2. We have plotted 8*BD, since the first term of the BD is $d^2/8$. (**b**) Comparison of linear and quadratic classifiers. These comparisons are for the corresponding accuracy. (**c**) Linear decision boundary (*grey line*) for a case in which BD predicted a large benefit of allowing unequal covariances, but the actual classification performance was the same for linear and quadratic decoders. The covariance ellipses are indicated in *black*. The *dots* indicate individual responses for two targets. In the linear case, the covariances are forced to be identical. (**d**) Same plot for quadratic decision boundary (*grey ellipse*). In the quadratic case, the *black ellipse* for target 1 is smaller than the response marker. It is not visible, but it is within the grey decision boundary ellipse

generated by the linear and quadratic classifiers, they resulted in similar classification rules. In the illustrated example (Fig. 4c–d), both classifiers operate according to the rule: if neither neuron fires a spike, classify the response as target 1, if either neuron fires 1 or more spikes, classify the response as target two. From the figures it can be seen that there were two reasons for this. First, neural responses are discrete, and second, neural responses are nonnegative. Thus, the Gaussian assumption does not hold up well when we examine the performance of the quadratic Gaussian classifier.

At this point, we wanted to extend our analyses further, by relaxing the Gaussian assumption, and using a very general classifier. Although it is always worth using a very general approach when estimating information, there are two drawbacks. The first is that it often does not allow insight to be gained into why correlations affect information coding, and the second more methodological reason is that these more general models often require many more trials of data for effective parameter estimation. Thus, they may perform poorly on small datasets. We did, however, find

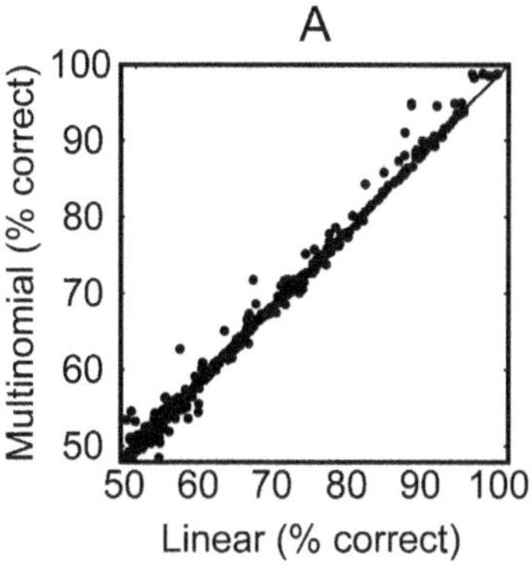

Fig. 5 Comparison of multinomial and linear classifiers for pairs of neurons

that we had enough data to fit a multinomial model, often referred to as a direct model when it is used to estimate Shannon information [36]. To fit the multinomial model, we simply calculated the fraction of times that each different pattern of responses occured, where the patterns are the spike counts in the pair of neurons for each target. Thus, we tabulated the fraction of times that we observe zero spikes in both neurons (00), one spike in neuron 1 and zero spikes in neuron 2 (01), as well as (02), (03), (11), etc. Formally, we estimate the frequencies as:

$$p\left(r_i | t = j\right) = \frac{n_{ij}}{N_j}, \tag{10}$$

which is the number of times response pattern i occurs for target j. We can then take any individual response, plug it into the table for each target, and see whether or not the response occurred more often for target 1 or target 2. We then classify the target using maximum likelihood, as the one which most likely gave rise to the response.

We did this and compared the performance of the multinomial classifier to the linear classifier. Again, we found that the multinomial classifier had essentially identical performance to the linear classifier, although it slightly outperformed it, as would be expected (Fig. 5). Thus, even if we completely drop the Gaussian assumption, we get the same classification performance as we do with the linear classifier, which is directly related to d^2.

Effects of Noise Correlations on Information Encoding and Decoding

Now that we have examined the performance of various classifiers under various assumptions, we can move to examining the effect of noise correlations on information encoding and decoding. The noise correlations we observed, in our experiment, were on average near zero (Fig. 6). However, some of the correlations could be as

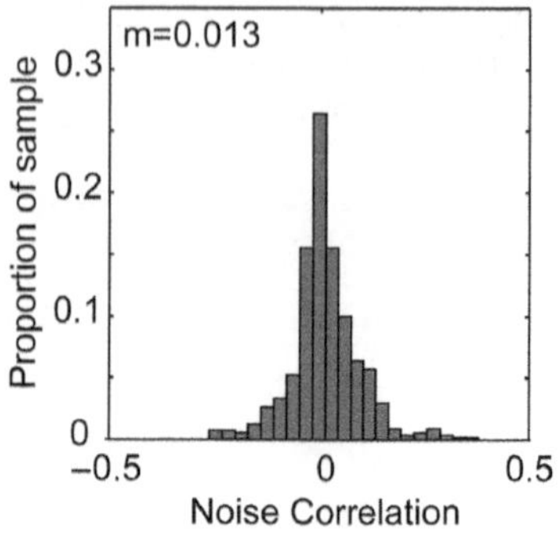

Fig. 6 Distribution of noise correlation values measured in our data

large as 0.1 or 0.2. Thus, correlations were moderate compared with what has been seen in other datasets. These values were dependent on the bin size [4]. The values reported here are for a bin size of 66 ms, as we found that this bin size maximized our decoding performance.

We examined the effect of these correlations on classification accuracy, by looking at classification metrics that paralleled our information metrics. Specifically, we looked at both:

$$\Delta A_{\text{shuffled}} = A - A_{\text{shuffled}}$$

and

$$\Delta A_{\text{diag}} = A - A_{\text{diag}}.$$

In general, there were three salient findings from these analyses. First, as with the individual classification performance, the predictions of the accuracy obtained by d^2 were similar to the actual classification performance, for these metrics (Fig. 7a, b). Second, the effects of noise correlations are quite small in general, although slightly larger in ensembles (Fig. 7b). Finally, as with classification performance, the effects of noise correlations on information encoding and decoding were similar, whether we used the linear Gaussian model, or the multinomial model (Fig. 7c). Thus, overall, we were able to accurately predict the effects of noise correlations on information encoding and decoding using d^2, d^2_{shuffled} and d^2_{diag}, the effects were quite small in pairs of neurons and only slightly larger in ensembles of 3–8 neurons and this was true independent of whether we used a linear Gaussian or a multinomial classifier.

Population Effects of Noise Correlations

The next question of relevance is whether or how we can extrapolate the effects we have seen in pairs of neurons to the population level. In other words, do small effects of correlations in pairs of neurons imply small effects of correlations in large populations of neurons? This question has not been directly addressed in experimental data. We can, however, examine population models, which were reviewed earlier,

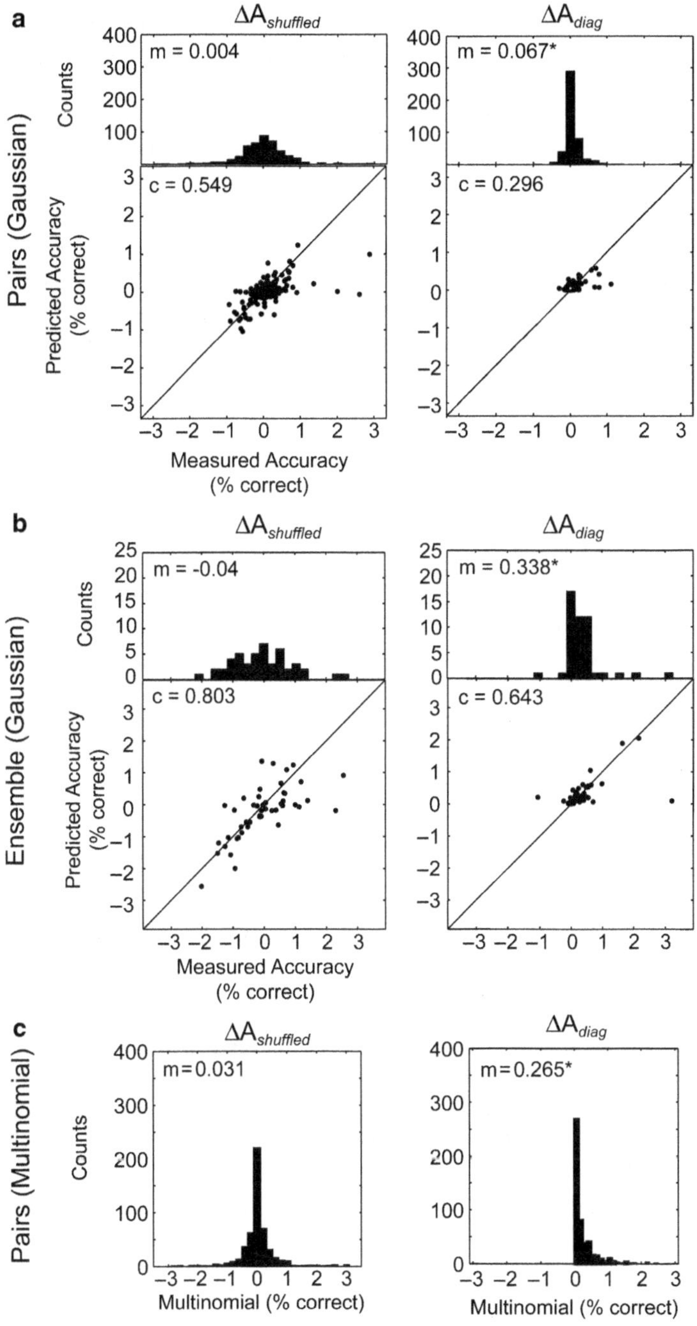

Fig. 7 Effects of noise correlations on information encoding and decoding. (**a**) Effects of noise correlations on information encoding (*left panel*) and decoding (*right panel*) for pairs of neurons. Comparison between effects predicted by d^2 and effects measured in data for both. *Histogram* shows distribution of measured effects, obtained by carrying out decoding analyses. (**b**) Same for ensembles. (**c**) Same analysis, carried out using multinomial classifier on pairs

and see what predictions they make. Specifically, do these models predict small effects in pairs that grow to larger effects in populations? If so, we cannot conclude that small effects in pairs lead to small effects at the population level. They may, but they may not.

In the empirical data in pairs of neurons, we found that the effects of noise correlations were small whether or not we considered signal-dependent correlations. With theoretical models we can also address the question of whether or not correlations have an effect, with or without signal-dependent correlations. Models have been developed which either do or do not include the effects of signal-dependent correlations. Indeed, when Fisher information is used to assess the effect of noise correlations on information coding, the effect of signal-dependent noise is a separate term, as it was with the Battacharyya distance used above, and as such, its independent contribution can be assessed. Specifically, Fisher information for Gaussian noise [10] is given by:

$$I(\theta) = f'(\theta)^{\mathrm{T}} Q^{-1}(\theta) f'(\theta) + \mathrm{tr}\left(Q'(\theta)Q(\theta)^{-1}Q'(\theta)Q(\theta)^{-1}\right), \tag{11}$$

where the first term assesses the information in the neural responses that can be extracted linearly and the second term assesses the information due to signal-dependent correlations that would have to be extracted nonlinearly. It can be seen that the first term is similar to d^2. We also note that Fisher information can vary with the value of the encoded variable, in this case θ, and therefore it is defined as a function of θ. Furthermore, like d^2, the Fisher information constrains the performance of estimators applied to data. Specifically, the Fisher information puts a lower bound on the inverse of the variance of any unbiased estimator of the encoded variable as:

$$\sigma^2 \geq \frac{1}{I(\theta)}. \tag{12}$$

Here we replicate models originally developed by Sompolinsky and colleagues [33, 35]. Similar results would be obtained with other models which have been proposed. If we begin by considering only the first term in the Fisher information equation, we can examine how information scales with the number of neurons in the population for different values of the correlation (Fig. 8a). We can see that, for uncorrelated neurons ($c = 0$) the information scales linearly with the number of neurons. However, for positive correlations, the information quickly saturates. This is the effect discussed earlier, originally reported by Zohary et al. [40]. This suggests that the effects of correlations will be more apparent in larger populations, because the difference between the uncorrelated and correlated populations increases as the size of the population increases. To look at that directly, we calculated both of our information metrics, which assess the effects of correlations on encoding and decoding, using the data from the model. It can be seen that the effects of correlations become much more pronounced in larger populations (Fig. 8b–c). Furthermore, the effects are generally larger on information encoding than they are on information decoding, but even for decoding one can lose 30% of the information if correlations are ignored by a decoding algorithm for even a modestly large population. Thus,

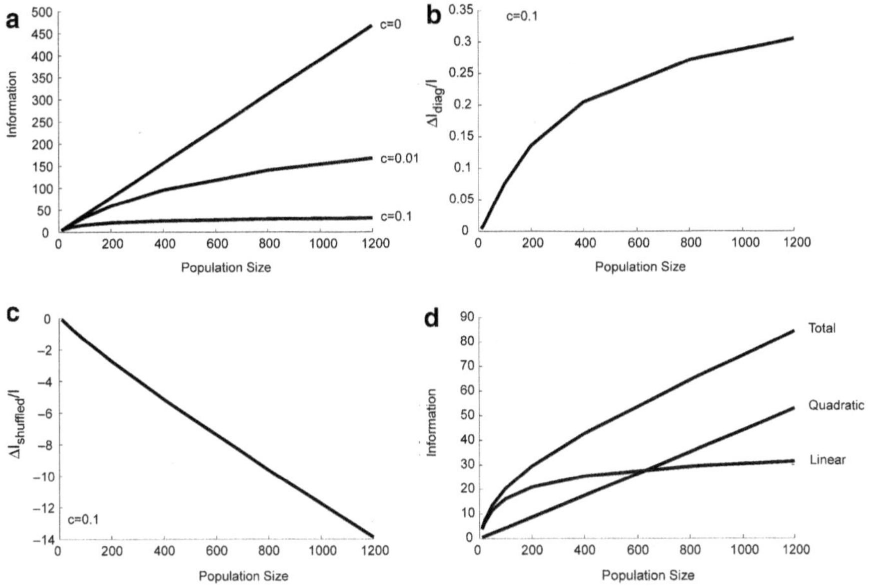

Fig. 8 Population effects of noise correlations. Panels (**a–c**) are from the original Sompolinsky model [35] and panel (**d**) is from the model of Shamir and Sompolinsky [33], which adds signal dependent noise. (**a**) Linear effects of noise correlations on information encoding. (**b**) Normalized effects of noise correlations on information decoding. (**c**) Normalized effects of noise correlations on information encoding. (**d**) Information in signal dependent noise. Line legend: Linear, information in first term of (11), Quadratic, information in second term, Total, the sum of linear and quadratic

extrapolating from pairs to populations is nontrivial. Small effects in pairs can become large effects in populations. Results are similar if we consider the potential role of signal-dependent noise, which is captured by the second, trace term in (11). For the model which has been developed by Shamir and Sompolinsky, the linear information saturates. However, the information in the signal-dependent noise (quadratic), which is small for a small number of neurons, begins to dominate as the size of the population increases (Fig. 8d). Thus, as is the case with the effect of noise correlations on linear information, the effect of signal-dependent noise also increases as the size of the population increases, growing linearly with population size.

As discussed earlier, information cannot grow to infinity, because the amount of information in the brain is constrained by the data processing inequality, and as such, information about a sensory variable cannot exceed the information in the peripheral receptor. This points to a new question: is population size in the brain constrained by the amount of information available? In other words, do populations of neurons operate in the saturated regime, where in effect correlations are having a large effect, or do they operate in the nonsaturated regime, where correlations are not constraining the information? Is this the same in all systems?

Answering this question with accuracy in an awake, behaving monkey is going to require technical advancements. There are two problems. First, it is difficult to record from even moderately large populations of single neurons simultaneously. However, methods are becoming available to record from 10s to maybe 100 or so neurons simultaneously, so that part of the problem should be tractable. The next question is: how much data will we have to collect to accurately measure information in a large population? For some models, under certain assumptions, we can estimate the amount of data that would be required. For example, for a linear Gaussian estimator given as:

$$\begin{aligned} y &= X\beta + \varepsilon \\ \varepsilon &\sim N\left(0, \sigma^2\right), \end{aligned} \tag{13}$$

one can show that the estimated variance using cross validation is related to the actual variance as:

$$<\hat{\sigma}^2> = \sigma^2\left(1 + \frac{M}{N}\right). \tag{14}$$

In this case $\hat{\sigma}^2$ is the estimate of the variance using cross validation, σ^2 is the true variance, M is the number of neurons being used to estimate y (i.e., the number of columns of X) and N is the number of trials available to estimate the model (i.e., the number of rows of X). When cross validation is not used, the sign in the parentheses is switched. The variable y would be the encoded parameter that we were trying to estimate using the neural responses. A plot of this for $M = 10$ and $M = 100$ shows that, whereas maybe 100 trials would be satisfactory for estimating information with 10 neurons, 1,000 trials might be necessary with 100 neurons, if we were extracting linear information (Fig. 9). Thus, the data requirements for even simple linear estimators become quite large. This is even more problematic if one begins to consider quadratic estimators, which one would use to try to extract information in signal-dependent noise, as discussed earlier. In this case, the number of parameters in the model scales as the square of the number of neurons, and thus the data requirements grow also as the square of the number of neurons. For quadratic estimators,

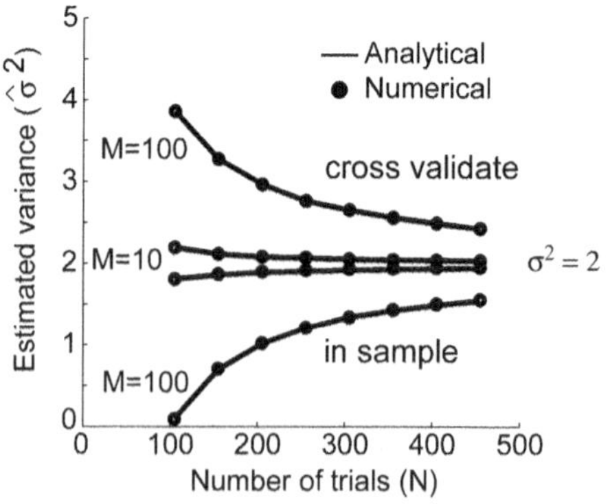

Fig. 9 Estimated variance as a function of the number of trials available for a Gaussian linear estimator. Results are shown for 10 and 100 predictors, with (*top two lines*) and without (*bottom two lines*) cross validation

one might require 100,000 trials to get accurate information estimates for 100 neurons. There is some hope that regularization techniques may be useful to reduce the amount of data necessary to estimate the models, but these will have to be used with much care, as they make rather strong assumptions about the data and if these assumptions are not met, comparisons between models can be problematic.

Conclusion

Considerable progress has been made on the question of whether or not correlated neuronal activity carries information. Much of the progress has been in generating well-defined questions. Furthermore, many groups have carried out analyses of the effects of correlations in pairs of neurons. Generally, they have shown that the effects are quite small. Theory, however, suggests that this is necessarily the case, and that the interesting effects of correlations will only manifest in large populations of neurons. The route for future progress will be in determining how we can accurately and directly test models which look for these effects in real neural data.

References

1. Abbott LF, Dayan P (1999) The effect of correlated variability on the accuracy of a population code. Neural Comput 11:91–101
2. Averbeck BB, Crowe DA, Chafee MV, et al. (2003) Neural activity in prefrontal cortex during copying geometrical shapes II. Decoding shape segments from neural ensembles. Exp Brain Res 150:142–153
3. Averbeck BB, Latham PE, Pouget A (2006) Neural correlations, population coding and computation. Nat Rev Neurosci 7:358–366
4. Averbeck BB, Lee D (2003) Neural noise and movement-related codes in the macaque supplementary motor area. J Neurosci 23:7630–7641
5. Averbeck BB, Lee D (2004) Coding and transmission of information by neural ensembles. Trends Neurosci 27:225–230
6. Averbeck BB, Lee D (2006) Effects of noise correlations on information encoding and decoding. J Neurophysiol 95:3633–3644
7. Basseville M (1989) Distance measures for signal processing and pattern recognition. Signal Process 18:349–369
8. Britten KH, Shadlen MN, Newsome WT, et al. (1993) Responses of neurons in macaque MT to stochastic motion signals. Vis Neurosci 10:1157–1169
9. Buracas GT, Zador AM, DeWeese MR, et al. (1998) Efficient discrimination of temporal patterns by motion-sensitive neurons in primate visual cortex. Neuron 20:959–969
10. Casella G, Berger RL (1990) Statistical Inference. Duxbury Press, Belmont, CA
11. Cover TM, Thomas JA (1991) Elements of Information Theory. Wiley, New York
12. Dan Y, Alonso JM, Usrey WM, et al. (1998) Coding of visual information by precisely correlated spikes in the lateral geniculate nucleus. Nat Neurosci 1:501–507
13. Gawne TJ, Kjaer TW, Hertz JA, et al. (1996) Adjacent visual cortical complex cells share about 20% of their stimulus-related information. Cereb Cortex 6:482–489
14. Gawne TJ, Richmond BJ (1993) How independent are the messages carried by adjacent inferior temporal cortical neurons? J Neurosci 13:2758–2771

15. Georgopoulos AP, Schwartz AB, Kettner RE (1986) Neuronal population coding of movement direction. Science 233:1416–1419
16. Golledge HD, Panzeri S, Zheng F, et al. (2003) Correlations, feature-binding and population coding in primary visual cortex. Neuroreport 14:1045–1050
17. Hinton GE, McClelland JL, Rumelhart DE (1986) Distributed Representations. In: Rumelhart DE and McClelland JL (ed) Parallel Distributed Processing. Explorations in the microstructure of cognition. Volume 1: Foundations. The MIT Press, Cambridge, MA
18. Johnson RA, Wichern DW (1998) Applied Multivariate Statistical Analysis. Prentice Hall, Saddle River, NJ
19. Maynard EM, Hatsopoulos NG, Ojakangas CL, et al. (1999) Neuronal interactions improve cortical population coding of movement direction. J Neurosci 19:8083–8093
20. Montani F, Kohn A, Smith MA, et al. (2007) The role of correlations in direction and contrast coding in the primary visual cortex. J Neurosci 27:2338–2348
21. Nirenberg S, Carcieri SM, Jacobs AL, et al. (2001) Retinal ganglion cells act largely as independent encoders. Nature 411:698–701
22. Oram MW, Hatsopoulos NG, Richmond BJ, et al. (2001) Excess synchrony in motor cortical neurons provides redundant direction information with that from coarse temporal measures. J Neurophysiol 86:1700–1716
23. Panzeri S, Pola G, Petroni F, et al. (2002) A critical assessment of different measures of the information carried by correlated neuronal firing. Biosystems 67:177–185
24. Panzeri S, Schultz SR (2001) A unified approach to the study of temporal, correlational, and rate coding. Neural Comput 13:1311–1349
25. Panzeri S, Schultz SR, Treves A, et al. (1999) Correlations and the encoding of information in the nervous system. Proc R Soc Lond B Biol Sci 266:1001–1012
26. Perkel DH, Bullock TH (1969) Neural Coding. In: Schmitt FO, Melnechuk T, Quarton GC and Adelman G (eds) Neurosciences Research Symposium Summaries. The MIT Press, Cambridge, MA 3: 405–527
27. Petersen RS, Panzeri S, Diamond ME (2001) Population coding of stimulus location in rat somatosensory cortex. Neuron 32:503–514
28. Petersen RS, Panzeri S, Diamond ME (2002) Population coding in somatosensory cortex. Curr Opin Neurobiol 12:441–447
29. Pola G, Thiele A, Hoffmann KP, et al. (2003) An exact method to quantify the information transmitted by different mechanisms of correlational coding. Network 14:35–60
30. Poor HV (1994) An Introduction to Signal Detection and Estimation. Springer, New York
31. Sejnowski TJ (1988) Neural populations revealed. Nature 332:308
32. Seriès P, Latham PE, Pouget A (2004) Tuning curve sharpening for orientation selectivity: coding efficiency and the impact of correlations. Nat Neurosci 7:1129–1135
33. Shamir M, Sompolinsky H (2004) Nonlinear Population Codes. Neural Comput 16:1105–1136
34. Shamir M, Sompolinsky H (2006) Implications of neuronal diversity on population coding. Neural Comput 18:1951–1986
35. Sompolinsky H, Yoon H, Kang K, et al. (2001) Population coding in neuronal systems with correlated noise. Phys Rev E 64:051904
36. Strong SP, Koberle R, De Ruyter Van, Steveninck RR, et al. (1998) Entropy and information in neural spike trains. Phys Rev Lett 80:197–200
37. Tolhurst DJ, Movshon JA, Thompson ID, et al. (1981) The dependence of response amplitude and variance of cat visual cortical neurones on stimulus contrast. Exp Brain Res 41:414–419
38. Wilke SD, Eurich CW (2002) Representational accuracy of stochastic neural populations. Neural Comput 14:155–189
39. Wu S, Nakahara H, Amari S (2001) Population coding with correlation and an unfaithful model. Neural Comput 13:775–797
40. Zohary E, Shadlen MN, Newsome WT (1994) Correlated neuronal discharge rate and its implications for psychophysical performance. Nature 370:140–143

Stochastic Synchrony in the Olfactory Bulb

Bard Ermentrout*, Nathaniel Urban, and Roberto F. Galán

Abstract Oscillations in the 30–100 Hz range are common in the olfactory bulb (OB) of mammals. The principle neurons (mitral cells) of the OB are believed to be responsible for these rhythms. We suggest that the mitral cells, which prefer to fire in a limited range could be synchronized by receiving correlated statistically random inputs (stochastic synchrony). We explore the mechanisms of stochastic synchrony using a combination of experimental, computational and theoretical methods.

Introduction

From the earliest recordings of brain electrical signals, synchronized oscillatory activity of large populations of neurons has been seen as a prominent feature of brain activity [8]. This synchronized activity occurs in a variety of brain areas and across a wide range of frequencies. The oscillations are particularly prominent at certain areas of the brain and in certain frequency bands [9]. The vertebrate olfactory bulb [1] generates several different prominent oscillations including low-frequency oscillations that are related to respiration, and also much higher-frequency oscillations in the 30–100 Hz range. Oscillations in the range of 40–80 Hz are observed in many brain areas and are known as gamma oscillations. These signals have attracted considerable interest because of their potential role in cognitive function and/or dysfunction. Here we describe some recent work that led us to propose a novel mechanism in which synchronization in the gamma frequency band can be caused by correlations of random, noise-like fluctuations and to apply this mechanism to

B. Ermentrout (✉)
Department of Mathematics, University of Pittsburgh, Pittsburgh, PA 15260
e-mail: bard@pitt.edu

*Supported by NIMH, NSF, and NIH CRCNS

K. Josić et al. (eds.), *Coherent Behavior in Neuronal Networks*, Springer Series in Computational Neuroscience 3, DOI 10.1007/978-1-4419-0389-1_12,

the understanding of oscillatory synchrony in the mouse olfactory bulb. We also discuss the possibility that similar mechanisms may account for gamma band synchronization across other brain areas, particularly across areas that are not tightly coupled, but which may receive correlated fluctuations.

Basic Circuitry of the Olfactory Bulb Mediates Recurrent and Lateral Inhibition

The main features of main olfactory bulb circuitry are indicated in Fig. 2a, b and have been recently reviewed [16,30,48]. The principal cells of the olfactory bulb, the mitral cells, receive many excitatory inputs from olfactory receptor neurons in the nose. These inputs are made onto the highly branched tuft of the primary dendrite of the mitral cell. These cells in turn provide output to higher brain areas. Mitral cell activity is modulated by several circuits intrinsic to the bulb, most notably by dendrodendritic recurrent and lateral inhibition mediated by olfactory bulb granule cells [3, 10, 25, 49–51, 60]. These circuits are believed to refine the spatial pattern of activity across bulbar neurons [2, 59] and also are known to play an important role in altering the timing of mitral cell activity [46].

Activity of granule cells triggers release of glutamate containing vesicles in mitral cell dendrites [25, 26, 33, 43]. This glutamate binds to NMDA and AMPA receptors on the dendritic spines of postsynaptic granule cells, depolarizing them. In some cases this depolarization is localized to a particular spine, resulting in release of GABA from only that spine, back onto the mitral cell [14]. Such local release mediates a form of recurrent inhibition. In other cases, stronger depolarization may cause the granule cell to fire an action potential [15, 33] which propagates throughout the dendritic tree of the granule cell, and may cause widespread release of GABA onto the dendrites of many mitral cells. Such global activation of granule cells is believed to cause a form of lateral inhibition.

Slow Kinetics of Lateral Inhibition are Incompatible with Synchronization of Fast Oscillations

Networks coupled by recurrent and lateral inhibition have been widely studied as generators of gamma oscillations [61, 63]. However, recent physiological data, mostly from in vitro preparations [40] indicate that olfactory bulb circuitry is more complicated and more dynamic than previously believed [3, 4, 13, 24, 48, 49, 54, 60]. Of particular relevance to discussions of high-frequency oscillatory synchrony is the observation that recurrent [25, 33, 47] and lateral inhibition [61] in vitro and in vivo [33] have decay times of approximately 350 ms. These long decay times are not due to slow kinetics of individual synaptic currents, but rather because the overall IPSC is made up of a prolonged barrage of small synaptic currents, and the rate of events in this barrage decays over several hundred milliseconds. These barrages of synaptic events are probably caused by long latency and repeated firing of olfactory bulb granule cells [27]. Each of these individual currents has a short time constant (10-ms decay) and the entire current is blocked by application of GABAA receptor antagonists. Thus, time course of granule cell-mediated inhibition spans more than 10 average gamma cycles, but is made up of many fast events. This slow time constant of lateral inhibition is incompatible with the synchronization of gamma oscillations [11, 61, 62]. However, we have described and propose further study of a mechanism whereby the fast fluctuating divergent outputs from single granule cells to multiple mitral cells provides a mechanism for synchronizing fast oscillations in mitral cells. To understand the mechanisms that may lead to this synchronization, we consider the known properties of olfactory bulb neurons and circuits.

Gamma Oscillations are Intrinsic to Olfactory Bulb and to Mitral Cells

Gamma oscillations have been observed in recordings from the olfactory bulb for many years [1, 41]. These oscillations can be readily observed by field potential recordings both in awake behaving and in anaesthetized animals [39, 53]. In vivo recordings in anaesthetized animals in which connections from cortex to the olfactory bulb were severed have shown that olfactory bulb gamma oscillations are generated intrinsically in the bulb, not requiring feedback connections from cortex [39]. These oscillations do depend on inhibition as they are not seen during pharmacological blockade of GABAA receptors and they are altered by genetic manipulation of GABAA receptors [41]. Recent work has further shown that gamma frequency oscillation can even be induced in acute olfactory bulb slices [18, 29] clearly indicating that they can be generated by the intrinsic bulbar circuitry [18,29]. This in vitro synchronization is prevented by blockade of GABAA receptors [29], and also of gap junctions, sodium current, and glutamate receptors [18].

Mitral Cells are Oscillators with a Preferred Frequency of 40 Hz

The biophysical properties of mitral cells have been investigated in detail in recent years. Mitral cells tend to have rather depolarized resting membrane potentials (−50 to −60 mV) and fire rather narrow action potentials (1-ms half width). Subthreshold current steps generate 25–50 Hz oscillations of subthreshold membrane potential in mitral cells [12] indicating that these cells have subthreshold resonance in the low gamma frequency range. Further depolarization by current steps of moderate amplitude result in long slow depolarizations which eventually generate high-frequency spiking [5]. These periods of spiking are interrupted by pauses that last hundreds of milliseconds, during which subthreshold oscillations in the membrane potential are again observed [12]. Increasing the amplitude of the current step results in shorter pauses in firing without a large effect on the frequency of firing during the spiking period [5]. Thus, increasing the amplitude of DC current injection causes a change in the average firing rate without much change in the most common interspike interval [5]. A similar phenomenon has been observed during in vivo whole cells recordings. In these recordings, many mitral cells show fluctuations in membrane potential that track the respiration cycle, which occurs in the theta frequency range (2–5 Hz) [33]. When these cells fire action potentials, these spikes generally occur during the peak of the theta cycle. A single theta cycle can be associated with multiple spikes and these spikes occur with an instantaneous frequency of approximately 40 Hz, independent of whether the cell fire as few as 2 or as many as 6 spikes in the single theta cycle. Thus a tripling of the average firing rate can occur even when the spikes that are generated have an interspike interval of 25 ms.

These observations show that individual mitral cells are strongly biased to fire in the gamma frequency range and that firing in this frequency range occurs across a wide range of steady state current values and/or a wide range of in vivo input strengths. Given that cells with similar firing rates are more easily synchronized, such dominance of gamma frequency spiking may be important for generating oscillatory synchrony across mitral cells. Our basic hypothesis is that odor inputs result in depolarization of the mitral cells and induce them to fire. Their tendency (due to intrinsic properties) to fire in a narrow frequency range (even if the inputs slowly vary) means that they can be treated as mathematical oscillators whose phase (but not frequency) is modulated by locally correlated inhibitory input from the granule cells. Interestingly, the firing rate of mitral cells does not vary much over several orders of magnitude of odor concentration when the granule cells are present [55]. Thus, we suggest that the long-lasting strong recurrent inhibition from the granule cells serves as a brake to mitral cell activity and keeps the firing rate in a retricted range. The fast correlated noisy transients that ride on the inhibition will serve to synchronize a local population of mitral cells as we will see in the next section.

Noise-Induced Oscillatory Synchrony

In the olfactory system, gamma frequency oscillations (20–80 Hz) have been observed since the earliest recordings [1] and are enhanced during certain states and olfactory behaviors [28, 44]. The mechanisms by which olfactory bulb gamma oscillations are generated and synchronized are not, however, well understood. Some fast oscillations are intrinsic to the bulb circuitry [39] even being observed in slice preparations [18, 29], suggesting that the intrinsic connectivity can give rise to synchronization. One long-standing hypothesis has been that recurrent and lateral inhibition mediated by dendrodendritic mitral cell-granule cell synapses (reviewed by [48]) are critical for the generation and synchronization, respectively, of high frequency oscillations in the olfactory bulb [6,34,52]. According to this hypothesis, mitral cell activity leads to recurrent inhibition which in turn stops mitral cell firing for some period. Synchronization is then achieved via lateral inhibition between mitral cells. That is, when one mitral cell inhibits its own firing, it also inhibits other mitral cells. Thus, the timing of the pauses in firing will be similar across mitral cells [12, 29]. Decaying inhibition then allows resumption of firing which again evokes recurrent and lateral inhibition. Several variants of this model have been proposed to explain olfactory bulb fast field potential oscillations [28,29,35,39,42,52]. However, little direct evidence showing that this mechanism can account for synchronous fast oscillations in the olfactory system has been provided. Alteration of inhibition changes fast field potential oscillations in vivo and in vitro [18, 29, 41], but this is consistent with other mechanisms (see below).

As described above, the kinetics of lateral inhibition in the olfactory bulb find it to be inconsistent with this proposed mechanism of gamma oscillations. We then use experimental and computational approaches to investigate the possibility that the olfactory bulb is using a different mechanism to generate synchronous oscillations. Specifically, we have shown that a mechanism that has been described theoretically [38, 58] but not previously applied to real oscillating neurons accounts for synchronization of fast olfactory bulb oscillations. According to this mechanism, mitral cells firing in a roughly oscillatory pattern are synchronized by correlated, but aperiodic inputs received from common granule cells. Such a mechanism of generating synchronous oscillations has not been observed experimentally in neural systems, though it may explain some previously observed phenomena [23, 45].

Stochastic Synchrony

Based on the above considerations, we suppose that the mitral cells can be regarded as noisy oscillators which receive some common (and thus correlated) input from surrounding granule cells. We can now ask if this is sufficient to cause some degree of synchronization and if so, what properties of the noise, correlation, and oscillators are necessary for this synchrony. First consider N identical nonlinear oscillators sharing a common signal:

$$\frac{\mathrm{d}X_j}{\mathrm{d}t} = F(X_j) + q\Xi(t) + \sqrt{1-q^2}\,\Xi_j(t) \tag{1}$$

where $\Xi(t)$ is a common noise term and $\Xi_j(t)$ are independent uncorrelated noise terms. (The noise could be colored, white, Poisson, etc). We assume that $X' = F(X)$ admits a stable limit cycle oscillator. If $q = 0$, then the intrinsic uncorrelated noise will drive the oscillators apart, however, for non-zero q, there is some shared signal which could lead to partial synchrony of the oscillators. Figure 1 shows an example simulation of 50 Hodgkin–Huxley oscillators (standard HH model with $10\,\mu\mathrm{A/cm}^2$ current injected so that they oscillate regularly)

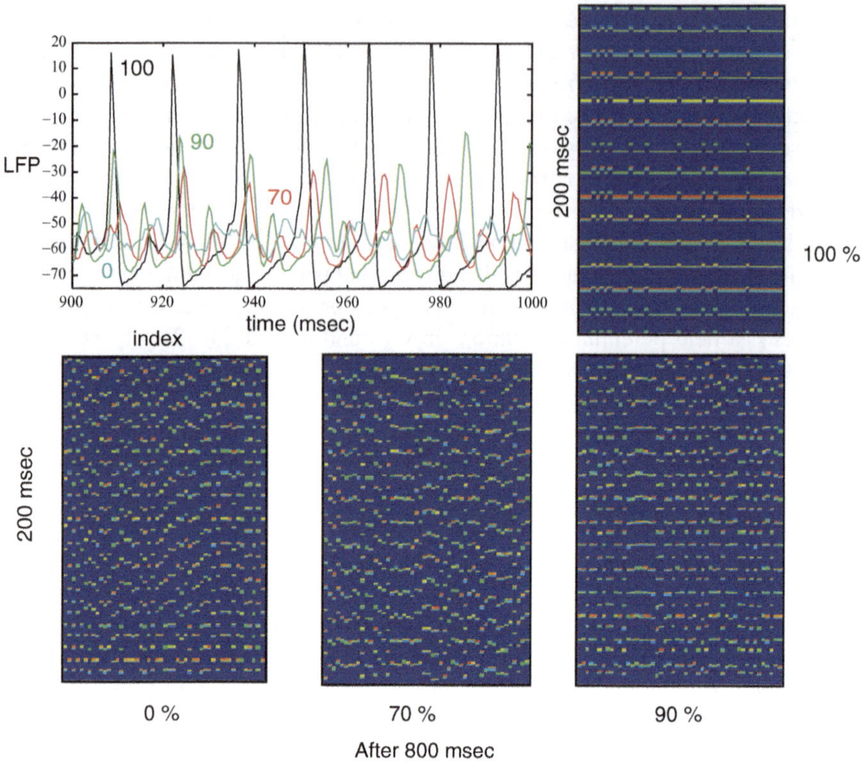

Fig. 1 Shared white noise between 50 HH oscillators. *Upper left* shows the averaged potentials of all 50 neurons as a function of the degree of correlation in the inputs. Remaining plots show potential as a function of time for these correlations.

with various values of q. The top left of the figure shows the average potential of all 50 cells: as the degree of shared input increases, a strong periodic rhythm emerges. Figure 2 shows that this behavior is not restricted to neural models. Fig. 2a, b show the underlying anatomy and membrane dynamics underlying stochastic synchrony in the olfactory bulb. In [19], we injected partially correlated input currents into a mitral cell and recorded the resulting potential. Figure 2c1 shows two trials of current injection (red and black curves) with 0% and 80% correlation. The potential traces of the mitral cells to these currents are shown in Fig. 2c2. There are clearly many more overlapping spikes when the correlation is high. To quantify this, we computed the cross-spectral density for different levels of correlation. As seen in figure 2c3,4, this grows with increased correlation and shows a peak in the 15–40 Hz range commonly found in the OB. These two figures demonstrate that common noise could play a large role in determining the synchrony between neurons.

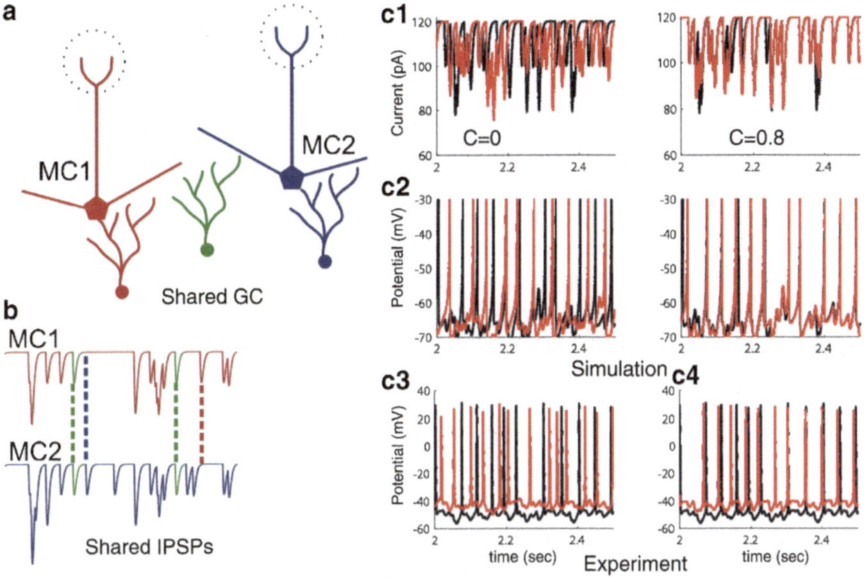

Fig. 2 Stochastic synchrony in mitral cells. (**a**) Diagram showing two uncoupled mitral cells with common granule cell; (**b**) Shared IPSPs between motral cells; (**c1**) Input currents 0% correlation and 80% correlation shown; (**c2**) Mitral cell responses to these two stimuli; (**c3**) Cross spectral density with different correlations; (**c4**) Power boost in the 15–40 Hz range due to correlation. (We depict the area under the curves in figure **c3** in the 15–40 Hz range divided by the total area under each curve.).

Phase Reduction and Lyapunov Exponents

To mathematically quantify the mechanism underlying stochastic synchrony, we apply the theory of phase reduction to (1). Since the oscillators are uncoupled and independent, we need only consider a pair of them to understand the phenomena. To consider the most general scenario, we assume that the oscillators can be slightly different and that the noise they receive is small. Furthermore, since we are interested in the role of shared currents, we assume that the only component of the oscillator which is perturbed is the somatic compartment and that the component of the vectors, Ξ, Ξ_j are ξ, ξ_j, respectively. Then, (see [58]) a pair of oscillators reduces to the pair

$$\theta_1' = \omega_1 + \Delta(\theta_1)[q\xi(t) + \sqrt{1-q^2}\xi_1(t)] \tag{2}$$

$$\theta_2' = \omega_2 + \Delta(\theta_2)[q\xi(t) + \sqrt{1-q^2}\xi_2(t)]. \tag{3}$$

In absence of stimuli, these oscillators fire at frequencies, ω_j and if the oscillators are identical, $\omega_1 = \omega_2$. The crucial function in this model is $\Delta(\theta)$, the phase resetting curve of the oscillator. Mathematically, $\Delta(\theta)$ is proportional to the voltage component of the solution, $Y(t)$ to the linear adjoint equation:

$$Y'(t) = -D_X F(U(t))^T Y(t), \quad Y^T(t)U'(t) = 1$$

where $U'(t) = F(U(t))$ is a stable limit cycle solution. Heuristically and experimentally, $\Delta(\theta)$ is computed as follows. Let us define the phase of the oscillator to be the time since it has last produced an action potential. Thus, $0 \le \theta < P$ where P is the period of the oscillator. Suppose that we inject a brief current pulse at phase, θ of the oscillation. This will cause an action potential to occur at a time $\hat{P}$ which is not generally the same time as the time, P when it would normally occur. The phase resetting curve (PRC) for the stimulus is:

$$\mathrm{PRC}(\theta, a) = P - \hat{P}$$

where, a parameterizes the magnitude of the perturbation (for example, the total charge delivered to the neuron). The quantity, $\Delta(\theta) := \lim_{a\to 0} \mathrm{PRC}(\theta, a)/a$ defines the infinitesimal PRC or the voltage component of the adjoint.

Figure 3 shows PRCs from both model and real neurons. One point that we want to make is that there are two qualitatively different types of PRCs: those which have both a negative and positive component and those which are strictly nonnegative. The PRCs on the left have a negative and positive component.

Given the PRC, $\Delta(\theta)$, the noise, ξ_j and the heterogeneity, ω_j we can now quantify the degree of synchronization for the uncoupled pair, (2–3). Let us first assume that they are identical and the noise is completely correlated. We can ask if solutions which start near synchrony will converge to synchrony and if so, how fast they will converge. (For simplicity, we will assume that Δ has period 1 without loss of generality.) Subtract the two equations and let $\phi = \theta_2 - \theta_1$. Then for ϕ small

$$\phi' = [\Delta(\theta_1 + \phi) - \Delta(\theta_1)]\xi(t) \approx \Delta'(\theta_1)\phi\xi(t).$$

We can study how ϕ varies over time when $\xi(t)$ is white by applying Ito's lemma. Let $y = \log\phi$. Then

$$y' = -\Delta'(\theta_1)^2\frac{\sigma^2}{2} + \Delta'(\theta)\xi(t),$$

where σ^2 is the variance of the noise, $\xi(t)$. y undergoes Brownian motion with a negative drift term. Since $\phi = \exp(y)$, on average, $\phi(t)$ will decay like $\exp(\lambda t)$ where λ is the average drift:

$$\lambda = \lim_{T\to\infty}\frac{1}{T}\int_0^T -\Delta'(\theta_1(t))^2\frac{\sigma^2}{2}\,\mathrm{d}t = -\frac{\sigma^2}{2}\int_0^1 \Delta'(\theta)^2 P(\theta)\,\mathrm{d}\theta, \tag{4}$$

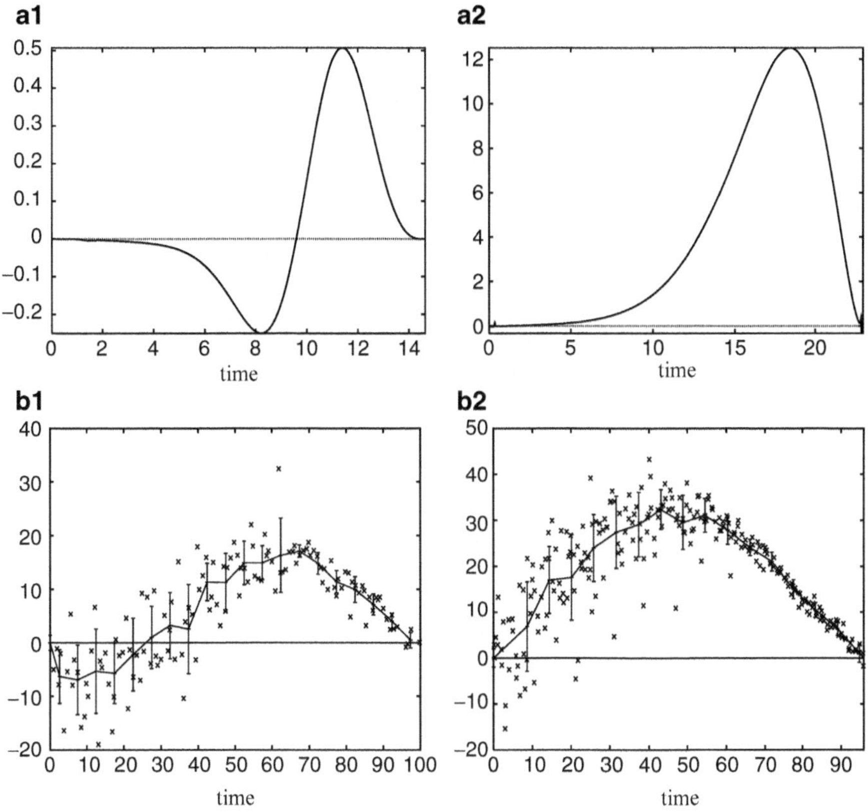

Fig. 3 Phase resetting curves. (**a1,2**) The Hodgkin–Huxley model and the Traub model with a calcium-dependent potassium current. (**b1,2**) PRCs from hippocampal neurons under different dynamic clamp scenarios. (Data provided by Theoden Netoff.)

where we have used the ergodicity of the noisy process to derive the last equality and where $P(\theta)$ is the invariant density of the phase

$$0 = -\omega \frac{\mathrm{d}P}{\mathrm{d}\theta} + \frac{\sigma^2}{2} \frac{\mathrm{d}\Delta(\theta)}{\mathrm{d}\theta} \left(\frac{\mathrm{d}\Delta(\theta) P}{\mathrm{d}\theta} \right). \tag{5}$$

If, instead of continuous noise processes, there are Poisson inputs, then the system of equations reduces to a pair of discrete maps [36]:

$$\begin{aligned} \theta_{n+1} &= \theta_n + \omega \tau_n + \sigma \Delta(\theta_n) \\ \phi_{n+1} &\approx [1 + \sigma \Delta'(\theta_n)] \phi_n, \end{aligned}$$

where σ is the magnitude of the pulse. For this model

$$\lambda = \int_0^1 \log\left(1 + \sigma\Delta'(\theta)\right) P(\theta)\,\mathrm{d}\theta, \tag{6}$$

$$P(\theta) = \int_0^1 Q(\theta - x - \sigma\Delta(x))P(x)\,\mathrm{d}x \tag{7}$$

where $Q(x)$ is the periodized density for an exponential distribution. Note that if σ is small, then we can expand (6) and we obtain the same equation as in (4) for the parameter λ which is called the Lyapunov exponent. In both cases, it is clearly a negative quantity, so that common noise will always cause nearby oscillators to converge to synchrony. The rate at which they converge is proportional to λ, so that more negative values of λ correspond to greater stochastic synchrony. In two papers, Tateno and Robinson [56,57] explored the Lyapunov exponent in model neurons and in cortical slices.

In recent work, Abouzeid and Ermentrout (in preparation) consider the pair of equations (4) and (5) along with a constraint

$$\int_0^1 a_0[\Delta(t)]^2 + a_1[\Delta'(t)]^2 + a_2[\Delta''(t)]^2\,\mathrm{d}t = 1$$

as an optimization problem in which one tries to minimize λ. They find using the Euler–Lagrange equations and perturbation methods that the optimal PRC is close to a sine wave, $\Delta(t) = C\sin 2\pi t$. One can also use specific parameterizations of $\Delta(t)$ which are close to the shapes of biological PRCs and then treat the optimization as a standard calculus problem. Indeed, consider $\Delta(t) = [\sin(2\pi t + a) - \sin(a)]/N(a)$ where $N(a)^2 = 3/2 - \cos(a)^2$ is chosen so the L^2 norm of $\Delta(t)$ is 1. If we assume weak noise, then $P(\theta)$ is roughly 1 and

$$\lambda \approx -\sigma^2 \frac{2\pi^2}{3 - 2\cos a}$$

which clearly has a maximum at $a = 0$. On the other hand, if we fix the square of $\Delta'(t)$ or $\Delta''(t)$ then the parameter a can be arbitrary. Similar approaches can be applied to the discrete Poisson case of equations (6) and (7). We remark that Tateno and Robinson found a similar result, PRCs with both negative and positive lobes have a more negative value for λ.

Noise Color and Reliability

Related to stochastic synchrony is the question of reliability of spikes. That is, given the same stimulus over and over again, how reliably times are the spikes of a neuron (or alternatively, how well correlated are the voltage traces). In a groundbreaking

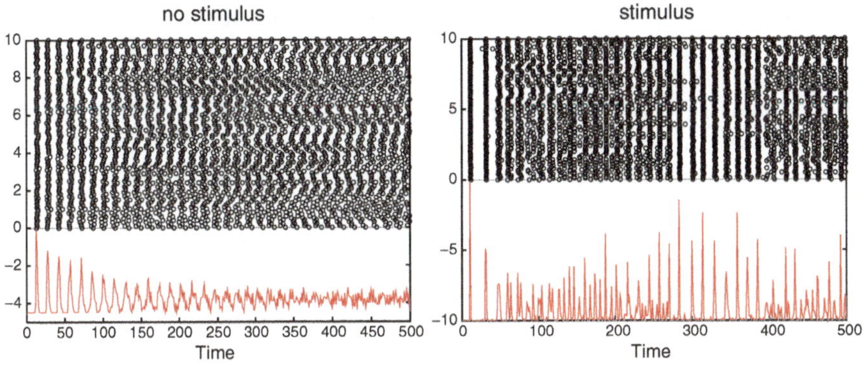

Fig. 4 Reliability in the HH model. *Left panel* shows spike rasters for 100 trials in which a constant current is applied and there is independent white noise. *Right panel* shows the same with an additional frozen noise current applied

paper, Bryant and Segundo [7] showed that a frozen white noise stimulus could produce very reliable spikes in a molluscan neuron. This was later applied by Mainen and Sejnowski [31] to cortical neurons. Figure 4 shows an example of this phenomena in the HH model. 100 trials are shown in which a constant current is applied at $t = 0$ which lasts for 500 ms. Below the spike rasters, we have binned the number of cells firing in a short time window. The initially reliable spikes degrade over time. On the other hand, if a small frozen noise signal is added on top of the constant current, then a large fraction of spikes can be reliably maintained over the duration of the stimulus. The implications for this in coding are reviewed in Ermentrout et al (2008).

We can quantify reliability as

$$R \equiv \lim_{T\to\infty} \frac{(1/T)\int_0^T s_1(t)s_2(t)\,\mathrm{d}t}{(1/T)\int_0^T s_1(t)^2\,\mathrm{d}t},$$

where $s_j(t)$ is a measure of the spike times, for example, a narrow Gaussian centered at the spike times of two signals. Note that this is like the normalized correlation. In an unpublished calculation, we show that when a stimulus is presented repeatedly in the presence of independent white noise, the reliability is related to the Lyapunov exponent:

$$R = \frac{\sqrt{-2\lambda b}}{\sqrt{-2b\lambda + \sigma_E^2}}, \qquad (8)$$

where b is a parameter related to the Gaussian smoothing of the spike times and σ_E is the extrinsic (independent) noise. Note that as $\sigma_E \to 0$, $R \to 1$. Since reliability is a monotonic function of $-\lambda$, this means that the maximal reliability occurs when $-\lambda$ is largest.

Conventional wisdom is that white noise is the best stimulus for reproducible spikes. However, if the frozen noise is small and the neurons are (noisy) oscillators, then it turns out that colored noise can produce greater reliability. Galán et al. [22] demonstrate this for both real and model neurons. Equation (8) shows that the reliability is related to the Lyapunov exponent, so that we will try to calculate this in the presence of colored noise generated by the Ornstein-Uhlenbeck process:

$$d\xi = -\beta\xi dt + \sqrt{\beta} dW,$$

where β determines the autocorrelation of the noise, $C(t) := \langle\xi(0)\xi(t)\rangle = \exp(-\beta|t|)$ and $W(t)$ is delta-correlated noise with variance, $\sigma^2/2$. Recall that the Lyapunov exponent satisfies

$$\lambda = \lim_{T\to\infty} \frac{1}{T} \int_0^T \Delta'(\theta(t))\xi(t)\, dt, \tag{9}$$

where $\theta(t)$ satisfies

$$\theta'(t) = 1 + \delta(\theta(t))\xi(t). \tag{10}$$

For small noise $\sigma \ll 1$, we can approximate the phase,

$$\theta(t) = t + \int_0^t \Delta(s)\xi(s)\, ds$$

and thus (9) is

$$\lambda \approx \lim_{T\to\infty} \frac{1}{T} \int_0^T \Delta''(t) \int_0^t \Delta(s)C(t-s)\, ds.$$

For example, if $\Delta(\theta) = a \sin 2\pi\theta$ and $C(t) = \exp(-\beta|t|)$, then

$$\lambda \approx -K \frac{\beta}{\beta^2 + 4\pi^2}$$

where K is a positive constant dependent on the magnitude of the noise, but not β. In this case, clearly the most negative λ occurs when $\beta = 2\pi$.

Figure 5a shows that there is a clear peak in the value of reliability as a function of $\tau := 1/\beta$ for model and real neurons driven to fire at about 40 Hz. Figure 5b shows the approximate value of the Lyapunov exponent from the calculations for a sinusoidal PRC along with the values obtained from a Monte Carlo simulation. For PRCs dominated by the first Fourier mode, the rule of thumb is that the optimal time constant for colored noise is $\tau_{opt} \approx P/2\pi$ where P is the period of the oscillator.

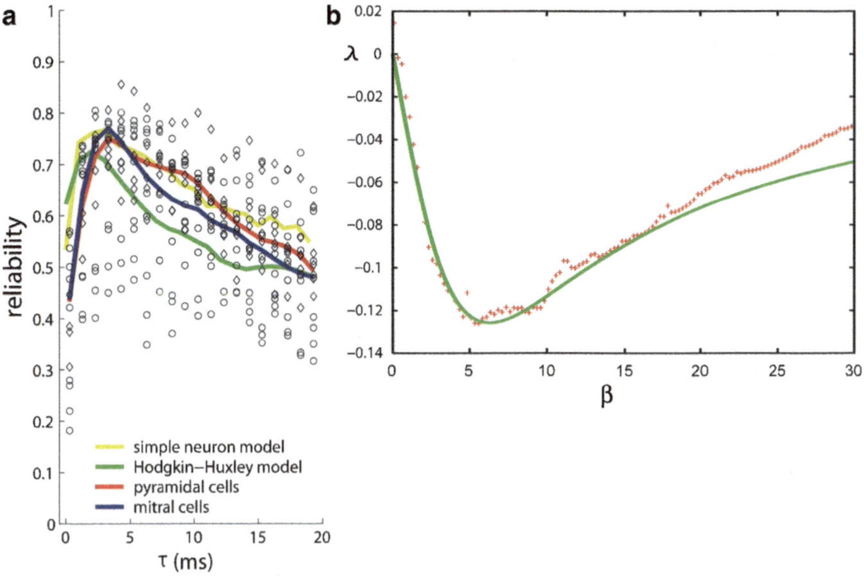

Fig. 5 Reliability as a function of the noise color. (**a**) Experiments and simulations show the peak of reliability for a 40-Hz oscillator at about 4 ms time constant. (**b**) Montecarlo and theory for a sinusoidal PRC.

Input/Output Correlations

To study the output vs. input correlations, we consider (2) and (3) where $\omega_1 = \omega_2$ but $q < 1$. Nakao et al. [37] studied this problem for white noise and Marella and Ermentrout [32] for Poisson inputs. In both cases, one is interested in the stationary density for the phase-differences between the two oscillators, that is, the random variable, $\phi := \theta_2 - \theta_1$. It turns out that through a series of perturbation expansions, one obtains the same result no matter whether the noise is white or Poisson:

$$P(\phi, c) = \frac{K}{1 - c\frac{h(\phi)}{h(0)}}, \tag{11}$$

where

$$h(\phi) = \int_0^1 \Delta(s)\Delta(s + \phi),$$

$c = 2q/(1 + q)$ is the correlation, and K is a normalization so that the integral is 1. Note that as $q \to 1$, this distribution approaches a delta function corresponding to perfect synchrony. One way to characterize the degree of synchrony is the "order parameter"

$$z(c) := \int_0^1 \cos 2\pi s P(s, c)\, \mathrm{d}s$$

which is zero for a uniform P and 1 for a delta function. Another way to characterize the degree of synchrony is to look at the deviation of the peak from the uniform distribution $P(0,c) - 1$. Marella and Ermentrout [32] explore the dependence of z and other order parameters on the shape of the PRC. In particular, they examine $z(q)$ with different PRC shapes and find that if the L^2 norm of the PRC is kept constant, PRCs which have the smallest DC component maximize this function. As a final example, for small correlation, c, Marella and Ermentrout obtain a simple expression for the deviation of the peak

$$P(0,c) - 1 \approx c\left[1 - \frac{\langle\Delta\rangle^2}{\langle\Delta^2\rangle}\right]$$

where $\langle x\rangle = \int_0^1 x(s)\,\mathrm{d}s$. If we keep the L^2 norm of the PRC constant, say 1, then the denominator is 1 and the peak deviation is maximal when Δ has zero mean – that is, no DC component.

Summary

Oscillations are ubiquitous in the nervous system and in the olfactory bulb in particular. In order for there to be large macroscopic local field potentials, there must be a good deal of synchrony in the rhythmic behavior of the principle cells, here, the mitral cells. Inhibition persists for too long and there is no direct coupling between these neurons. Thus, we have posited that a primary mechanism for synchronization is shared "noisy" inhibitory postsynaptic currents (inhibitory miniature events) from the interneurons, granule cells. We have characterized the degree of this so-called stochastic synchronization by reducing complex neuronal networks to dynamics of the phases of each neural oscillator. This has allowed us to derive some expression for the degree of synchrony as a function of properties of the noise and properties of the underlying oscillations. Furthermore, our simple theories have been put to experimental tests both in complicated membrane models and in real central nervous system neurons.

References

1. Adrian ED (1942) Olfactory reactions in the brain of the hedgehog. J Physiol 100:459–473.
2. Arevian AC, Kapoor V, Urban NN (2008) Activity-dependent gating of lateral inhibition in the mouse olfactory bulb. Nat Neurosci 11:80–87.
3. Aroniadou-Anderjaska V, Ennis M, Shipley MT (1999) Dendrodendritic recurrent excitation in mitral cells of the rat olfactory bulb. J Neurophysiol 82:489–494.
4. Aungst JL, Heyward PM, Puche AC, Karnup SV, Hayar A, Szabo G, Shipley MT (2003) Centre-surround inhibition among olfactory bulb glomeruli. Nature 426:623–629.
5. Balu R, Larimer P, Strowbridge BW (2004) Phasic stimuli evoke precisely timed spikes in intermittently discharging mitral cells. J Neurophysiol 92:743–753.

6. Bressler SL, Freeman WJ (1980) Frequency analysis of olfactory system EEG in cat, rabbit, and rat. Electroencephalogr Clin Neurophysiol 50:19–24.
7. Bryant, HL, Segundo, JP (1976) Spike initiation by transmembrane current: a white-noise analysis. J. Physiol. 260:279–314.
8. Buzsaki G (2006) Rhythms of the Brain. Oxford: Oxford University Press.
9. Buzsaki G, Draguhn A (2004) Neuronal oscillations in cortical networks. Science 304:1926–1929.
10. Carlson GC, Shipley MT, Keller A (2000) Long-lasting depolarizations in mitral cells of the rat olfactory bulb. J Neurosci 20:2011–2021.
11. Chow CC, White JA, Ritt J, Kopell N (1998) Frequency control in synchronized networks of inhibitory neurons. J Comput Neurosci 5:407–420.
12. Desmaisons D, Vincent JD, Lledo PM (1999) Control of action potential timing by intrinsic subthreshold oscillations in olfactory bulb output neurons. J Neurosci 19:10727–10737.
13. Didier A, Carleton A, Bjaalie JG, Vincent JD, Ottersen OP, Storm-Mathisen J, Lledo PM (2001) A dendrodendritic reciprocal synapse provides a recurrent excitatory connection in the olfactory bulb. Proc Natl Acad Sci USA 98:6441–6446.
14. Egger V, Svoboda K, Mainen ZF (2003) Mechanisms of lateral inhibition in the olfactory bulb: efficiency and modulation of spike-evoked calcium influx into granule cells. J Neurosci 23:7551–7558.
15. Egger V, Svoboda K, Mainen ZF (2005) Dendrodendritic synaptic signals in olfactory bulb granule cells: local spine boost and global low-threshold spike. J Neurosci 25:3521–3530.
16. Egger V, Urban NN (2006) Dynamic connectivity in the mitral cell-granule cell microcircuit. Sem Cell Develop Biol 17.
17. Ermentrout GB, Galán RF, Urban NN. (2008) Reliability, synchrony and noise. Trends Neurosci 31(8):428–434.
18. Friedman D, Strowbridge BW (2003) Both electrical and chemical synapses mediate fast network oscillations in the olfactory bulb. J Neurophysiol 89:2601–2610.
19. Galán, RF, Ermentrout, GB, Urban, NN (2005) Efficient estimation of phase-resetting curves in real neurons and its significance for neural-network modeling. Phys Rev Lett 94, 158101.
20. Galán, RF, Fourcaud-Trocme, N, Ermentrout, GB, Urban, NN (2006) Correlation-induced synchronization of oscillations in olfactory bulb neurons. J Neurosci 26, 3646–3655.
21. Galán, RF, Ermentrout, GB, Urban, NN (2007) Stochastic dynamics of uncoupled neural oscillators: Fokker-Planck studies with the finite element method. Phys Rev E Stat Nonlin Soft Matter Phys 76, 056110.
22. Galán, RF, Ermentrout, GB, Urban, NN (2008) Optimal time scale for spike-time reliability: theory, simulations and experiments. J Neurophysiol 99, 277–283.
23. Hasenstaub A, Shu Y, Haider B, Kraushaar U, Duque A, McCormick DA (2005) Inhibitory postsynaptic potentials carry synchronized frequency information in active cortical networks. Neuron 47:423–435.
24. Isaacson JS (1999) Glutamate spillover mediates excitatory transmission in the rat olfactory bulb [see comments]. Neuron 23:377–384.
25. Isaacson JS, Strowbridge BW (1998) Olfactory reciprocal synapses: dendritic signaling in the CNS. Neuron 20:749–761.
26. Jahr CE, Nicoll RA (1980) Dendrodendritic inhibition: demonstration with intracellular recording. Science 207:1473–1475.
27. Kapoor V, Urban NN (2006) Glomerulus-specific, long-latency activity in the olfactory bulb granule cell network. J Neurosci 26:11709–11719.
28. Kay LM, Laurent G (1999) Odor- and context-dependent modulation of mitral cell activity in behaving rats. Nat Neurosci 2:1003–1009.
29. Lagier S, Carleton A, Lledo PM (2004) Interplay between local GABAergic interneurons and relay neurons generates gamma oscillations in the rat olfactory bulb. J Neurosci 24:4382–4392.
30. Lowe GD, Woodward M, Rumley A, Morrison CE, Nieuwenhuizen W (2003) Associations of plasma fibrinogen assays, C-reactive protein and interleukin-6 with previous myocardial infarction. J Thromb Haemost 1:2312–2316.

31. Mainen, ZF, Sejnowski, TJ (1995) Reliability of spike timing in neocortical neurons. Science 268:1503–1506.
32. Marella, S, Ermentrout, GB (2008) Class-II neurons display a higher degree of stochastic synchronization than class-I neurons. Phys Rev E Stat Nonlin Soft Matter Phys 77, 041918.
33. Margrie TW, Sakmann B, Urban NN (2001a) Action potential propagation in mitral cell lateral dendrites is decremental and controls recurrent and lateral inhibition in the mammalian olfactory bulb. Proc Natl Acad Sci USA 98:319–324.
34. Mori K, Nowycky MC, Shepherd GM (1981) Electrophysiological analysis of mitral cells in the isolated turtle olfactory bulb. J Physiol (Lond) 314:281–294.
35. Mori K, Takagi SF (1978) An intracellular study of dendrodendritic inhibitory synapses on mitral cells in the rabbit olfactory bulb. J Physiol (Lond) 279:569–588.
36. Nagai, K et al. (2005) Synchrony of neural oscillators induced by random telegraphic currents. Phys Rev E Stat Nonlin Soft Matter Phys 71, 036217.
37. Nakao H, Arai K, Kawamura Y (2007) Noise-Induced Synchronization and Clustering in Ensembles of Uncoupled Limit-Cycle Oscillators, Phys. Rev. Lett. 98, 184101.
38. Nakao H, Arai K-S, Nagai K, Tsubo Y, Kuramoto Y (2005) Synchrony of limit-cycle oscillators induced by random external impulses. Phys Rev E 72.
39. Neville KR, Haberly LB (2003) Beta and gamma oscillations in the olfactory system of the urethane-anesthetized rat. J Neurophysiol 90:3921–3930.
40. Nickell WT, Shipley MT, Behbehani MM (1996) Orthodromic synaptic activation of rat olfactory bulb mitral cells in isolated slices. Brain Res Bull 39:57–62.
41. Nusser Z, Kay LM, Laurent G, Homanics GE, Mody I (2001) Disruption of GABA(A) receptors on GABAergic interneurons leads to increased oscillatory power in the olfactory bulb network. J Neurophysiol 86:2823–2833.
42. Powell KR, Koppelman LF, Holtzman SG (1999) Differential involvement of dopamine in mediating the discriminative stimulus effects of low and high doses of caffeine in rats. Behav Pharmacol 10:707–716.
43. Rall W, Shepherd GM, Reese TS, Brightman MW (1966) Dendrodendritic synaptic pathway for inhibition in the olfactory bulb. Exp Neurol 14:44–56.
44. Ravel N, Chabaud P, Martin C, Gaveau V, Hugues E, Tallon-Baudry C, Bertrand O, Gervais R (2003) Olfactory learning modifies the expression of odour-induced oscillatory responses in the gamma (60-90 Hz) and beta (15-40 Hz) bands in the rat olfactory bulb. Eur J Neurosci 17:350–358.
45. Reyes AD (2003) Synchrony-dependent propagation of firing rate in iteratively constructed networks in vitro. Nat Neurosci 6:593–599.
46. Schaefer AT, Angelo K, Spors H, Margrie TW (2006) Neuronal Oscillations Enhance Stimulus Discrimination by Ensuring Action Potential Precision. PLoS Biol 4:e163.
47. Schoppa NE, Kinzie JM, Sahara Y, Segerson TP, Westbrook GL (1998) Dendrodendritic inhibition in the olfactory bulb is driven by NMDA receptors. J Neurosci 18:6790–6802.
48. Schoppa NE, Urban NN (2003) Dendritic processing within olfactory bulb circuits. Trends Neurosci 26:501–506.
49. Schoppa NE, Westbrook GL (2001a) Glomerulus-specific synchronization of mitral cells in the olfactory bulb. Neuron 31:639–651.
50. Schoppa NE, Westbrook GL (2001b) NMDA receptors turn to another channel for inhibition. Neuron 31:877–879.
51. Schoppa NE, Westbrook GL (2002) AMPA autoreceptors drive correlated spiking in olfactory bulb glomeruli. Nat Neurosci 5:1194–1202.
52. Segev I (1999) Taming time in the olfactory bulb. Nat Neurosci 2:1041–1043.
53. Singer W, Gray C, Engel A, Konig P, Artola A, Brocher S (1990) Formation of cortical cell assemblies. Cold Spring Harb Symp Quant Biol 55:939–952.
54. Smith TC, Jahr CE (2002) Self-inhibition of olfactory bulb neurons. Nat Neurosci 5:760–766.
55. Stopfer M, Jayaraman V, Laurent G. (2003) Intensity versus identity coding in an olfactory system. Neuron 39:991–1004.
56. Tateno T, Robinson HP (2007a) Quantifying noise-induced stability of a cortical fast-spiking cell model with Kv3-channel-like current. Biosystems 89(1-3):110–116.

57. Tateno T, Robinson HP (2007b) Phase resetting curves and oscillatory stability in interneurons of rat somatosensory cortex. Biophys J 92(2):683–695.
58. Teramae JN, Tanaka D (2004) Robustness of the noise-induced phase synchronization in a general class of limit cycle oscillators. Phys Rev Lett 93:204103.
59. Urban NN (2002) Lateral inhibition in the olfactory bulb and in olfaction. Physiol Behav 77:607–612.
60. Urban NN, Sakmann B (2002) Reciprocal intraglomerular excitation and intra- and inter-glomerular lateral inhibition between mouse olfactory bulb mitral cells. J Physiol 542:355–367.
61. Wang XJ, Buzsaki G (1996) Gamma oscillation by synaptic inhibition in a hippocampal interneuronal network model. J Neurosci 16:6402–6413.
62. White JA, Chow CC, Ritt J, Soto-Trevino C, Kopell N (1998) Synchronization and oscillatory dynamics in heterogeneous, mutually inhibited neurons. J Comput Neurosci 5:5–16.
63. Whittington MA, Traub RD, Kopell N, Ermentrout B, Buhl EH (2000) Inhibition-based rhythms: experimental and mathematical observations on network dynamics. Int J Psychophysiol 38:315–336.

Stochastic Neural Dynamics as a Principle of Perception

Gustavo Deco and Ranulfo Romo

Abstract Typically, the neuronal firing activity underlying brain functions exhibits a high degree of variability both within and between trials. The key question is: are these fluctuations just a concomitant effect of the neuronal substrate without playing any computational role or do they have a functional relevance? In this chapter, we first review the theoretical framework of stochastic neurodynamics that allows us to investigate the roles of noise and neurodynamics in the computation of probabilistic behavior. The relevance of this framework for neuroscience will be demonstrated by focusing on the simplest type of perceptual task, namely sensory detection. We focus on the following remarkable observation in a somatosensory task: when a near-threshold vibrotactile stimulus is presented, a sensory percept may or may not be produced. These perceptual judgments are believed to be determined by the fluctuation in activity of early sensory cortices. We show, however, that the behavioral outcomes associated with near-threshold stimuli depend of the neuronal fluctuations of more central areas to early somatosensory cortices. The theoretical analysis of the behavioral and neuronal correlates of sensation will show how variability at the neuronal level in those central areas can give rise to probabilistic behavior at the network level and how these fluctuations influence network dynamics.

Introduction

In this chapter, we consider how the noise contributed by the probabilistic spiking times of neurons (spiking noise) plays an important and advantageous role in brain function. We go beyond the deterministic noiseless description of the dynamics of cortical networks, and show how the properties of the system are influenced by the spiking noise. We show that the spiking noise has a significant contribution to the outcome that is reached, in that this noise is a factor in a network with a finite

G. Deco (✉)
Institució Catalana de Recerca i Estudis Avançats (ICREA), Universitat Pompeu Fabra, Passeig de Circumval.lació, 8 08003 Barcelona, Spain
e-mail: gustavo.deco@upf.edu

K. Josić et al. (eds.), *Coherent Behavior in Neuronal Networks*, Springer Series in Computational Neuroscience 3, DOI 10.1007/978-1-4419-0389-1_13,

(i.e., limited) number of neurons. This spiking noise can be described by introducing statistical fluctuations into the finite-size system. It is important that the outcome that is reached, and not just its time course, is influenced on each trial by these statistical fluctuations.

In particular, we will use integrate-and-fire models with spiking neurons to model the actual neuronal data that are obtained from neurophysiological experiments. The integrate-and-fire simulations capture the stochastic nature of the computations. However, we show that to understand analytically (mathematically) the stable points of the network, for example what decisions may be reached, it is helpful to incorporate a mean field approach that is consistent with the integrate-and-fire model. The mean field approach allows one to determine, for example, the synaptic strengths of the interconnected neurons that will lead to stable states of the network, each of which might correspond to a different decision, or no decision at all. The spiking simulations then examine which fixed points (or decisions) are reached on individual trials, and how the probabilistic spiking of the neurons influences these outcomes.

More specifically, we will show that both neurodynamics and stochastic fluctuations matter, in the sense that both have an essential computational role for a complete explanation of perception. To this purpose, we will take as a prototypical example the most elemental and historical task of perceptual detection. By constructing and analyzing computational models, we will establish the link that accounts for measurements both at the cellular and behavioral level. In particular, we show that the behavioral correlate of perceptual detection is essentially given by a noise driven transition in a multistable neurodynamical system. Thus, neuronal fluctuations can be an advantage for brain processing, as they lead to probabilistic behavior in decision-making in this and other sensory tasks. For example, decisions may be difficult without noise. In the choice dilemma described in the medieval Duns Scotus paradox, a donkey who could not decide between two equidistant food rewards might suffer the consequences of the indecision. The problem raised is that with a deterministic system, there is nothing to break the symmetry, and the system can become deadlocked. In this situation, the addition of noise can produce probabilistic choice, which is advantageous, as will be described in this paper.

Brain Dynamics: From Spiking Neurons to Reduced Rate-Models

The computation underlying brain functions emerges from the collective dynamics of spiking neuronal networks. A spiking neuron transforms a large set of incoming input spike trains, coming from different neurons, into an output spike train. Thus, at the microscopic level, neuronal circuits of the brain encode and process information by spatiotemporal spike patterns. We assume that the transient (nonstationary) dynamics of spiking neurons is properly captured by one-compartment, point-like models of neurons, such as the leaky integrate-and-fire (LIF) model [38]. In the LIF model, each neuron i can be fully described in terms of a single internal variable, namely the depolarization $V_i(t)$ of the neural membrane. The basic circuit of a LIF model consists of a capacitor C in parallel with a resistor R driven by a synaptic current (excitatory or inhibitory postsynaptic potential, EPSP or IPSP, respectively). When the voltage across the capacitor reaches a threshold θ, the circuit is shunted to a reset potential V_{reset}, and a δ-pulse (spike) is generated and transmitted to other neurons. The subthreshold membrane potential of each neuron evolves as a simple RC-circuit, with a time constant $\tau = RC$ given by the following equation:

$$\tau \frac{\mathrm{d}V_i(t)}{\mathrm{d}t} = -[V_i(t) - V_{\mathrm{L}}] + \tau \sum_{j=1}^{N} J_{ij} \sum_{k} \delta(t - t_j^{(k)}), \tag{1}$$

where V_{L} is the leak potential of the cell in the absence of external afferent inputs and the total synaptic current flow into cell i is given by the sum of the contributions of δ-spikes produced at presynaptic neurons, with J_{ij} the efficacy of synapse j and $t_j^{(k)}$ the emission time of the kth spike from the jth presynaptic neuron.

In the brain, local neuronal networks comprise a large number of neurons which are massively interconnected. The set of coupled differential equations (1) above describe the underlying dynamics of such networks. Direct simulations of these equations yield a complex spatiotemporal pattern, covering the individual trajectory of the internal state of each neuron in the network. This type of direct simulation is computationally expensive, making it very difficult to analyze how the underlying connectivity relates to various dynamics. One way to overcome these difficulties is by adopting the population density approach, using the Fokker–Planck formalism [21, 22, 28]. We will follow here a derivation done by Stefano Fusi (private communication). In this approach, individual integrate-and-fire neurons are grouped together into populations of statistically similar neurons. A statistical description of each population is given by a probability density function that expresses the distribution of neuronal states (i.e., membrane potential) over the population. In general, neurons with the same state $V(t)$ at a given time t have a different history because of random fluctuations in the input current $I(t)$. The main source of randomness is from fluctuations in the currents. The key assumption in the population density approach is that the afferent input currents impinging on neurons in one population are uncorrelated. Thus, neurons sharing the same state $V(t)$ in a population are indistinguishable. The population density $p(\upsilon, t)$ expresses the fraction of neurons at time t that have a membrane potential $V(t)$ in the interval $[\upsilon, \upsilon + d\upsilon]$. The evolution of the population density is given by the Chapman–Kolmogorov equation

$$p(\upsilon, t + \mathrm{d}t) = \int_{-\infty}^{+\infty} p(\upsilon - \varepsilon, t)\rho(\varepsilon|\upsilon - \varepsilon)\mathrm{d}\varepsilon, \tag{2}$$

where $\rho(\varepsilon|\upsilon) = \mathrm{Prob}\{V(t + \mathrm{d}t) = \upsilon + \varepsilon|V(t) = \upsilon\}$ is the conditional probability that generates an infinitesimal change $\varepsilon = V(t+\mathrm{d}t) - V(t)$ in the infinitesimal interval $\mathrm{d}t$. The temporal evolution of the population density can be reduced to a simpler differential equation by the mean-field approximation. In this approximation, the currents impinging on each neuron in a population have the same statistics, because, as mentioned above, the history of these currents is uncorrelated. The mean-field approximation entails replacing the time-averaged discharge rate of individual cells with a common time-dependent population activity (ensemble average). This assumes ergodicity for all neurons in the population. The mean-field technique allows us to discard the index denoting the identity of any single neuron. The resulting differential equation describing the temporal evolution of the population density is called the Fokker–Planck equation, and reads

$$\frac{\partial p(\upsilon,t)}{\partial t} = \frac{1}{2\tau}\sigma^2(t)\frac{\partial^2 p(\upsilon,t)}{\partial \upsilon^2} + \frac{\partial}{\partial \upsilon}\left[\left(\frac{\upsilon - V_{\mathrm{L}} - \mu(t)}{\tau}\right) p(\upsilon,t)\right]. \tag{3}$$

In the particular case that the drift is linear and the diffusion coefficient $\sigma^2(t)$ is given by a constant, the Fokker–Planck equation describes a well-known stochastic process called the Ornstein–Uhlenbeck process [31]. The Ornstein–Uhlenbeck process describes the temporal evolution of the membrane potential $V(t)$ when the input afferent current is $\mu(t) + \sigma\sqrt{\tau}w(t)$, with $w(t)$ a white noise process. This can be interpreted, by means of the Central Limit Theorem, as the case in which the sum of many Poisson processes becomes a normal random variable with mean $\mu(t)$ and variance σ^2.

The nonstationary solutions of the Fokker–Planck equation (3) describe the dynamical behavior of the network. However, these simulations, as the direct simulation of the original network of spiking neurons (1), are computationally expensive and their results probabilistic, which makes them unsuitable for systematic explorations of parameter space. On the other hand, the stationary solutions of the Fokker–Planck equation (3) represent the stationary solutions of the original integrate-and-fire neuronal system. The stationary solution of the Fokker–Planck equation satisfying specific boundary conditions (see [3, 23, 30]) yields the *population transfer function* of Ricciardi (ϕ):

$$\nu = \left[t_{\mathrm{ref}} + \tau\sqrt{\pi}\int_{\frac{V_{\mathrm{reset}} - V_{\mathrm{L}} - \mu}{\sigma}}^{\frac{\theta - V_{\mathrm{L}} - \mu}{\sigma}} \mathrm{e}^{x^2}\{1 + \mathrm{erf}(x)\}\mathrm{d}x \right]^{-1} = \phi(\mu,\sigma), \tag{4}$$

where $\mathrm{erf}(x) = 2/\sqrt{\pi}\int_0^x \mathrm{e}^{y^2}\mathrm{d}y$. In last equation t_{ref} is the refractory time.

The population transfer function gives the average population firing rate as a function of the average input current. For more than one population, the network is partitioned into populations of neurons whose input currents share the same statistical properties and fire spikes independently at the same rate. The set of stationary, self-reproducing rates ν_i for different populations i in the network can be found by solving a set of coupled self-consistency equations, given by:

$$\nu_i = \phi(\mu_i, \sigma_i), \tag{5}$$

This reduced system of equations allows a thorough investigation of the parameters. In particular, one can construct bifurcation diagrams to understand the nonlinear mechanisms underlying equilibrium dynamics and in this way solve the "inverse problem," i.e., the selection of the parameters that generate the attractors (steady states) that are consistent with the experimental evidence. This is the crucial role of the mean-field approximation: to simplify analyses through the stationary solutions of the Fokker–Planck equation for a population density under the diffusion approximation (Ornstein–Uhlenbeck process) in a self-consistent form. After that, one can perform full nonstationary simulations using these parameters in the integrate-and-fire scheme to generate *true dynamics*. The mean field approach ensures that these dynamics will converge to a stationary attractor that is consistent with the steady-state dynamics we require [3, 16]. The stochastic (random) firing times of neurons introduces noise into neuronal networks, and it is the consequences of this randomness expressed in a finite (limited) sized network of such neurons with which we are concerned in this review. We show that the noise in such systems not only helps us to understand many aspects of decision-making as implemented in the brain, but is in fact beneficial to the operation of decision-making processes.

The mean-field approach has been applied to model single neuronal responses, fMRI activation patterns, psychophysical measurements, and the effects of pharmacological agents and of local cortical lesions [4, 5, 8, 10–15, 32, 37].

Perceptual Detection and Stochastic Dynamics

Neurophysiology

The detection of sensory stimuli is among the simplest perceptual experiences and is a prerequisite for any further sensory processing. A fundamental problem posed by the sensory detection tasks is that repeated presentation of a near-threshold stimulus might unpredictably fail or succeed in producing a sensory percept. Where in the brain are the neuronal correlates of these varying perceptual judgments? Pioneering studies on the neuronal correlates of sensory detection showed that, in the case of vibrotactile stimuli, the responses of S1 neurons account for the measured psychophysical accuracy [27]. However direct comparisons between S1 responses and detection performance were not directly addressed and, therefore, it is not clear whether the activity of S1 accounts for the variability of the behavioral responses. Psychophysical performance was measured in human observers and S1 recordings were made in anesthetized monkeys.

This problem has been recently addressed [6, 7]. These authors trained monkeys to perform a detection task. In each trial, the animal had to report whether the tip of a mechanical stimulator vibrated or not. Stimuli were sinusoidal, had a fixed frequency of 20 Hz and were delivered to the glabrous skin of one fingertip. Crucially, they varied in amplitude across trials. Stimulus-present trials were interleaved with an equal number of stimulus-absent trials in which no mechanical vibrations were delivered. Depending on the monkeys' responses, trials could be classified into four types of responses: hits and misses in the stimulus-present condition, and correct rejections and false alarms in the stimulus-absent condition. Stimulus detection thresholds were calculated from the behavioral responses. Thus an important issue in this and similar tasks is to determine the neuronal correlates that account for these behavioral reports.

De Lafuente and Romo [6] simultaneously characterized the activity of S1 neurons (areas 3b and 1) and the monkey's psychophysical performance by recording the extracellular spike potentials of single S1 neurons while the monkeys performed the detection task. Figure 1 shows the experimental design and main results. To test whether the responses of S1 neurons accounted for the monkey's psychophysical performance, [6] calculated neurometric detection curves and compared them with the psychometric curves. The proportion of "yes" responses for neurometric curves was defined, for a given amplitude, as the proportion of trials in which the neuron's firing rate reached or exceeded a criterion value [6, 18]. For each neuron, this criterion was chosen to maximize the number of correct responses. Pairwise comparisons of detection thresholds obtained from logistic fits to the simultaneously obtained neurometric and psychometric data showed that the detection thresholds of individual S1 neurons were not significantly different from the animals' psychophysical thresholds, and the two thresholds measures highly covaried. In addition, the shape

Fig. 1 The detection task. (**a**) Drawing of monkey working in the detection task. (**b**) The sequence of events during the detection trials. Trials began when the stimulation probe indented the skin of one fingertip of the left, retrained hand (probe down, PD). The monkey then placed its right, free hand on an immovable key (key down, KD). On half of the randomly selected trials, after a variable prestimulus period (Prestim, 1.5–3.5 s), a vibratory stimulus (Stim, 20 Hz, 0.5 s) was presented. Then, after a fixed delay period (Delay, 3 s), the stimulator probe moved up (probe up, UP), indicating to the monkey that it could make the response movement (MT) to one of the two buttons. The button pressed indicated whether or not the monkey felt the stimulus. Henceforth referred to as yes and no responses, respectively. (**c**) Depending on whether the stimulus was present or absent and on the behavioral response, trial outcome was classified as a hit, miss, false alarm (FA), or correct reject (CR). Trials were pseudo-randomly chosen: 90 trials were stimulus absent (amplitude 0), and 90 trials were stimulus present with varying amplitudes (9 amplitudes with 10 repetitions each). (**d**) Classical psychometric detection curve obtained by plotting the proportion of yes responses as a function of the stimulus amplitude. (**e**) Mean firing rate of hit trials for S1 ($n = 59$) and MPC ($n = 50$) neurons. (**f**) Comparison of normalized neuronal population activity of S1 neurons during hits and misses for near-threshold stimuli, and during correct rejections and false alarms in stimulus-absent trials. Normalized activity was calculated as a function of time, using a 200 ms window displaced every 50 ms. This was calculated by substractng the mean activity and dividing by the standard deviation of the activity from a 200 ms window of the prestimulus period. *Lower panels* show the choice probability index as a function of time. This quantity measures the overlap between two response distributions: in this case, between hits and misses and between correct rejection and false alarm trials. *Dotted lines* mark significance levels. (**g**) The same as in **f**, but for a neuronal population activity of MPC neurons. Adapted with permission from De LaFuente and Romo, 2006

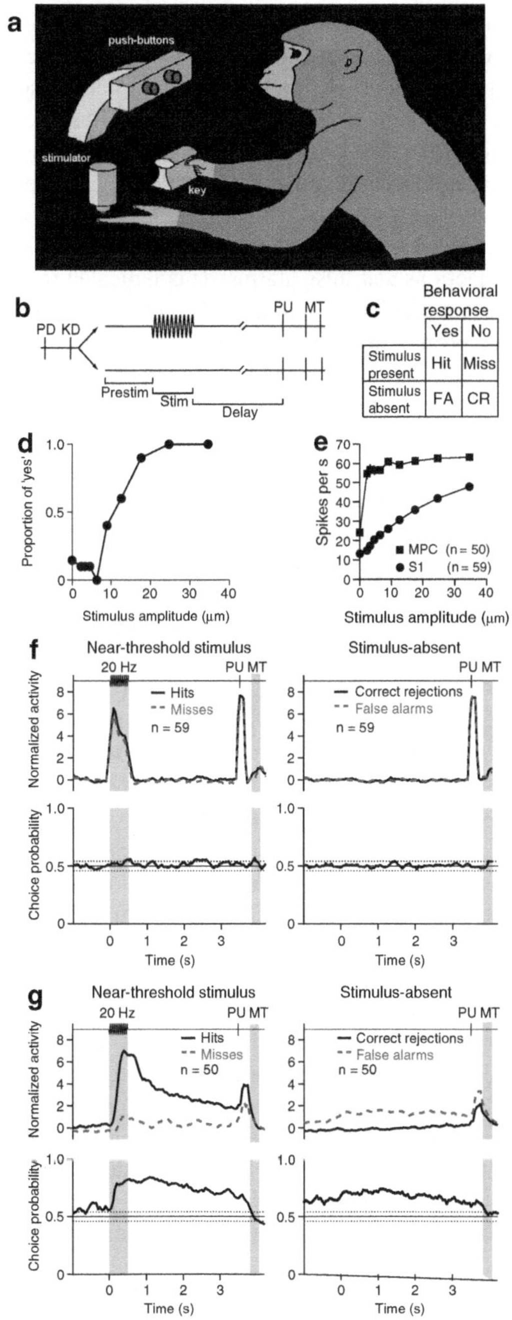
a
push-buttons
stimulator
key
b
PD KD
PU MT
Prestim
Stim
Delay
c
Behavioral response
Yes
No
Stimulus present
Hit
Miss
Stimulus absent
FA
CR
d
Proportion of 'yes'
Stimulus amplitude (μm)
e
Spikes per s
MPC (n = 50)
S1 (n = 59)
Stimulus amplitude (μm)
f
Near-threshold stimulus
Stimulus-absent
20 Hz
PU MT
Hits
Misses
n = 59
Correct rejections
False alarms
n = 59
Normalized activity
Choice probability
Time (s)
g
Near-threshold stimulus
Stimulus-absent
20 Hz
PU MT
Hits
Misses
n = 50
Correct rejections
False alarms
n = 50
Normalized activity
Choice probability
Time (s)

of the mean neurometric curve resulting from the activity of the S1 neurons showed close correspondence with the shape of the mean psychometric curve.

An important question addressed in this study is whether the activity of S1 neurons covaries with the perceptual "yes"–"no" judgments that the monkeys made on a trial-by-trial basis [6]. To test this, these authors compared the activity during hit and miss trials for the near-threshold stimulus, as well as for the corresponding activity in correct reject and false alarm trials in the stimulus-absent condition. They found no significant differences in the activity of S1 neurons between hits and misses, nor between correct rejections and false alarms. This indicated that activity of individual S1 neurons did not predict the monkey's behavior. To further quantify this, [6] calculated a choice probability index, which estimates the probability with which the behavioral outcome can be predicted from the neuronal responses [2, 20]. Again they found no significant differences between hits and misses, or between correct rejections and false alarm trials.

The low choice probability values are consistent with a detection model in which the activity of S1 serves as input to an additional processing stage(s) that determines whether a stimulus has occurred or not. Under this hypothesis, the correlation between S1 activity and the final decision about the stimulus presence or absence is highly dependent on the amount of correlated-noise among sensory neurons [39]. Indeed, [6] found that the mean noise correlation coefficient across pairs of S1 neurons was 0.16 ± 0.02. This amount of correlated-noise is similar to that reported in previous studies [1, 35, 39], and is also consistent with the near chance choice probability values reported in the study of [6]. These results further support a detection model in which a central area(s) must be observing the activity of S1 neurons to judge about the stimulus presence or absence. Therefore, the functional role of S1 in this and other perceptual tasks may be mainly to generate a neural representation of the sensory stimulus for further processing in areas central to it [19, 33–36]. However, a previous study found that fMRI signals in primary visual cortex (V1) reflected the percepts of human subjects, rather than the encoded stimulus features [29]. This result suggests that, in V1, top-down signals (nonsensory inputs delivered to visual cortex via feedback projections) can be combined with bottom-up (sensory) information and contribute to sensory percepts [29]. S1 data did not show evidence for this type of neural interaction; rather, it indicated that S1 represents the physical properties of stimuli and contributes little to near-threshold percepts [6]. The discrepancy could be due to fundamentally different organizations across sensory cortices, or to differences between species. Another possibility to consider is that the modulation revealed through fMRI may have an effect that is invisible from the point of view of single neurons. This would happen if, for instance, such modulation acted only to synchronize the spikes of multiple target neurons [17].

To test whether the neuronal correlates of the perceptual decisions associated with detection might reside outside S1, [6] recorded neurons from the medial premotor cortex (MPC), a frontal cortical area known to be involved in the evaluation of sensory information and in decision-making processes [20, 24]. They found that, in contrast to the graded dependence on stimulus amplitude observed in S1, MPC neurons responded in an all-or-none manner that was only weakly modulated by the stimulus amplitude, but that closely correlated with "yes" and "no" behavioral responses. The mean normalized activity across the MPC neurons was strong and sustained, and with near-threshold stimuli it was clearly different for hit and miss trials. Moreover, almost 70% of the false alarm trials were predicted from increases in neuronal activity in stimulus-absent trials. de Lafuente and Romo [6] also found that the MPC activity preceding stimulus onset was higher during hits than during misses. These early increases in activity predicted detection success significantly above chance levels, as is shown by the choice probability plots. Although de Lafuente and Romo (2005, 2006) do not know the role of this increased prestimulus activity, they speculate that it might be associated with trial history during a run. To support this conjecture, [6] wondered about the behavioral responses on trials previous to false alarm responses. They found that the probability of an "yes" response was increased in those trials preceding a false alarm, supporting the notion that monkeys were biased toward "yes" responses. de Lafuente and Romo [6] speculated that given that "yes" responses to the three subthreshold amplitudes were rewarded, this could have encouraged the monkeys to respond "yes" in the next trial, producing a false alarm response. The results indicate that the responses of all MPC neurons studied were associated with stimulus presence or with false alarms; that is, with "yes" responses. They did not find neurons whose increases in their activities were associated with "no" responses. [6] do not know the reason for this but they speculate that "no" is the default response that is installed from trial beginning and that the stimulus presentation overrides this default response.

The close association between neuronal responses and behavioral responses, and the weak relationship between neuronal activity and stimulus amplitude, supported the interpretation that MPC neurons do not code the physical attributes of stimuli, but rather represent perceptual judgments about their presence of absence. As the monkeys reported their decisions by a motor act, a key question needed to be answered: was the MPC activity truly related to stimulus perception, or was it simply reflecting the different motor actions associated with the two response buttons? To test this, [6] designed a control task in which the correct response button was illuminated at the beginning of every trial. In this variant of the detection task, the monkeys simply had to wait until the end of the trial to push the illuminated button, without the need to attend to the presence or absence of the mechanical vibration. It is important to note that in this test condition all-or-none activity was still observed in response to the near-threshold stimulus, and the probability of activation depended on the stimulus amplitude, similar to that observed in the standard detection task. Given that in the control test the monkeys did not have to choose a response button based on the vibratory stimulus, the results are consistent with the interpretation that the activity of MPC neurons is related to the subjective perception of sensory stimuli, rather than to the selection of the motor plan. These results therefore favor the hypothesis that this MPC population reflects the failure or success of the near-threshold stimulus in triggering a sensory percept.

A Computational Model of Probabilistic Detection

An aim of this chapter is to show how stochastic dynamics helps to understand the computational mechanisms involved in perceptual detection. The computational analysis of detection focuses on the paradigm and experimental results of [6] described above. In summary, they used a behavioral task where trained awake monkeys report the presence or absence of a mechanical vibration applied to their fingertips by pressing one of two pushbuttons. They found that the activity of MPC neurons was only weakly modulated by the stimulus amplitude, and covaried with the monkeys' trial-by-trial reports. On the contrary, S1 neurons did not covary with the animals' perceptual reports, but their firing rate did show a monotonically increasing graded dependence on the stimulus amplitude (see Fig. 1d and e). The fact that MPC neurons correlate with the behavioral performance, with a high firing rate for an "yes" report and a low firing rate for a "no" report, suggests an underlying bistable dynamic in an attractor framework.

A minimal network model is now described that captures the computation involved in perceptual detection and is consistent with the neurophysiological and behavioral evidence described [9]. The main idea of the model is to establish a neurodynamics that shows two possible *bistable* decision states associated with the two possible behavioral responses: "stimulus detection" and "no stimulus detection." The computation underlying perceptual detection is then understood as a fluctuation-driven, probabilistic transition to one of the two possible bistable decision states.

A patch of MPC neurons in the frontal lobe is modeled by a network of interacting neurons organized into a discrete set of populations. Populations are defined as groups of excitatory or inhibitory neurons sharing the same inputs and connectivities. Some of the excitatory population of neurons have a selective response, which reflects the sensitivity to an external applied vibrotactile stimulus (note that for simplicity in 1A only one selective population is shown for the single specific vibrotactile frequency utilized in the experiment). All other excitatory neurons are grouped in a "Non-selective" population. There is also one inhibitory population of local inhibitory neurons that regulates the overall activity by implementing competition in the network. Neurons in the networks are connected via three types of receptors that mediate the synaptic currents flowing into them: AMPA and NMDA glutamate receptors, and GABA receptors. Neurons within a specific excitatory population are mutually coupled with a strong weight ω_+. Neurons between two different selective populations have anticorrelated activity, which results in weaker connections ω_-.

In this model, activity in a selective excitatory population corresponds to the detection of a percept associated with an external applied vibrotactile stimulus. The strength of the input (λ) impinging on that excitatory population is proportional to the strength of the presented vibrotactile stimulus (as for example encoded in S1, i.e., the input to MPC is transmitted from S1). When a stimulus is presented, there is just one population sensitive to it. To model this characteristic, we use a network composed of two selective populations, but only one will be selective to the stimuli applied. The relevant bistability is therefore given by the state where the excitatory populations have low activity (corresponding to no detection of a percept, i.e., a "no" response), and the state where the excitatory population sensitive to the presented vibrotactile stimulus is highly activated (corresponding to the detection of the percept, i.e., an "yes" response). We refer to this model as "Non-Competing Yes-Neurons" (NCYN) (Fig. 2a). Just the selective population sensitive to the applied vibrotactile stimulation used in the experiment is represented by a specific excitatory population. (A full specification of the whole connectivity is provided in [9].)

The characteristics of the network in the stationary conditions were studied with the mean-field approach reviewed above. Using this approximation, the relevant parameter space given by the population cohesion ω_+ vs. the external input λ was scanned. The mean-field results for the NCYN-model are illustrated in a phase diagram (Fig. 2b) that shows different regimes of the network. For small values of λ and for a weak population cohesion, the network has one stable state where all populations are firing at a weak level (spontaneous state). This spontaneous state encodes the "no" response in the NYCN model. For higher population cohesion and higher values of λ, a state corresponding to strong activation of the selective population sensitive to the applied vibrotactile stimulation emerges. We call this excited state encoding the "yes" response, the "yes" state. Between these two regions, there is a bistable region where the state corresponding to weak ("no" response) or strong ("yes" response) activation states of the selective population sensitive to the applied vibrotactile stimulation are both stable.

To study the probabilistic behavior of the neuronal dynamics of the network, the spiking simulations of the configurations corresponding to the region of bistability were analyzed with methods similar to those used for the neurophysiological data by [6].

The results are presented of the nonstationary probabilistic analysis calculated by means of the full spiking simulations averaged over several trials. In all cases, the aim was to model the behavior of the MPC neurons which are shown in Fig. 1d and e, which reflect detection of the percept [6]. It is proposed that the perceptual response results from a neurodynamical bistability [9]. In this framework, each of the stable states corresponds to one possible perceptual response: "stimulus detected" or "stimulus not detected." The probability of detecting the stimulus is given by the transitions between these two states. In fact, the probabilistic character of the system results from the stochastic nature of the networks. The source of this stochasticity is the approximately random spiking of the neurons in the finite-size network. We note that there are two sources of noise in such spiking networks: the randomly arriving external Poissonian spike trains and the fluctuations due to the finite size of the network. Here we refer to finite-size effects due to the fact that the populations are described by a finite number N of neurons. In the mean-field framework, (see [25,26]) "incoherent" fluctuations due to quenched randomness in the neurons' connectivity and/or to external input are already taken into account in the variance, and "coherent" fluctuations give rise to new phenomena. In fact, the number of spikes emitted by the network in a time interval $[t, t + \mathrm{d}t)$ is a Poisson variable with mean and variance $N\nu(t)\mathrm{d}t$. The estimate of $\nu(t)$, is then a stochastic process $\nu_N(t)$, well

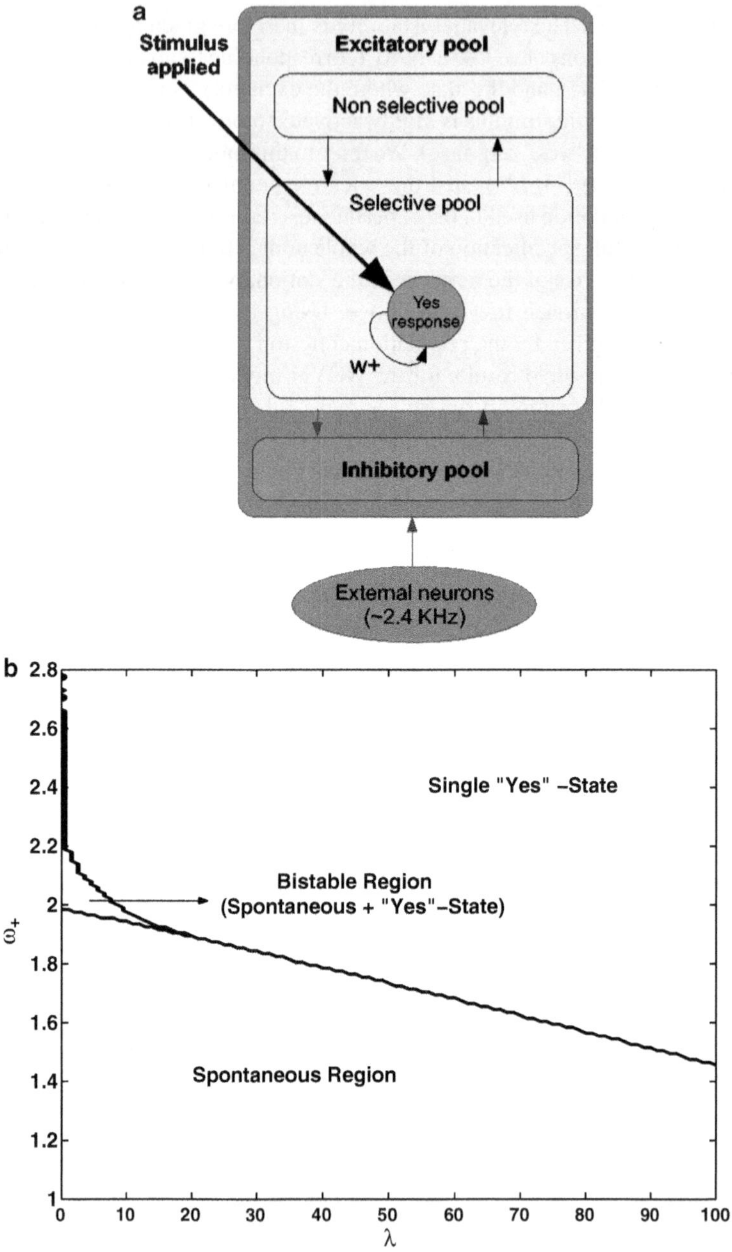

Fig. 2 (**a**) The perceptual detection model (NCYN) has excitatory populations selective to the applied vibrotactile stimulation. A "no" response is given when the selective population has low activity and an "yes" response when it has high activity. The *arrows* indicate the recurrent connections between the different neurons in a pool. (See text for more details). (**b**) Phase diagrams for the NYCN model for parameter exploration of the attractor states (stationary states) of the underlying dynamical system. The diagrams show the different attractor regions as a function of the stimulus input (λ) and the level of coupling ω_+ within the neurons of the same selective population (cohesion)

described in the limit of large $N\nu$ by $\nu_N(t) \simeq \nu(t) + \sqrt{\nu(t)/N}\xi(t)$, where $\xi(t)$ is Gaussian white noise with zero mean and unit variance, and $\nu(t)$ is the probability of emitting a spike per unit time in the infinite-size network. Such finite-N fluctuations, which affect the global activity ν_N, are coherently felt by all neurons in the network and lead to an additive Gaussian noise corrections in the mean-field equations.

To compare the theoretical results with the experimental results, the characteristics of the bistable neurodynamical model NCYN were studied. The behavior of the relevant populations encoding the different bistable states corresponding to the two alternative choices is shown in Fig. 3. Figure 3a plots the proportion of "yes" responses as a function of the intensity of the applied vibrotactile stimulation, i.e., as a function of the strength λ of the stimulus presented. The figure shows that the proportion of "yes" responses (hits) increases as the intensity of the stimulus applied grows. The model is consistent with the experimental results of Lafuente and Romo shown in Fig. 1. Hence, the model shows a probabilistic behavior that emulates the real behavior of subjects detecting a vibrotactile stimulus [6]. Let us now concentrate on the level of firing activity observed in MPC neurons that covary with the behavioral responses. Figure 3b shows the activity of the neurons encoding the "yes" response (selective excitatory population sensitive to the applied vibrotactile stimulus) averaged over trials that reported a percept (hits). In the model, the mean firing activity is almost constant and is not linearly related to the stimulus amplitude, as reflected in the experimental results. The fact that neurons encoding the "yes" response present a relatively constant level of activation on trials that report a detected percept, whereas on trials that fail to detect a percept these neurons have low activity (spontaneous level), is consistent with an attractor network. Therefore the transitions driven by the spiking-related statistical fluctuations are consistent with the behavioral data.

Deco et al. [9] studied also a second different bistable network model called "Competing Yes-No-Neurons" (CYNN). Both models (NCYN and CYNN) are consistent with the existing single cell recordings, but they involve different types of bistable decision states, and consequently different types of computation and neurodynamics. By analyzing the temporal evolution of the firing rate activity of neurons on trials associated with the two different behavioral responses, they were able to produce evidence in favor of the CYNN model. Specifically, the CYNN model predicts the existence of some neurons that encode the "no" response, and other neurons that encode the "yes" response. The first set of neurons slightly decrease their activity at the end of the trial, whereas the second group of neurons increase their firing activity when a stimulus is presented. Thus in this case, the simulations indicate that the CYNN model fits the experimental data better than the NCYN model.

In conclusion, computational stochastic neurodynamical models provide a deeper understanding of the fundamental mechanisms underlying perceptual detection and how these are related to experimental neuroscience data. We argue that addressing such a task is a prerequisite for grounding empirical neuroscience in a cogent theoretical framework.

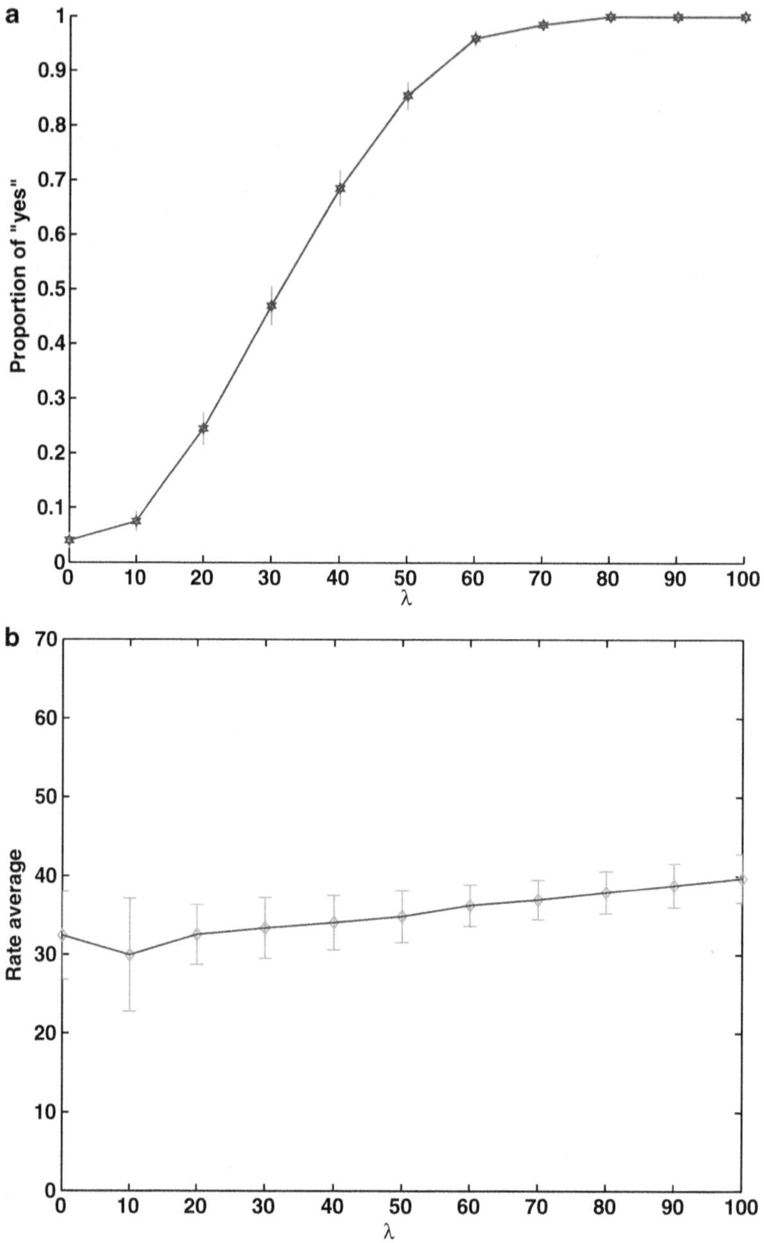

Fig. 3 Simulated results plotting the detection curves resulting from 200 trials (overall performance) and the mean rate activity of hit trials at a function of the input strength λ for the MPC neurons for the experimental design of de Lafuente and Romo (2005). (**a**) Probability of an "yes" response (hit). (**b**) Mean firing rate activity of neurons in the "yes" population on "yes" trials. The simulations of the nonstationary and probabilistic behavior of the neurodynamical activity were performed by a full spiking and synaptic simulation of the network

References

1. W. Bair, E. Zohary, and W. Newsome. Correlated firing in macaque in visual mt: Time scales and relationship to behavior. *Journal of Neuroscience*, 21:1676–1697, 2001.
2. K. Britten, W. Newsome, M. Shadlen, S. Celebrini, and J. Movshon. A relationship between behavioral choice and the visual responses of neurons in macaque mt. *Visual Neuroscience*, 13:87–100, 1996.
3. N. Brunel and X. J. Wang. Effects of neuromodulation in a cortical network model of object working memory dominated by recurrent inhibition. *Journal of Computational Neuroscience*, 11:63–85, 2001.
4. S. Corchs and G. Deco. Large-scale neural model for visual attention: integration of experimental single cell and fMRI data. *Cerebral Cortex*, 12:339–348, 2002.
5. S. Corchs and G. Deco. Feature based attention in human visual cortex: simulation of fMRI data. *Neuroimage*, 21, 36–45, 2004.
6. V. de Lafuente and R. Romo. Neuronal correlates of subjective sensory experience. *Nature Neuroscience*, 8:1698–1703, 2005.
7. V. de Lafuente and R. Romo. Neural correlates of subjective sensory experience gradually builds up accross cortical areas. *Proceedings of the National Academy of Science USA*, 103:14266–14271, 2006.
8. G. Deco and T. S. Lee. A unified model of spatial and object attention based on inter-cortical biased competition. *Neurocomputing*, 44–46:775–781, 2002.
9. G. Deco, M. Perez-Sanagustin, V. de Lafuente, and R. Romo. Perceptual detection as a dynamical bistability phenomenon: a neurocomputational correlate of sensation. *Proceedings of the National Academy of Sciences USA*, 104:20073–20077, 2007.
10. G. Deco, O. Pollatos, and J. Zihl. The time course of selective visual attention: theory and experiments. *Vision Research*, 42:2925–2945, 2002.
11. G. Deco and E. Rolls. Neurodynamics of biased competition and cooperation for attention: a model with spiking neurons. *Journal of Neurophysiology*, 94:295–313, 2005.
12. G. Deco and E. T. Rolls. Object-based visual neglect: a computational hypothesis. *European Journal of Neuroscience*, 16:1994–2000, 2002.
13. G. Deco and E. T. Rolls. Attention and working memory: a dynamical model of neuronal activity in the prefrontal cortex. *European Journal of Neuroscience*, 18:2374–2390, 2003.
14. G. Deco and E. T. Rolls. A neurodynamical cortical model of visual attention and invariant object recognition. *Vision Research*, 44:621–644, 2004.
15. G. Deco, E. T. Rolls, and B. Horwitz. 'What' and 'where' in visual working memory: a computational neurodynamical perspective for integrating fMRI and single-neuron data. *Journal of Cognitive Neuroscience*, 16:683–701, 2004.
16. P. Del Giudice, S. Fusi, and M. Mattia. Modeling the formation of working memory with networks of integrate-and-fire neurons connected by plastic synapses. *Journal of Physiology Paris*, 97:659–681, 2003.
17. P. Fries, J. Reynolds, A. Rorie, and R. Desimone. Modulation of oscillatory neuronal synchronization by selective visual attention. *Science*, 291:1560–1563, 2001.
18. D.M. Green and J.A. Swets. *Signal Detection Theory and Psychophysics*. Wiley, New York, 1966.
19. A. Hernandez, A. Zainos, and R. Romo. Neuronal correlates of sensory discrimination in somatosensory cortex. *Proceedings of the National Academy of Sciences USA*, 97:6191–6196, 2000.
20. A. Hernandez, A. Zainos, and R. Romo. Temporal evolution of a decision-making process in medial premotor cortex. *Neuron*, 33:959–972, 2002.
21. B. Knight. Dynamics of encoding in neuron populations: some general mathematical features. *Neural Computation*, 12(3):473–518, 2000.
22. B.W. Knight, D. Manin, and L. Sirovich. Dynamical models of interacting neuron populations. In E.C. Gerf, ed. *Symposium on Robotics and Cybernetics: Computational Engineering in Systems Applications*. Cite Scientifique, Lille, France, 1996.

23. P. Lansky, L. Sacerdote, and F. Tomassetti. On the comparison of Feller and Ornstein-Uhlenbeck models for neural activity. *Biological Cybernetics*, 73:457–465, 1995.
24. L. Lemus, A. Hernandez, R. Luna, A. Zainos, V. Nacher, and R. Romo. Neural correlates of a postponed decision report. *Proceedings of the National Academy of Sciences USA*, 104:17174–17179, 2007.
25. M. Mattia and P.D. Giudice. Population dynamics of interacting spiking neurons. *Phys Rev E*, 66(5):051917, 2002.
26. M. Mattia and P.D. Giudice. Finite-size dynamics of inhibitory and excitatory interacting spiking neurons. *Phys Rev E*, 70:052903, 2004.
27. V. Mountcastle, W. Talbot, H. Sakata, and J. Hyvarinen. Cortical neuronal mechanisms in flutter-vibration studied in unanesthetized monkeys. neuronal periodicity and frequency discrimination. *Journal of Neurophysiology*, 32:452–484, 1969.
28. A. Omurtag, B.W. Knight, and L. Sirovich. On the simulation of large populations of neurons. *Journal of Computational Neuroscience*, 8:51–53, 2000.
29. D. Ress and D. Heeger. Neuronal correlates of perception in early visual cortex. *Nature Neuroscience*, 6:414–420, 2003.
30. L. Ricciardi and L. Sacerdote. The Ornstein-Uhlenbeck process as a model for neuronal activity. I. mean and variance of the firing time. *Biological Cybernetics*, 35:1–9, 1979.
31. H. Risken. *The Fokker-Planck Equation: Methods of Solution and Applications*. Springer Verlag, Berlin, 1996.
32. E. T. Rolls and G. Deco. *Computational Neuroscience of Vision*. Oxford University Press, Oxford, 2002.
33. R. Romo, C. Brody, A. Hernandez, and L. Lemus. Neuronal correlates of parametric working memory in the prefrontal cortex. *Nature*, 339:470–473, 1999.
34. R. Romo, A. Hernandez, A. Zainos, L. Lemus, and C. Brody. Neural correlates of decision-making in secondary somatosensory cortex. *Nature Neuroscience*, 5:1217–1225, 2002.
35. R. Romo, A. Hernandez, A. Zainos, and E. Salinas. Correlated neuronal discharges that increase coding efficiency during perceptual discrimination. *Neuron*, 38:649–657, 2003.
36. E. Salinas, A. Hernandez, A. Zainos, and R. Romo. Periodicity and firing rate as candidate neural codes for the frequency of vibrotactile stimuli. *Journal of Neuroscience*, 20:5503–5515, 2000.
37. M. Szabo, G. Deco, S. Fusi, P. Del Giudice, M. Mattia, and M. Stetter. Learning to attend: Modelling the shaping of selectivity in inferotemporal cortex in a categorization task. *Biological Cybernetics*, 94:351–365, 2006.
38. H. Tuckwell. *Introduction to Theoretical Neurobiology*. Cambridge University Press, Cambridge, 1988.
39. E. Zohary, M. N. Shadlen, and W. T. Newsome. Correlated neuronal discharge rate and its implications for psychophysical performance. *Nature*, 370:140–143, 1994.

Large-Scale Computational Modeling of the Primary Visual Cortex

Aaditya V. Rangan, Louis Tao, Gregor Kovačič, and David Cai

Abstract This chapter reviews our approach to large-scale computational modeling of the primary visual cortex (V1). The main objectives of our modeling are to (1) capture groups of experimentally observed phenomena in a single theoretical model of cortical circuitry, and (2) identify the physiological mechanisms underlying the model dynamics. We have achieved these objectives by building parsimonious models based on minimal, yet sufficient, sets of anatomical and physiological assumptions. We have also verified the structural robustness of the proposed network mechanisms. During the modeling process, we have identified a particular operating state of our model cortex from which we believe that V1 responds to changes in visual stimulation. This state is characterized by (1) high total conductance, (2) strong inhibition, (3) large synaptic fluctuations, (4) an important role of NMDA conductance in the orientation-specific, long-range interactions, and (5) a high degree of correlation between the neuronal membrane potentials, NMDA-type conductances, and firing rates. Tuning our model to this operating state in the absence of stimuli, we have used it to identify and investigate model neuronal network mechanisms underlying cortical phenomena including (1) spatiotemporal patterns of spontaneous cortical activity, (2) cortical activity patterns induced by the Hikosaka line-motion illusion stimulus paradigm, (3) membrane potential synchronization in nonspiking neurons several millimeters apart, and (4) neuronal orientation tuning in V1.

Introduction

The primary visual cortex (V1) is one area of the brain where computational modeling has been successfully used to investigate the link between physiological mechanisms and cortical function. Our approach to large-scale computational modeling of V1 [1–3] is reviewed in this chapter. Clearly, to gain a deeper theoretical

D. Cai (✉)
Courant Institute of Mathematical Sciences, New York University, 251 Mercer Street, New York, NY 10012, USA
e-mail: cai@cims.nyu.edu

K. Josić et al. (eds.), *Coherent Behavior in Neuronal Networks*, Springer Series in Computational Neuroscience 3, DOI 10.1007/978-1-4419-0389-1_14,

understanding of even the simplest brain functions, modeling must strike a careful balance between mathematical abstraction and physiological detail. Moreover, to achieve a good understanding of network dynamics through computational modeling, fast and efficient computational methods for simulating large-scale neuronal networks have become a necessity.

Modeling Objectives In our view, the main objectives of computational neuronal network modeling are to (1) capture groups of experimentally observed cortical phenomena in a single theoretical model of cortical circuitry, and (2) identify the physiological mechanisms underlying the resulting model dynamics. In this way, computational modeling with sufficient realism may help to pick from among a number of theoretical scenarios those that could be truly realized in nature. This objective makes it necessary to build "parsimonious" models based on a minimal, yet sufficient, set of anatomical and physiological assumptions that allow such models to qualitatively and quantitatively reproduce a given set of distinct physiological effects within a unified dynamical regime and with a single realistic cortical architecture. The spatial and temporal scales resolved by the model must be chosen judiciously for the model to faithfully reflect the described phenomena. A clear advantage of a large-scale computational model over more idealized models is that it is broad enough to explore a large number of possible dynamical regimes, all within a single framework, and thus identify those regimes that are physiologically relevant. For example, a computational model can explore the cortical operation of both correlated and uncorrelated firing activity, whereas an idealized firing-rate model can usually describe the latter only. Finally, one of the ultimate goals of large-scale computational neuronal network modeling is to use the network mechanisms identified in computational modeling for guiding the design of new experiments, as well as to contrast these mechanisms with experimental results, so that we can reach a better understanding of the underlying physiological phenomena.

In contrast to statistical physics that provides general principles governing large-scale equilibrium systems, no unifying law has so far been found that would govern large-scale network systems in neuroscience. We must therefore strive to extract the general governing features of the investigated neuronal assemblies in a robust way that is insensitive to the insignificant details of both the computational model used and the parameter regime it operates in. However, unlike in analytical considerations where this simply means discussing a sufficiently general model family, computational models are very concrete in terms of their specifications, such as the values of parameters used in the model. Therefore, special attention must be paid to the "structural robustness" of the discovered network mechanisms. This again means that the models must be able to capture multiple phenomena in a single dynamical regime, within *broad parameter ranges*, and also that the models should capture bifurcations wherever they exist and reproduce their correct dynamical behavior as observed experimentally. These requirements constrain the models structurally. Additionally, in a stronger sense, one can only be reasonably convinced that the network mechanisms discovered via this modeling process are robust structurally when physiologically reasonable variations of the network architecture all reproduce the studied phenomena and confirm the discovered mechanisms in similar parameter regimes and with comparable accuracy. Moreover, when we study phenomena that are hypothesized to be network induced instead of being controlled by the cellular dynamics of particular neurons, we can demonstrate the robustness of the hypothesized network mechanism by replacing the underlying Hodgkin–Huxley-type equations with, for example, a simpler integrate-and-fire (I&F) neuron model. If the presumed network-induced mechanism is indeed at work, the particular choice of the neuronal equations (Hodgkin–Huxley; or linear, quadratic, or exponential I&F) should make no essential difference. In fact, a study by comparing a number of such related models can be systematically employed for examining the robustness of the hypothesized mechanisms.

Due to the large scales of the modeled cortical areas, and the correspondingly large numbers of neurons involved, our computational models are large in terms of the numbers of neurons included and physical areas covered. Modeling of large cortical areas allows us to theoretically explain results of large-scale modern neurophysiological experiments, such as imaging with voltage-sensitive dyes [4, 5] and multielectrode array measurements [6]. It is important to emphasize that the numbers of model neurons must have approximately the same orders of magnitude as the numbers of the neurons *in vivo* or *in vitro*, since some of the most important dynamical properties of neuronal networks, such as the level of fluctuations, are often controlled by the numbers of neurons and connectivity (sparse or dense) of their synaptic couplings. Inappropriate choice of the number of neurons and synaptic connectivity in the model may result in dynamical behavior inconsistent with physiological effects. For example, too many weak pre-synaptic connections per neuron tend to lead to mean-driven dynamics and prevent spiking that is induced by subthreshold fluctuations as often observed in experiments [7]. The large numbers of neurons in our network model, dictated by the cortical phenomena we study, require us to devise fast algorithms that take advantage of the cortical architecture properties. This need is particularly acute because, in order to capture the sought-for phenomena and confirm their robustness, many simulations (up to 10^3–10^4) are necessary to sufficiently explore considerable portions of the parameter space.

Overview of Results We have been implementing a parsimonious modeling strategy in the modeling of the primary visual cortex (V1). Using this strategy, we have investigated network mechanisms underlying cortical phenomena including spatiotemporal patterns of spontaneous cortical activity [1], cortical activity patterns induced by the Hikosaka line-motion illusion stimulus paradigm [2], and neuronal orientation tuning [3]. Our large-scale computational model contains $\sim 10^6$ conductance-based, I&F, point neurons, and the simulations involve a range of spatiotemporal scales pertaining to cortical processing in V1. The time scales involved span from the fastest network time scale ($\sim$2 ms) [8, 9] to the slow scale of the NMDA receptor ($\sim$50−200 ms) [10]. The spatial scales involve both local, isotropic ($<$0.5 mm) and long-range, orientation-specific ($\sim$1–8 mm) lateral cortico-cortical connections in V1, distributed according to a realistic cortical architecture [11, 12]. We assume these cortico-cortical connections to be sparse.

We first addressed V1 dynamics in the absence of any stimuli [1], to isolate cortical phenomena that are independent of the LGN input dynamics. At this modeling stage, we reproduced spatiotemporal cortical patterns of spontaneous cortical activity [13, 14], a striking example of cortical processing in V1 observed by real-time optical imaging based on voltage sensitive dyes [15]. Far from being the expected featureless noise [16], these patterns are highly correlated on millimeter scales, appear to become activated in multiple areas of iso-orientation preference, and tend to migrate to nearby such areas after about ~80 ms. We identified a specific cortical operating state (see IDS state below) that we believe to underlie this and many other cortical response properties in V1.

To study stimulus-driven phenomena, we incorporated an LGN model. We have successfully modeled the spatiotemporal V1 neuronal activity that is associated with the Hikosaka line-motion illusion [2]. This illusion is induced by showing a small stationary square followed by a long stationary bar to create the illusory motion perception of the square "growing" to become the bar [17]. In V1, it was observed that actual subthreshold cortical activity in response to the Hikosaka stimulus is very similar to that induced by real moving stimuli [18]. Finally, we have also successfully modeled the orientation tuning dynamics of V1 neurons within the same computational model [3].

The parsimony of our models manifests itself in how we restrict our assumptions underlying the cortical phenomena we investigate. For example, for the study of neuronal orientation tuning, we choose to consider only small-sized stimuli, so that only the local network with its short-range cortico-cortical connections is activated. Thus, we were able to restrict our model to a local version with all the long-range effects lumped in an effective uniform inhibition. Both our large-scale model and this local model can successfully model the orientation tuning dynamics. Yet more markedly so, parsimony is reflected in the modeling of the line-motion illusion. In particular, from the experiments, it is not clear whether the spatiotemporal V1 activity associated with this illusion emerges in V1 or from cooperative effects with strong feedback from higher cortical areas. In the model neuronal network, we have only considered the dynamics in V1 and explicitly excluded any structured feedback from the higher cortical areas, yet this model was sufficient to capture the cortical activity corresponding to the line-motion illusion. This allows us to hypothesize that the cortical activity observed in voltage-sensitive-dye imaging is mainly produced by the V1 circuitry.

We have found a single dynamical regime of our model cortex that captures a number of experimental phenomena at once. In addition to the three mentioned above, these include membrane potential synchronization in nonspiking neurons several millimeters apart [19], and the "similarity index" dynamics caused by drifting grating stimuli turned on and off periodically [13], both described in section "Patterns of Spontaneous Cortical Activity." In all these cases, the theoretical advantage of our parsimonious model is clear: It highlights the most sharply delineated proposed mechanisms, while simultaneously demonstrating their realizability in a single cortical state in our model.

We refer to the dynamical regime in our model that can capture all the phenomena mentioned above as the "intermittent de-suppressed" (IDS) operating state [1, 2]. This IDS state is characterized by (1) high total conductance (cf. [9, 20–23]), (2) strong inhibition, (3) large synaptic fluctuations (cf. [24–26]), (4) an important role of NMDA conductance in the orientation-specific, long-range interactions (cf. [27]), and (5) a high degree of correlation between the neuronal membrane potentials, NMDA-type conductances, and firing rates. It is from this type of operating state that we believe V1 responds to changes in visual stimulation. The IDS state appears to be highly stochastic and sensitive to external stimuli. It is not an attractor in the sense of, say, the marginal state, which is very stable, strongly locked-in, and largely insensitive to external stimuli [28, 29].

Computational Approach The dynamical properties of the IDS operating state, together with the cortical architecture of (1) strong isotropic, nonspecific local connections and (2) weak, orientation-specific, long-range cortico-cortical connections, guide us in the design of our computational scheme. In addition, as in the corresponding experiments, the nature of observables is statistical, so the appropriate computational aim is to achieve statistical rather than trajectory-wise accuracy of our simulations. In particular, to obtain efficiency using large time-steps we had to address the following computational issues: (1) stiffness due to the high-conductance state of the network in the IDS state, (2) correctly accounting for the influence of each cortical spike, computed within a large numerical time-step, on the dynamical variables and other cortical spikes within the same time-step, and (3) efficiently accounting for the influence of strong, spatially local interactions and relatively weak, modulational, long-range interactions arising in the V1 cortical architecture.

To address these issues, we have developed a novel, highly effective and efficient computational method for evolving V1 dynamics [30], which allows us to minimize the computational overhead associated with each spike, and evolve the I&F network model of V1 with N neurons so that each neuron fires approximately once in $\mathcal{O}(N)$ operations. Our method not only evolves the system with trajectory-wise accuracy when the time steps used are sufficiently small, but also evolves the system with statistical accuracy when the time steps in the simulation are quite large, 1–2 ms, approaching the smallest physiologically relevant time scale in the model. More precisely, such simulations still render accurate network firing rates; distributions of interspike intervals, conductances, and voltages; and spatiotemporal patterns formed by conductances and voltages. Finally, we stress that our computational strategy should be applicable to neuronal assemblies other than V1. In fact, the issues addressed by this strategy are fairly general, and our scheme may easily be adapted to simulate other areas of the brain.

The remainder of this chapter is organized as follows. In section "Physiological Background," we discuss some of the anatomical and physiological background that we use in our computational model. In section "The Large-Scale Computational Model," we describe the model and its mathematical and physiological components. The model V1 dynamics are presented in section "Dynamics of the Primary Visual

Cortex." In particular, spontaneous cortical activity patterns are discussed in section "Patterns of Spontaneous Cortical Activity," line-motion illusion in section "Line-Motion Illusion," and orientation tuning of V1 neurons in section "Orientaton Tuning." Structural robustness of the model is discussed in section "Discussion."

Physiological Background

The primary visual cortex (V1) is a thin sheath of densely packed and highly interconnected neuronal cells (neurons), located at the back of the skull. Along the "visual pathway," *Retina* $\rightarrow$ *LGN* $\rightarrow$ *V1* $\rightarrow$ *And Beyond*, it is in V1 where neuronal responses are first simultaneously selective to elementary features of visual scenes, including a pattern's orientation. For example, *orientation tuning* is the selective response of a single neuron to some orientations of a simple visual pattern (say a bar or grating), but not to other orientations [31].

The primary visual cortex is several cm^2 in lateral area and 1–2 mm in depth. It has a complex, layered substructure (layers 1, 2/3, $4B$, $4C\alpha$, $4C\beta$, 5, and 6, labeled from the cortical surface inwards). Each layer is anatomically distinct, containing excitatory and inhibitory neurons with dense lateral connectivity, augmented by specific feed-forward and feed-back projections between different layers. Visual input first arrives at V1 through (excitatory) axons from the Lateral Geniculate Nucleus (LGN) primarily into the layers $4C\alpha$ ("magno pathway") and β ("parvo pathway").

Neurons in V1 are roughly divided into "simple" and "complex" cells. This division dates back to [31]. The responses of simple cells to visual stimuli tend to be approximately linear, while those of complex cells tend to be nonlinear. For instance, if the stimulus is a drifting grating, the spiking rate of a simple cell will be modulated at the frequency with which the grating's peaks and troughs pass through the cell's receptive field; the spiking rate of a complex cell will increase with the onset of the stimulus, but then stay approximately constant in time for its duration. For a standing, contrast-reversing stimulus, simple-cell firing rates are sensitive to its spatial phase and modulate at the stimulus frequency, while complex-cells are spatial-phase insensitive and modulate at double the stimulus frequency [32, 33]. The theoretical model of [31] proposes that simple cells receive LGN input and pool their output to drive complex cells, with evidence for excitatory connections from simple to complex cells found in [34]. Phase sensitivity is lost in this pooling. However, most V1 neurons are neither completely simple nor completely complex [35]. Complex cell also receives strong input from other complex cells [34] and the LGN [36–38], not just from simple cells, and can be excited without strongly exciting simple cells [39–43]. Therefore, an alternative hypothesis is that the amount of excitatory LGN input varies from one V1 neuron to the next (indirect evidence for this is given in [44–46]), and is compensated by the amount of cortical excitation, so that each V1 neuron receives roughly the same amount of excitation [47], as suggested by cortical development theories [48,49] and experiments [50,51]. We adopt this hypothesis in our model, as described below. We note that simple cell properties were recovered in a model of V1 neurons that all received equal amount of LGN drive [52]. In this model network, strong cortical inhibition cancels the nonlinearity in the LGN drive to produce linear response properties of the simple cells.

Optical imaging experiments [53–55] reveal orientation preference as organized into millimeter-scale "orientation hypercolumns" that tessellate the cortical surface, with orientation preference laid out in spokes emanating from "pinwheel centers," with ocular dominance arranged in left-eye/right-eye stripes. Orientation preference selectivity appears to be well correlated even between single pairs of nearby cortical neurons, whereas preferred spatial phase does not, indicating the possibility that spatial phase preference may be mapped across V1 in a disordered fashion [56]. The exact nature of the spatial frequency preference distribution across V1 is still somewhat in dispute: Interpretations of experiments have ranged from domains with only high or low spatial frequency preference [57] to continuous pinwheel patterns [58], but appear to have converged on disordered distributions [59–61].

Anatomical, optical imaging, and electrophysiological studies suggest that lateral connectivity shows different types of organization on different spatial and temporal scales. At hypercolumn scales ($<500\,\mu$m), the pattern of connectivity appears isotropic, with monosynaptic inhibition at or below the range of excitation [62–65]. The excitatory short-range connections appear to be mostly mediated by the fast, AMPA, neurotransmitter [27] (with persistence time-scale $\sim$3 ms [66]), while the inhibitory connections are mediated by $GABA_A$ (with persistence time-scale $\sim$7 ms [66]). We use these facts in our model.

At longer scales, $\sim 1-5$ mm, the largely intralaminar and reciprocal lateral connections [67–72] (also referred to as horizontal connections) in V1 are much less isotropic. These horizontal connections arise purely from excitatory neurons, and terminate on both excitatory ($\sim$75%) and inhibitory ($\sim$25%) neurons [73–75]. They are only strong enough to elicit subthreshold responses in their postsynaptic neurons [76, 77], and are believed to only modulate their firing rates.

Long-range connections have patchy terminals in the superficial layers 1–3 [67–71]. They have been observed in the input layers 4B and upper 4Cα in primates, where they have bar-like terminals [69,71,72], and in layers 5 and 6 in primates and carnivores, where they tend to be more diffuse [69, 70, 78]. (They do not, however, seem to exist in the layer 4 of the tree shrew [79]). A clear tendency has been revealed for the patchy long-range projections of neurons in layer 2/3 to align with the bar-like projections of neurons in layer 4B that lie directly below them [72].

We here remark that, as described in section "Dynamics of Primary Visual Cortex," our model is used both as a model of the input layer 4Cα, and as an "effective" or "lumped" model which does not include the detailed laminar structure of V1. In the second case, we believe that it still renders an adequate description of the signal observed in the voltage-sensitive-dye-based imaging experiments whose results we address. This is because the signal in these experiments reflects bulk membrane potential variations in the imaged area, which, in turn, largely reflect subthreshold synaptic potentials and action potentials in all the dendrites that reach the superficial V1 layers, regardless of the layer in which their respective soma lies [80]. Therefore, in the model, we also use anatomical details of the horizontal connections that are more in line with those in the superficial layers.

Long-range projections have been found to connect sites of like preferences, such as orientation preference [81, 82], ocular dominance and cytochrome oxydase blobs [83], and direction preference [84]. The shapes of the cortical regions covered by horizontal projections of a given neuron differ from species to species, ranging from just barely elongated along the retinotopic axis in macaque [83] and new world monkeys [68] (anisotropy ratio ~1.5–1.8) to highly elongated in the tree shrew [67] (anisotropy ratio ~4).

In contrast to short-range connections, long-range connections in V1 appear to be mediated by both AMPA and NMDA. In particular, *in vitro* stimulation of white matter leads to the conclusion that firing by layer 3 pyramidal neurons may be driven and synchronized by long-range, horizontal connections, mediated in part by NMDA [27]. Additionally, long-range horizontal inputs to cells in layers 2 and 3 can sum nonlinearly [77], which is indicative of NMDA receptor involvement in long-range connections due to the voltage-dependent conductance of the NMDA channel [85]. Moreover, visual response in the superficial V1 layers 1–3 was observed to be in part mediated by NMDA receptors, both in cats [86] and the macaque [87]. This evidence should give credence to the claim that both AMPA and NMDA mediate synaptic transmission through long-range horizontal connections in V1.

The precise role of the long-range horizontal connections in V1 is as yet unknown, however, it appears that they contribute to spatial summation of stimuli and contextual effects from outside of a given neuron's classical receptive field [69, 88]. They may also contribute to synchronous firing of cells with similar orientation preferences, especially when those cells are separated by more than 0.4 mm [89, 90] (see also section "Patterns of Spontaneous Cortical Activity," especially Fig. 5), and the synchronization of fast, γ-band (25–90 Hz), oscillations present in the collective firing rates of neuronal populations over distances of ~5 mm [91, 92]. Simulations using our model [1, 2] suggest that particularly striking examples of the long-range connection contributions may be in millimeter-scale spatiotemporal patterns of spontaneous cortical activity [13, 14] and activity induced by the Hikosaka-motion-illusion stimulus [18], which have been observed in experiments using voltage-sensitive dyes (see sections "Patterns of Spontaneous Cortical Activity" and "Line-Motion Illusions").

In addition to the diverse spatial scales, the neuronal network in V1 also operates within a large range of temporal scales. The manifestations of selectivities such as orientation tuning are actually strongly dynamical, as revealed by reverse-time correlation experiments [93, 94], which reveal some of the time-scales operative in V1 cortical processing. These are: the LGN response time $\tau_{\text{lgn}} = \mathcal{O}(10^2)$ ms, reflecting the composition of retinal and genicular processing of visual stimulation; the various time-scales of synaptically mediated currents $\tau_{\text{syn}} = \mathcal{O}(3 - 200\,\text{ms})$, as described in the introduction; and $\tau_G = C/[G]$, where C is cellular capacitance and $[G]$ a characteristic size of total synaptic conductances. Recall that τ_G is the time-scale of response of a neuron, and is a property of network activity. The higher the activity, the shorter is τ_G, usually about $\mathcal{O}(2-5\,\text{ms})$. Intracellular measurements have shown that under visual stimulation, cellular conductances can become large, increasing by factors of two or three, and tend to be dominated by (cortico-cortical) inhibition [9, 20–23].

Some prior theoretical studies of cortical effects induced by short- and long-range horizontal connections in V1 include (but are by no means limited to) the following: An I&F computational model with an idealized architecture was studied in [95]. The largely analytical studies of [96–98] address the role of long-range connections, studying stationary cortical pattern formation and stability. The role of recurrent excitation in a network model was studied in [99]. A large-scale computational model of neuronal orientation tuning in V1 was presented in [100]. A detailed large-scale, highly realistic, local computational model of neurons in four orientation hypercolumns in the input layer $4C\alpha$ of macaque V1 [8, 52, 101], which included only short-range connections. Orientation selectivity of cells in this model was shown to be greatly enhanced by recurrent interactions [101]. In [47], the model was extended to include heterogeneity in LGN input.

The Large-Scale Computational Model

We model a patch ($\sim$25 mm^2) of primary visual cortex by a large network of $N \sim 5\times 10^5$ coupled, excitatory and inhibitory, simple and complex, integrate-and-fire (I&F) point neurons distributed uniformly over a two-dimensional lattice. Of these neurons, $\sim$75% are excitatory and $\sim$25% inhibitory. They are labeled by the index $i = (i_1, i_2)$, corresponding to their positions in the lattice $\mathbf{x}_i$. Their type $\mathcal{L}_i \in \{E, I\}$ (excitatory, inhibitory) is assigned randomly with the probability corresponding to its respective percentage in the population.

The intracellular membrane potentials of the neurons, $V_i(t)$, are driven by the changes in the conductances, $G_i^Q(t)$. Together, these variables evolve according to the set of equations

$$\frac{\mathrm{d}}{\mathrm{d}t}V_i(t) = -G^{\mathrm{L}}\left[V_i(t) - \varepsilon^{\mathrm{L}}\right] - \sum_Q G_i^Q(t)\left[V_i(t) - \varepsilon^Q\right], \tag{1a}$$

$$\frac{\mathrm{d}}{\mathrm{d}t}G_i^Q(t) = -\frac{G_i^Q(t)}{\sigma^Q} + \sum_j \sum_k S_{i,j}^Q \delta(t - T_{j,k}) + \sum_k F_i^Q \delta(t - T_{i,k}^F), \tag{1b}$$

except at the kth spike time, $T_{i,k}$, of the the ith neuron, which occurs when the membrane potential $V_i(t)$ reaches the firing threshold $V_i = \varepsilon^{\mathrm{T}}$. The spike time $T_{i,k}$ is recorded and the potential V_i reset to ε^{R}, where it is held for an absolute refractory period of τ_{ref} ms.

In (1a), G^{L} and ε^{L} are the leakage conductance and potential, respectively. The index Q in (1a) runs over the types of conductances used, which are characterized by their different decay time scales σ^Q and reversal potentials ε^Q. In our model, we consider three conductance types $Q \in \{\text{AMPA}, \text{NMDA}, \text{GABA}_A\}$, the first two of which are excitatory and the last inhibitory. Each spike from the jth neuron gives rise to a jump of magnitude $S_{i,j}^Q$ in the Q-type conductance of the ith neuron. By setting the coupling strengths $S_{i,j}^{\text{AMPA}}$ and $S_{i,j}^{\text{NMDA}}$ to zero whenever $\mathcal{L}_j = I$, and similarly setting $S_{i,j}^{\text{GABA}_A}$ to zero whenever $\mathcal{L}_j = E$, we achieve that the excitatory conductances G^{AMPA} and G^{NMDA} of any model neuron only jump when that neuron receives a spike from an excitatory neuron within the network, while its inhibitory conductance G^{GABA_A} only jumps when it receives a spike from an inhibitory neuron. Note that, the coupling strengths $S_{i,j}^Q$ can be chosen so as to encode many different types of network architecture. The system is also driven by external input from the LGN. Each external input spike makes that neuron's Q-type conductance jump by magnitude F_i^Q. We will further discuss the external input below.

In our model, we use reduced-dimensional units [101], in which only time retains dimension, with units of conductance being [ms^{-1}]. Typically, in these units, we set the conductance time-scales $\sigma^{\text{AMPA}} = 2$ ms, $\sigma^{\text{NMDA}} = 80$ ms, $\sigma^{\text{GABA}_A} = 7$ ms, the reversal potentials values $\varepsilon^{\text{L}} = 0$, $\varepsilon^{\text{AMPA}} = \varepsilon^{\text{NMDA}} = 14/3$, $\varepsilon^{\text{GABA}_A} = -2/3$, the threshold and reset voltages $\varepsilon^{\text{T}} = 1$ and $\varepsilon^{\text{R}} = 0$, the refractory period τ_{ref}=2 ms, and the leakage conductance $G^{\text{L}} = 0.05$. The voltage constants correspond to the physiological values $\varepsilon^{\text{L}} = -70$ mV, $\varepsilon^{\text{AMPA}} = \varepsilon^{\text{NMDA}} = 0$ mV, $\varepsilon^{\text{GABA}_A} = -80$ mV, $\varepsilon^{\text{T}} = -55$ mV. The physiological leakage conductance is $G^{\text{L}} = 50 \times 10^{-6}\,\Omega^{-1}\,\text{cm}^{-2}$.

According to (1b), the rise of each conductance G_i^Q upon receiving a spike is instantaneous, while its decay takes place on the time-scale σ^Q. However, our treatment can be readily extended to conductances in the form of an α-function with both rise and decay time-scales. In this case, we typically use the rise time-scale 0.05 ms for the AMPA and GABA$_A$ conductances, and 0.5 ms for the NMDA conductance.

Short-Range Connections Each model neuron is isotropically, randomly connected to other nearby neurons, with interaction strengths $S_{i,j}^{\text{SR},Q} = f_{Q,\mathcal{L}_j}\Delta_{i,j}\bar{S}_{\mathcal{L}_i}^{\text{SR},Q}$ $K^{\text{SR},Q}(|\mathbf{x}_i - \mathbf{x}_j|)$. The normalized spatial kernel $K^{\text{SR},Q}(\mathbf{r})$ is chosen to be the Gaussian

$$K^{\text{SR},Q}(\mathbf{r}) = \frac{1}{\pi(D^Q)^2}\exp\left[-\frac{|\mathbf{r}|^2}{(D^Q)^2}\right], \tag{2}$$

which decays on the spatial scale $D^Q \sim 0.3$ mm. In addition to being isotropic, this choice of the coupling kernel also makes the modeled short-range connections nonspecific in neurons' phase preference. The coefficients $f_{Q,\mathcal{L}_j}$ are chosen to reflect the fact that a spiking excitatory neuron can only increase excitatory conductances, and a spiking inhibitory neuron can only increase inhibitory conductances. The random connectivity matrix $\Delta_{i,j}$ indicates whether neuron j is connected to neuron i. The maximum strength $\bar{S}_{\mathcal{L}_i}^{\text{SR},Q}$ only depends on the type of conductance and the type of the postsynaptic neuron.

Long-Range Connections Orientation specific, long-range (LR) connections are anisotropic, excitatory, and project onto both excitatory and inhibitory cells. Their coupling coefficients are given by $S_{i,j}^{\text{LR},Q} = \Delta_{i,j}\bar{S}_{\mathcal{L}_i}^{\text{LR},Q} K^{\text{LR},Q}(\mathbf{x}_i, \theta_i, \mathbf{x}_j, \theta_j)$, with maximum coupling strengths $\bar{S}_{\mathcal{L}_i}^{\text{LR},Q}$ and long-range coupling kernels $K^{\text{LR},Q}$. We require that $\bar{S}_{\mathcal{L}_i}^{\text{LR,GABA}_A} = 0$, reflecting the purely excitatory nature of the connections. The long-range kernel $K^{\text{LR},Q}(\mathbf{x}_i, \theta_i, \mathbf{x}_j, \theta_j)$ connects neurons in different pinwheels if the difference in their preferred orientations $\theta_i - \theta_j \leq \Delta\theta \approx \pi/16$. It has a two-dimensional Gaussian shape on the spatial scale $D_{\text{LR}} \sim 1.5$ mm, with eccentricity 1~2 [11]. In addition, we take the coupling strengths of the form $\bar{S}_{\mathcal{L}_i}^{\text{LR,AMPA}} = \Lambda \bar{S}_{\mathcal{L}_i}^{\text{LR}}$ and $\bar{S}_{\mathcal{L}_i}^{\text{LR,NMDA}} = (1-\Lambda)\bar{S}_{\mathcal{L}_i}^{\text{LR}}$, where Λ denotes the percentage of NMDA receptor contribution to the total LR conductance [87, 102, 103].

Altogether, the coupling coefficients in (1a, 1b) are the sum of the short- and long-range coupling coefficients, $S_{i,j}^{Q} = S_{i,j}^{\text{SR},Q} + S_{i,j}^{\text{LR},Q}$.

Modeling the LGN Input and Background Noise The drive to our model cortex is provided by model background noise and a model LGN (mLGN). The background and mLGN spike trains arriving at a V1 neuron are modeled as independent Poisson trains, with the corresponding spike times denoted by $T_{i,k}^{\text{B}}$ and $T_{i,k}^{\text{LGN}}$, respectively. We assume the background to be homogeneous, firing with the uniform rate R_{B}. We choose R_{B} so that the spontaneous firing rate of the model V1 neurons due to both the model background and the mLGN spontaneous firing is $\lesssim$5 spikes per second.

We assume that our rectangular model cortical patch corresponds to a rectangular patch Ω of the visual space. (In our largest model so far, $\Omega = 9° \times 6°$.) To the ith neuron, located at the lattice point $\mathbf{x}_i$ in the cortical patch, we assign a corresponding lattice point $\mathbf{y}_i$ in Ω. We take the correspondence $\mathbf{x}_i \leftrightarrow \mathbf{y}_i$ to be linear, which ignores detailed retinotopic effects. This is reasonable provided the solid angle subtended by Ω is sufficiently small. The ith neuron's receptive field center $\mathbf{r}_i$ is chosen at a point scattered randomly within the solid angle $\alpha \lesssim 1°$ from $\mathbf{y}_i$.

In cat and macaque V1, both orientation preference and spatial phase preference are conferred on cortical cells from the convergence of output from many LGN cells [104]. In the simplest model, the details of these LGN cells are ignored, and for each V1 neuron only the center and size of its receptive field, and its feature preferences, are considered. Such a model describes the mLGN input rate to the ith neuron in the linear spatiotemporal convolution form

$$R_i^{\text{LGN}}(t) = \bar{R}_{\text{LGN}} + \int_0^\infty \mathrm{d}\tau \int_\Omega \mathrm{d}\boldsymbol{\rho}\, I(\mathbf{r}_i - \boldsymbol{\rho}, t - \tau)\, K^{\theta_i,\phi_i,\omega_i}(\boldsymbol{\rho})\, K^{\text{T}}(\tau), \qquad (3)$$

where $I(\mathbf{r}, t)$ is the visual stimulus in Ω, and θ_i, ϕ_i, and ω_i are this neuron's preferred orientation, spatial phase, and spatial frequency, respectively. We choose the spontaneous rate $\bar{R}_{\text{LGN}} = 170$ spikes/s. The Gabor spatial kernel $K^{\theta,\phi,\omega}(\boldsymbol{\rho}) = K^{rf}(\boldsymbol{\rho})\cos\left[\omega(\rho_x \cos\theta + \rho_y \sin\theta) + \phi\right] - \bar{K}^{\theta,\phi,\omega}(\boldsymbol{\rho})$ has a Gaussian spatial envelope K^{rf} with receptive field size $\sigma_{rf} \approx 1°$, as given by (2). The time kernel is given by $K^{\text{T}}(\tau) = \bar{K}^{\text{T}} t \left(\tau_r^{-2} \mathrm{e}^{-t/\tau_r} - \tau_f^{-2} \mathrm{e}^{-t/\tau_f}\right)$, with rise and decay times $\tau_r = 10 \sim 16$ ms, $\tau_f = 40 \sim 64$ ms [105]. The radial kernel $\bar{K}^{\theta,\phi,\omega}(\boldsymbol{\rho})$ and the constant $\bar{K}^{\text{T}}$ are chosen so that $K^{\theta,\phi,\omega}$ has 0 mean and K^{T} has a maximum of 340 spikes/s. The LGN input model (3) is valid provided the stimulus contrast $\max(I) - \min(I)$ is sufficiently small in our simulation that the firing rate R_i^{LGN} never drops below zero. A more detailed LGN model, valid for larger contrast values, is described in [47].

From neuron to neuron, the preferred orientation is laid out in pinwheel patterns [53–55, 106–108], the preferred phase is randomly chosen with uniform distribution in $[-\pi, \pi]$ [56], and the preferred frequency is randomly chosen with the average $\omega \approx 1$ cycle per degree [59–61].

As only the AMPA conductances receive feedforward input from the LGN, we put the strengths $F_{i,\mathrm{LGN}}^{\mathrm{NMDA}} = F_{i,\mathrm{LGN}}^{\mathrm{GABA}_A} = 0$. The strengths $F_{i,B}^{Q}$ and $F_{i,\mathrm{LGN}}^{\mathrm{AMPA}}$ of the background and LGN spikes are chosen to reflect the fact that, in addition to cortico-cortical couplings, simple cells are driven by the mLGN and complex neurons are driven by stimulus-independent background, with a continuum of "mixed type" cells in-between [47, 49–51].

Computational Method As mentioned in the introduction, we have developed a new, highly effective and efficient computational method for solving (1a and 1b). This method has the following properties: (1) We employ an integrating factor exploiting the near-slaving of the neuronal membrane potential to the effective reversal potential in a high-conductance state, and write the solution in the form of a numerically tractable integral equation. In this way, we can take large time-steps even in the high-conductance state in which the I&F equations are stiff. (2) Within each time-step, we sort the approximated spike-times and apply an iterated correction procedure to account for the effects of each spike on all future spikes within this time-step. We can thus account for the spike–spike interactions within a large time-step of duration 1–2 ms. (This is very different from spike-time interpolation of [109, 110].) (3) We divide the network up into local clusters of approximately the same size as the spatial scale of the local interactions, and treat these clusters in such a way as to minimize the computational work associated with computing the cortico-cortical interactions. The details of our computational method are described in [30].

Dynamics of the Primary Visual Cortex

In our large-scale model, by adjusting the strengths of the local and long-range cortico-cortical connections, we identify an "intermittent de-suppressed" (IDS) cortical operating state, characterized by (1) high conductance, (2) strong inhibition, (3) large fluctuations, and (4) strong correlations among the neuronal membrane potentials, NMDA conductances, and firing rates. Fluctuations in the IDS state arise from intermittent firing events which are strongly correlated in time and in orientation domains, and whose correlation time is controlled by the decay time-scale of the NMDA conductance. In addition, this state is just below a "fluctuation controlled criticality," i.e., a bifurcation point above which there is hysteresis and bistability in the firing dynamics of complex cells.

We emphasize that the significance of this single IDS state manifests itself in the fact that the IDS state captures the dynamics of (1) large-scale, coarse-grained spatiotemporal activity patterns occurring in the spontaneous as well as periodically stimulated states in V1, as observed in the voltage-sensitive-dye imaging experiments [13, 14], (2) the spatiotemporal activity in V1 associated with the Hikosaka line-motion illusion [18], and (3) orientation tuning of V1 neurons.

Patterns of Spontaneous Cortical Activity

The cortical operating state in the absence of external stimuli still undergoes reorganization of information and so is expected to undergo rich spontaneous dynamics. This is indeed the case in V1, where experiments of [13, 14] on anaesthetized cats show that such activity forms highly structured and correlated coherent patterns on the scales of several millimeters, which appear in regions of like orientation preference over many orientation hypercolumns and persist over time scales of $\sim$80 ms. In our working hypothesis, this persistence time scale and the spatial-correlation structures implicate the combined effect of the orientation-specific, long-range cortico-cortical connections and the NMDA conductance time scale. In our view, these spontaneous activity patterns reflect the rich dynamics of cortical operating states and any theoretical V1 model must be constrained to successfully reproduce this spatiotemporal activity. To study this spontaneous cortical activity without the modeling complications of the LGN, and to examine our working hypothesis, we employ a network comprised of $\sim 5\times 10^5$ model neurons within a $\sim 16\,\text{mm}^2$ patch of 64 orientation hypercolumns, in which the long-range connections are orientation specific and the time-scale associated with these connections is of the NMDA type [1].

Figure 1a shows the orientation preference map of V1 neurons, conferred on them in our model by their afferent inputs from the LGN. Figure 1b, c shows two instantaneous patterns of the neuronal membrane potentials $V(\mathbf{x}, t)$, which are strongly correlated over like-orientation domains. Following [13, 14], we define two space-dependent measures of the model cortical state. The first is the *"preferred cortical state"* of a neuron, which is defined to be the spatial pattern of the average voltage $V_{\mathrm{P}}(\mathbf{x}; \theta_d) = \sum_i V\left(\mathbf{x}, t^i; \theta_d\right)/N$, evoked by a strong stimulus at this neuron's optimal orientation θ_d. Here, t^i is this neuron's ith spiking time and N is its total number of spikes. The second measure, for the same neuron, is the *"spike-triggered spontaneous activity pattern,"* $V_{\mathrm{st}}(\mathbf{x}) = \sum_j V\left(\mathbf{x}, t^j\right)/M$ of the network without stimulus, again triggered by and averaged over this neuron's spike times. At our IDS operating state, achieved with moderate strengths of long-range lateral connections, $V_{\mathrm{P}}(\mathbf{x}; \theta_d)$ and $V_{\mathrm{st}}(\mathbf{x})$ computed in our model are strongly correlated with one another, as clearly shown in Fig. 2a, d. In addition, from Fig. 1a, it can be seen that they also strongly resemble iso-orientation domains within the neuronal orientation preference map.

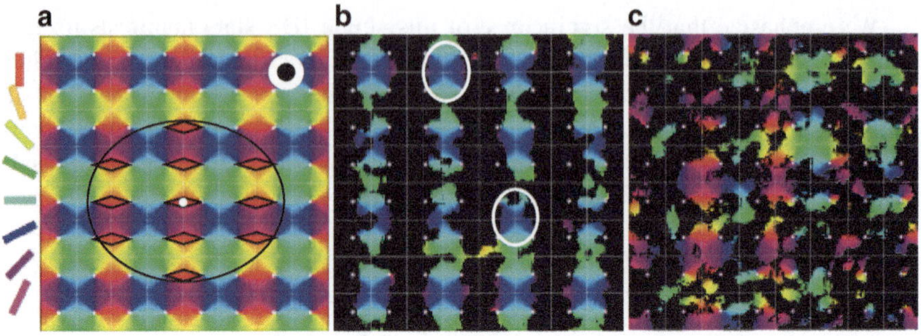

Fig. 1 (**a**) Model V1 cortical area with orientation hypercolumns. Preferred orientation is coded by color. Pinwheel centers are marked by small white dots. The lengthscale of local inhibitory/excitatory couplings is indicated by the inner/outer radius of the black/white annulus in the upper right corner. The extent of long-range connections is indicated by the *large ellipse*. Orientation domains that are coupled by the long-range connections to the neurons in the orientation domain in the middle of the ellipse are indicated by *rhombuses*. (**b**) Spontaneous activity pattern covering orientation domains of near-horizontal angles. The *two ovals* cover regions where voltages can be highly correlated in time. (**c**) Spontaneous activity pattern covering orientation domains of two orthogonal angles. [Reproduced with permission from [1], http://www.pnas.org (Copyright 2005, National Academy of Sciences, USA).].

The *"similarity index"* $\rho(\theta_d;t)$ is the spatial correlation coefficient between the preferred cortical state $V_P(\mathbf{x};\theta_d)$ and the membrane potential $V(\mathbf{x},t)$ of the neurons in the network. In particular,

$$\rho(\theta_d;t) = \frac{\int_{\mathbf{x}}\left[V_P(\mathbf{x};\theta_d) - \int_{\mathbf{y}} V_P(\mathbf{y};\theta_d)\,d\mathbf{y}\right]\left[V(\mathbf{x},t) - \int_{\mathbf{y}} V(\mathbf{y},t)d\mathbf{y}\right]d\mathbf{x}}{\sqrt{\int_{\mathbf{x}}\left[V_P(\mathbf{x};\theta_d) - \int_{\mathbf{y}} V_P(\mathbf{y};\theta_d)\,d\mathbf{y}\right]^2 d\mathbf{x}}\sqrt{\int_{\mathbf{x}}\left[V(\mathbf{x},t) - \int_{\mathbf{y}} V(\mathbf{y},t)d\mathbf{y}\right]^2 d\mathbf{x}}},$$

computed over an area of 4×4 pinwheels, which was roughly the area used in experiments. We use $\rho(\theta_d;t)$ to detect the evolution and persistence time scale of the cortical activity patterns in the IDS state. The time evolution of $\rho(\theta_d;t)$ in the three computed regimes (two of which are discussed below) is presented in Fig. 2g, i, k, and its time trace at $\theta_d = -60°$ in Fig. 2h, j, l. In the IDS regime, the evolution presented in Fig. 2g, h indicates the typical pattern duration, before it switches to other orientations, to be $\sim$ 80 ms. This scale can also be gleaned from the voltage and the similarity index temporal auto-correlation functions, as shown in Fig. 2b, c.

To further quantify our data in comparison with experiment, we compute the all-time similarity-index histogram by binning the similarity index $\rho(\theta_d;t)$, for all θ_d, sampled at the rate $\nu_S = 1$ frame/ms for a total duration $T = 256$ s. We also compute the spike-triggered similarity-index histogram by binning $\rho(\theta_d;t^i)$, sampled only on the spike times t^i of neurons that have the same orientation preference $\theta_{\text{op}} = \theta_d$, for the same duration $T = 256$ s, and average over all θ_d. The all-time and spike-triggered similarity index histograms for the IDS state are shown in Fig. 3d, e. The IDS firing rate as a function of the similarity index, computed as the ratio

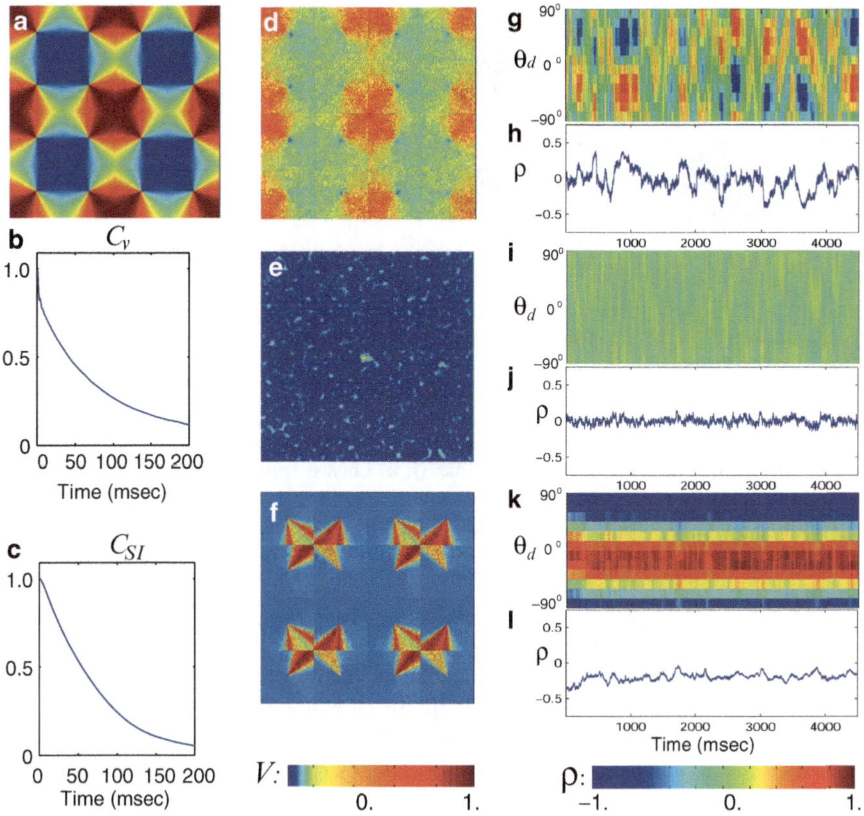

Fig. 2 (**a**) Preferred cortical state $V_P(\mathbf{x}; \theta_d)$ of the neuron in the center of the square. (**b, c**) Temporal autocorrelations of the membrane potential trace V and similarity index ρ in the IDS state, averaged over the cortical location and θ_d. (**d–l**) Spike-triggered activity pattern $V_{st}(\mathbf{x})$ for the neuron in (A) and time-evolution of the similarity index over all preferred orientations and for orientation preference $-60°$: (**d, g, h**) IDS, (**e, i, j**) uniform, and (**f, k, l**) locked state. [Reproduced with permission from [1], http://www.pnas.org (Copyright 2005, National Academy of Sciences, USA).].

between the distributions in panels E and D of Fig. 3, multiplied by the rate ν_S, is shown in Fig. 3f. Clearly, there is very good agreement with the corresponding experimental results of [13], as reproduced in Fig. 3a–c.

Computationally, we analyzed two more possible operating states of our model cortex. If the long-range connections are sufficiently weak, a neuron's spike-triggered spontaneous activity pattern, $V_{st}(\mathbf{x})$, becomes spatially homogeneous, as shown in Fig. 2e, i, j and Fig. 3g, h, i. This state closely resembles the "homogeneous phase" state investigated in an idealized mean-field model in [29]. If the long-range connections are very strong, $V_{st}(\mathbf{x})$ becomes locked to a specific pattern of orientations that tends not to correlate with the neuron's preferred cortical state $V_P(\mathbf{x}; \theta_d)$ for most of the simulation time, as shown in Fig. 2f, k, l and Fig. 3j, k.

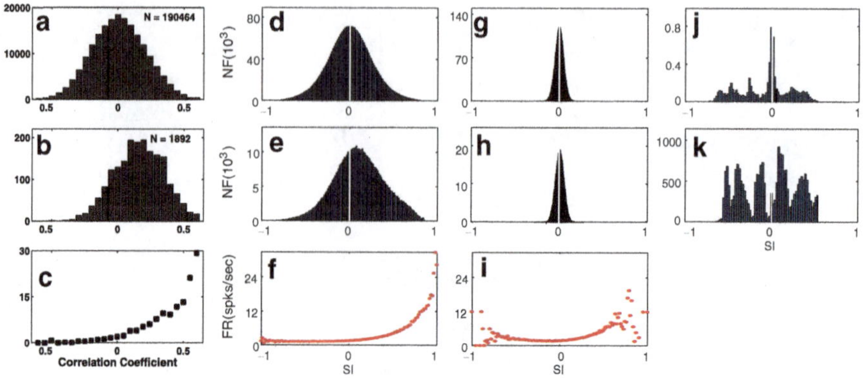

Fig. 3 *Top row*: all-time similarity-index histogram. *Middle row*: spike-triggered similarity-index histogram. *Bottom row*: firing rate as a function of the similarity index. (**a–c**) Experimental data. (**d–f**) IDS regime. (**g–i**) Homogeneous regime. (**j, k**) Locked regime. [Reproduced with permission from [13] (Copyright 1999, by the American Association for the Advancement of Science).] [Reproduced with permission from [1], http://www.pnas.org (Copyright 2005, National Academy of Sciences, USA).]

This state is similar to the "marginal phase" state in [29]. In neither of these last two cases is the spike-triggered spontaneous activity pattern correlated with the preferred cortical state of the neuron in question, which is at variance with [14], nor does the similarity index behavior capture experimental observations in [13].

All three of the analyzed model cortical operating states, in particular IDS, are structurally robust. The robustness of IDS is further discussed in section "Discussion."

Fixing the strengths of the long-range connections so that the model network without any stimulus is in the IDS state, we now drive the model with a drifting grating stimulus of given orientation θ_d, which is periodically turned on for 1 s and then off for 2 s. The activity patterns triggered on the spikes of a neuron with the preferred orientation θ_d are again similar to this neuron's preferred cortical state, $V_P(\mathbf{x};\theta_d)$. The time evolution of the similarity index $\rho(\theta_d;t)$ corresponding to the orientation θ_d is shown in the right panel of Fig. 4. When the stimulus is turned on, $\rho(\theta_d;t)$ tends to rapidly swing into positive values. This reflects the activity pattern strongly correlating with $V_P(\mathbf{x};\theta_d)$ for the evoked duration. When the stimulus is turned off, $\rho(\theta_d;t)$ tends to rapidly swing into negative values, and then slowly return closer to zero. These features, including the "negative-swing" phenomenon, are in good qualitative agreement with those observed experimentally in [13], reproduced in the left panel of Fig. 4.

Our analysis of the IDS state in our model V1 reveals that there is a chain of strong correlations among the voltages, conductances, and firing rates in the system. We find that the network operates in a state of high total conductance, $G^{\mathrm{T}} \gg G^{\mathrm{L}}$, which implies that the membrane potential $V(\mathbf{x},t)$ is slaved to the effective reversal potential $V^S(\mathbf{x},t)$ over the leakage-conductance time scale $(G^{\mathrm{L}})^{-1} \sim 10\,\mathrm{ms}$. (The total conductance is defined as $G^{\mathrm{T}} = \sum_Q G^Q$, and the effective reversal potential

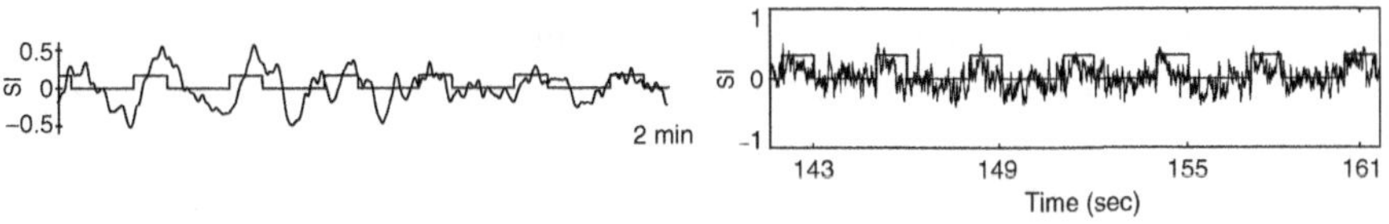

Fig. 4 Temporal dynamics of the similarity index $\rho(\theta_d; t)$. *Left*: Experiment. *Right*: IDS state simulation. [Reproduced with permission from [13] (Copyright 1999, by the American Association for the Advancement of Science).] [Reproduced with permission from Ref. [1], http://www.pnas.org (Copyright 2005, National Academy of Sciences, USA).]

as $V^S = \sum_Q G^Q \varepsilon^Q / G^T$, with $Q = L, I$, and E.) Our computations show that both these potentials stay well below the firing threshold on average, making the model operation "fluctuation-driven." The inhibitory conductances are much higher than the excitatory conductances, $G^I \gg G^E$, and the NMDA conductances are much higher than the AMPA conductances. Using the above conductance relations and definition of the effective reversal potential, it is easy to show that there is a chain of statistical correlations $V(\mathbf{x}, t) \sim V^S(\mathbf{x}, t) \sim G^{NMDA}/G^I$. Furthermore, our IDS state exhibits a correlation between G^{NMDA} and G^I in a sublinear fashion, i.e, a smaller increase in G^I is correlated with a larger increase in G^{NMDA}. Therefore, we have G^{NMDA} correlated with $V(\mathbf{x}, t)$ and $V^S(\mathbf{x}, t)$. In addition, a strong correlation of the firing rate $m(t)$ with the membrane potential $V(\mathbf{x}, t)$ and the NMDA conductance G^{NMDA} is observed in the IDS state. The fact that these physiologically reasonable correlations are observed only in the IDS state strongly constrains the model's cortico-cortical coupling strengths. Finally, we mention that only in the IDS state is the subthreshold activity consistent with physiologically realistic 2–10 spikes/s spontaneous and 10–60 spikes/s evoked firing rates.

Dynamically, the IDS operating state is an intermittent cycle: A small number of spontaneous firings of excitatory neurons increases (typically over a timecourse of $\leq$5 ms) the NMDA conductances and so also the membrane potentials of neurons in domains with like orientation preference within $\sim$1 mm via the long-range connections. These conductances and potentials become nearly statistically synchronized. When the inhibition in a region undergoes a transient drop, (i.e., the network becomes de-suppressed), a spontaneous firing of a single excitatory neuron will recruit many other neurons to fire. These correlated firing events occur within $\approx$10 ms. These patterns are then spread to other iso-orientation domains of like preference via a cascade of successive firing events. As a result, the induced spatial patterns of the voltage closely resemble the iso-orientation domains in the orientation preference map and the "preferred cortical state." Shortly thereafter, these excitatory recruitment events trigger strong inhibition mediated by local connections, which suppresses any further recruitment. The pattern then slowly drifts or decays on the NMDA conductance decay scale $\sigma^{NMDA} \sim 80$ ms, and the inhibition decays with it. Thus, a key mechanism underlying the spontaneous cortical activity patterns in the IDS operating state is this rapid correlated recruitment of excitatory neurons in iso-orientation domains.

To illustrate these features of the IDS, we return to Fig. 1b, c, which show two instantaneous patterns of spontaneous cortical activity at our IDS operating state. In both figures, instantaneous neuronal membrane potentials $V(\mathbf{x},t)$ are plotted, with low activity regions masked off by black color. The regions of high activity shown in Fig. 1b cover iso-orientation domains corresponding to near-horizontal angles. As discussed in the previous paragraph, even relatively distant parts of such high-activity regions become activated almost simultaneously, with the entire pattern then drifting slowly over the cortical surface, persisting for $\sim$80 ms. Multiangle patterns, such as the one shown in Fig. 1c, which corresponds to two orthogonal angles, can also occur. In this case, the activation regions corresponding to the two angles are mostly well separated, but parts of them may be present in the same orientation pinwheels. This pattern also persists over $\sim$80 ms, and is thus not a short transient. These drifting, multiangle patterns do not occur in the locked state described above and are not consistent with the theory of marginal phase [29].

Another manifestation of the above-described scenario and the role of the long-range connections in the IDS state is the near-synchronization of neuronal membrane potentials in pairs of neurons located within $\sim$0.5 mm from one-another in our model V1, as shown in the right panel of Fig. 5, which occurs even when the neurons in the pair are not spiking. This synchronization can be explained by the common synaptic inputs from long-range connections. These near-synchronized voltage traces capture well the corresponding experimental result [19], as shown in the left panel of Fig. 5.

The NMDA component present in the long-range cortico-cortical connections is crucial for capturing the spatiotemporal dynamics and persistence scale of the spontaneous cortical activity patterns. If there were no NMDA conductance, i.e., $\Lambda = 0$, the model cortex still can manifest an IDS-like state, but it exhibits a wrong persistence time scale for the patterns, that of only $\sim$20 ms. We note that the involvement of the NMDA conductances in the evolution of spontaneous activity patterns in the model cortex is consistent with the experimentally observed reduction of such activity in cats *in vivo* after blocking the NMDA receptors [86, 111].

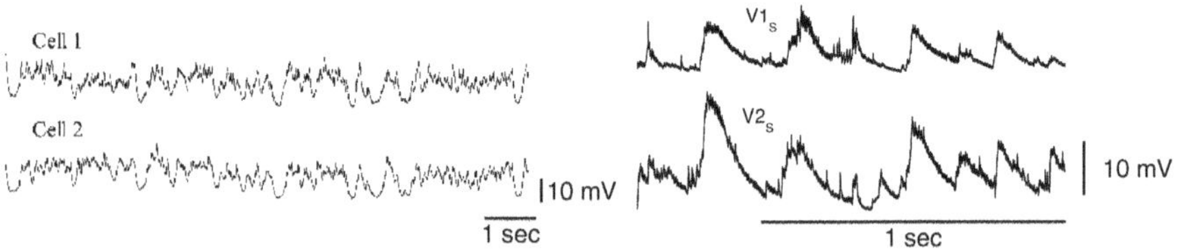

Fig. 5 Synchronized trace pairs of neuronal voltages. (The neurons in the pair are not firing.) *Left*: Experimental measurements in cat V1. *Right*: Simulation in the IDS state in our model V1. [Reprinted from [19], Copyright 1999, with permission from Elsevier.] [Reproduced with permission from [1], http://www.pnas.org (Copyright 2005, National Academy of Sciences, USA).]

Line-Motion Illusion

The stimulus paradigm that induces the Hikosaka line-motion illusion is the following: a cue of a small stationary square is flashed on a display for $\sim$50 ms, which is then followed by an adjacent stationary bar $\sim$10 ms after the removal of the square [17]. This stimulus creates the perception that the bar continuously grows out of the square, which is termed as the line-motion illusion. As revealed by optical imaging experiments using voltage-sensitive dye, this stimulus in fact creates in V1 a cortical activity pattern very similar to that created by a small, fast-moving square [18]. It was suggested that this similarity is the neurophysiological correlate associated with the pre-attentive perception of illusiory motion.

To investigate the mechanisms underlying the V1 activity patterns associated with the Hikosaka line-motion illusion, we examine the spatiotemporal cortical dynamics under the Hikosaka stimulus using our model cortex consisting of $\sim 10^6$ neurons, covering 96 orientation hypercolumns spanning an $\sim$6 mm $\times$ 4 mm area of V1.

In this modeling, we aim to examine the response properties of the IDS state, so, in the absence of external stimuli, we tune the model cortex to the IDS state [2]. We first discuss the cortical input sculpted by our model LGN, which is a spatiotemporal convolution of the external image. For the moving square and the Hikosaka stimulus, it is apparent that these cortical inputs from the model LGN bear little resemblance to one-another, as shown in Fig. 6.

We calibrate the strength of our model LGN using a flashed square as the stimulus. The input strength of the model LGN is the only additional parameter to the IDS

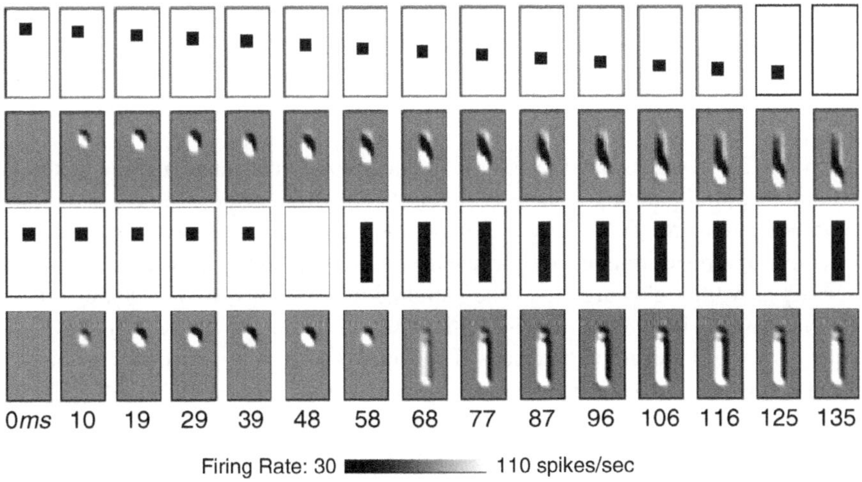

Fig. 6 Cortical input from the model LGN, which is a spatiotemporal convolution of the external stimulus. *Top*: the moving square. *Bottom*: the Hikosaka stimulus. *White rectangles*: stimulus. *Gray rectangles*: LGN input to V1. [Reproduced with permission from [2] http://www.pnas.org (Copyright 2005, National Academy of Sciences, USA).]

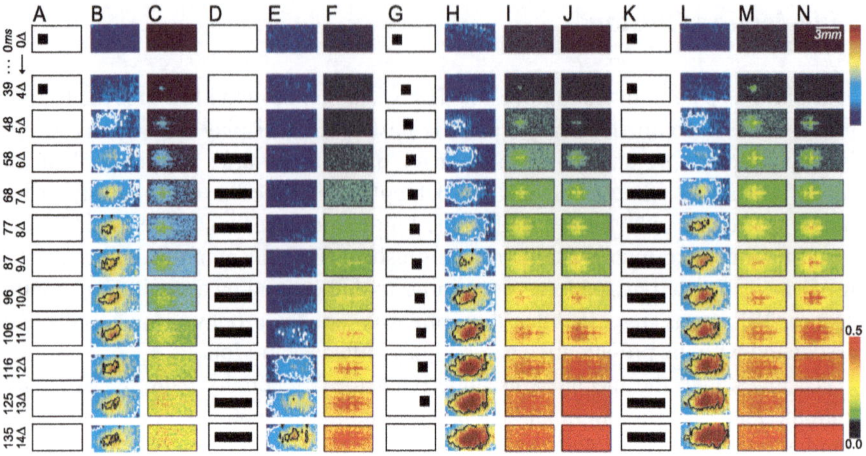

Fig. 7 Spatiotemporal pattern of the cortical activity corresponding to the (**a**) flashed and (**g**) moving square, (**d**) bar, and (**k**) Hikosaka stimulus. (**b, e, h, l**) Voltage-sensitive-dye signal measured in the experiments. (**c, f, i, m**) Effective reversal potentials in model V1. (**j, n**) NMDA conductances in model V1. [Adapted by permission from Macmillan Publishers Ltd. from [18], copyright 2004.] [Reproduced with permission from [2], http://www.pnas.org (Copyright 2005, National Academy of Sciences, USA).].

state and is chosen so as to ensure reasonable agreement between the spatiotemporal activity of our model V1 and the experimental signal in [18], as displayed in Fig. 7a–c. The square is flashed for 48 ms. The delayed surge of cortical activity evoked by the square appears at ~50 ms after the square is first presented, and eventually decays. After fixing the LGN strength, we next examine the cortical activity patterns evoked by a stationary bar. The delayed surge of activity due to the bar again appears ~50 ms after the presentation of the stimulus. The spatiotemporal dynamics exhibits good agreement between the real and model cortices, as shown in Fig. 7d–f.

In Fig. 7g–n, we display the activity due to the moving square and the Hikosaka stimuli. We find good agreement between the effective reversal potential V^S and NMDA conductance G^{NMDA} profiles in our model V1 and the experimental voltage-sensitive-dye measurement. In particular, the moving square produces an area of activity growing rightward out of the initial cortical response to the square, with an onset delay of about ~50 ms. This activity area fills out the cortical image of the square's path, and eventually decays. The Hikosaka stimulus – a square flashed for 48 ms followed 10 ms later by a bar – also produces an area of activity growing rightward out of the initial cortical response to the square. For both stimuli, there is a remarkable degree of correlation between the effective reversal potential V^S and NMDA conductance G^{NMDA} profiles. It should be stressed that such close correlation only occurs in the IDS operating state. Most importantly, both in the real and model cortex, the activity patterns corresponding to both stimuli are remarkably similar. As mentioned above, this similarity was suggested to be the neurophysiological correlate associated with the illusory motion perception [18].

What is the network mechanism underlying the cortical activity associated with the motion illusion? Our model V1 dynamics provides an intriguing scenario: In our V1 network, as soon as the square is flashed, it temporarily increases the model LGN input to the left side of the model cortex. Enough excitatory conductance builds up under this input during the 48 ms of the square's presentation that within ~50 ms the neurons in the impacted cortical area are recruited to begin spiking. Through the long-range connections, these spikes raise the NMDA conductances in a larger cortical region, reaching the middle and right of the model cortex. Due to the spatial decay of the long-range coupling strength, G^{NMDA} elevation is more pronounced closer to the trace of the square in the cortex. The elevated G^{NMDA} persists for about ~80 ms, and because of the $G^{\mathrm{NMDA}} - V^S$ correlation, the neuronal membrane potentials also rise in the same cortical area during the same time period. As the bar is shown on the screen and the model LGN input increases in the cortical footprint of the bar, the neurons more to the left have on average higher voltages and thus have higher probability to fire than those more to the right. (See [2] for details.)

Crucial in the scenario about the network mechanism underlying the spatiotemporal activity associated with the line-motion illusion is the abovementioned "priming" effect of the NMDA conductances raised through the long-range cortical connections of the model network in the IDS state. It is well known [76,77] (see also section "Physiological Background") that long-range connections are not strong enough to directly cause firing of their target neurons, but do increase activity through subthreshold modulation. This observation places a strong constraint on the long-range connection strength in our model cortex, since long-range connections that are too weak would not be able to properly "prime" the rest of the model cortex at all (see also below), and long-range connections that are too strong would cause direct neuronal spiking in large parts of the model cortex instead of priming the cortex via subthreshold modulation. In neither case would we be able to reproduce experimentally observed spatiotemporal activity associated with the line-motion illusion.

We note that this priming mechanism provides an explanation for yet more experimentally observed phenomena [18]: (1) The propagation speed of the cortical activity is independent of the contrast of the flashed square cue for a constant contrast of the bar. (2) The onset time of the growing spatiotemporal pattern does depend on the contrast. In particular, for a cue with less contrast, the activity emerges later. Figure 8 clearly demonstrates the agreement on these points between our modeling results and experimental observation [18]. The simple reason for this phenomenon is that the neurons in the middle of the model cortex are "primed" more by a higher-contrast stimulus, and thus can fire earlier when the bar is shown.

As emphasized above, our computational modeling approach affords us great flexibility to investigate cortical mechanisms. Here, to better understand the role of the NMDA-induced priming in the cortical dynamics under the Hikosaka stimulus, we perform numerical experiments on a number of hypothetical alternative cortices and/or cortical setups. First, we consider a model cortex with very weak long-range connections, i.e., in the homogeneous operating state, shown in Fig. 9a, b. In this state, the correlation between the effective reversal potential V^S and the NMDA

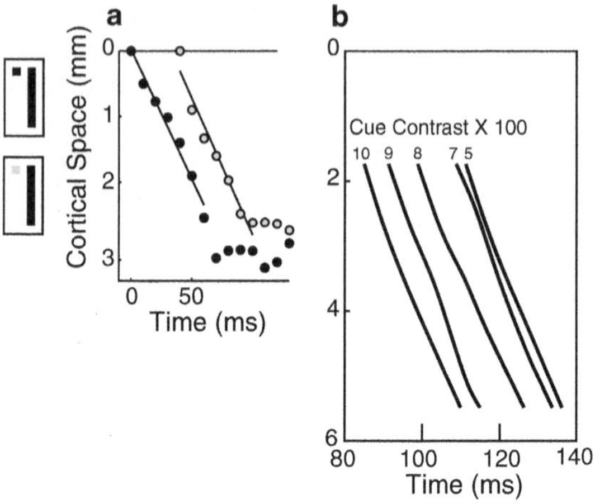

Fig. 8 Contrast dependence of the velocity and onset of the activity associated with the line-motion illusion. (**a**) Experiment [18]. (**b**) Simulation. [Adapted by permission from Macmillan Publishers Ltd. from [18], copyright 2004.] [Reproduced with permission from [2], http://www.pnas.org (Copyright 2005, National Academy of Sciences, USA).]

conductance G^{NMDA} is lost, and so is the priming effect of the square cue which the long-range connections are too weak to transmit. This cortex essentially reproduces the LGN input. A model cortex with the long-range connections dominated by the AMPA conductances (i.e., $\Lambda \approx 0$) exhibits similar, LGN-dominated behavior, shown in Fig. 9c, d. In this case, G^{AMPA} decays too fast for the priming effect of the long-range connections to last sufficiently long. In both these cases, the neuronal membrane potentials in the middle of the model cortex are not elevated when the LGN stimulus caused by the bar arrives, and, therefore, the neurons in the middle fire no sooner than the neurons on the right of the cortex.

Figure 9e, f shows a thought-experiment with the model cortex in the IDS state and driven by the Hikosaka stimulus, but with the G^{NMDA} values flipped at 68 ms from left to right. Thus, the right of the model cortex is more primed than the middle, and consequently the activity corresponding to the illusion now grows from right to left instead of from left to right. Figure 9g, h shows what happens if we instead block the NMDA release on the left side of the model cortex. As the square cue can no longer prime the middle of the model cortex, the $V^S - G^{\text{NMDA}}$ activity profile on its right side appears similar to the profile in the case when only the bar is flashed, shown in Fig. 7f. If the model cortex is very strongly locally inhibited, its right area recruits before its middle area, which is under suppression induced by the square cue, as shown in Fig. 9i, j.

If we stimulate the model cortex by a bar growing from left to right, the resulting cortical activity is illustrated in Fig. 9k, l. The spatial patterns of voltage and NMDA in this case strongly resemble those of the Hikosaka stimulus, shown in Fig. 7m, n and again in Fig. 9m, n.

In our priming mechanisms discussed above, the spatiotemporal activity associated with the line-motion illusion does not depend critically on the exact shape of the cue or the subsequent bar. In particular, any small object that gives rise to an NMDA profile with a spatial gradient can serve as a cue. Any other, longer object, whose position on the screen is close enough to that of the cue, can serve

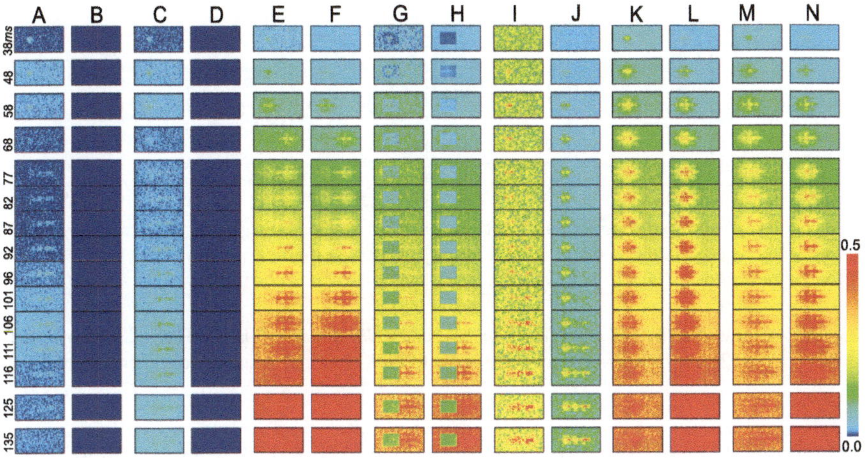

Fig. 9 Voltage and NMDA conductance activity patterns in alternative cortical operating states and parameter regimes: (**a** and **b**) Homogeneous state. (**c** and **d**) AMPA-dominated long-range connections. (**e** and **f**) G^{NMDA} values flipped from left to right at 68 ms. (**g** and **h**) NMDA release blocked on the left side of the model cortex. (**i** and **j**) Strongly locally inhibited model cortex. (**k** and **l**) Bar growing from left to right. (**m** and **n**) Activity due to the Hikosaka stimulus. [Reproduced with permission from [2] http://www.pnas.org (Copyright 2005, National Academy of Sciences, USA).].

instead of the bar by taking advantage of the NMDA gradient to induce recruitment near the cortical image of the cue faster than farther away from this image. If we interpret the spatiotemporal activity in V1 induced by the Hikosaka stimulus as the neurophysiological correlate associated with the line-motion illusion, then any stimuli satisfying the just-described attributes should produce illusory percept according to our priming mechanism. This is indeed the case: In Fig. 10, we show several Hikosaka-like stimuli that all induce the illusory motion sensation: These stimuli all have diffuse and/or curved edges, some are not connected in space (Fig. 10b, c), have a cue that is not a part of the "bar" (Fig. 10a, b), or have low or varying contrast (Fig. 10c, d), yet they all give a motion illusion sensation. (See the movie in the supplementary material of [2].) Since we are dealing with the preattentive line-motion illusion, the cue must be shown for ~40 ms, and then the "bar" almost immediately afterwards, to avoid the attention-induced line-motion illusion.

The main ingredients in our network mechanism that give rise to the spatiotemporal activity associated with the line-motion illusion are (1) the spatiotemporal input structure sculpted by the model LGN, (2) the cue-induced priming effect of the long-range NMDA cortico-cortical conductances, and (3) the strong $V^S - G^{\mathrm{NMDA}}$ correlation in the IDS operating state. We should stress that, in our model, the spatiotemporal activity associated with the line-motion illusion arises within V1 as a result of local and long-range cortico-cortical interactions without any feedback from other cortical levels (except spatiotemporally uniform Poisson noise to model the simple baseline activity of V1). The success of this parsimonious model

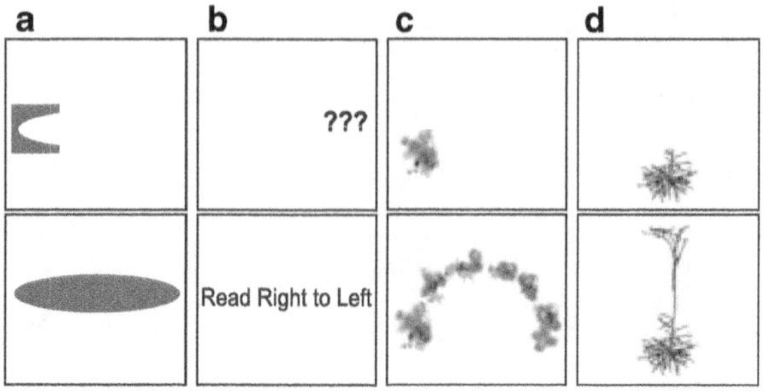

Fig. 10 Alternative Hikosaka stimuli. *Top row*: cues. *Bottom row*: "bars." [Reproduced with permission from [2], http://www.pnas.org (Copyright 2005, National Academy of Sciences, USA).]

in modeling the spatiotemporal activity associated with the line-motion illusion and its similarity to that induced by a moving square therefore makes the V1 cortical circuitry a likely candidate for the similarity observed in cat V1 between the spatiotemporal activity induced by a moving square and that induced by the Hikosaka stimulus in a preattentive line-motion illusion [18].

Orientation Tuning

To investigate the dynamics of orientation tuning in V1, we use both a large-scale model as in the previous section and a model incorporating only short-range cortico-cortical connections and representing a 1 mm^2 local patch, covering four orientation hypercolumns that contain $\mathcal{O}(10^4)$ model neurons [3]. After being constrained to operate in an IDS-like state, the model network reproduces a number of experimentally observed properties related to orientation tuning dynamics in V1, and reveals possible mechanisms responsible for orientation selectivity in V1.

The model network gives rise to a continuum of simple and complex cells, as characterized by the modulation ratio F1/F0 of the firing rate averaged over the cycle of the drifting-grating used as the stimulus. This is the ratio between its first Fourier component and its mean at preferred stimulus orientation, and is $>1/2$ for simple and $<1/2$ for complex cells. We found the distribution of the firing-rate F1/F0 across model cortex to be bimodal and broad, as in [35, 112]. We also found the distribution of modulation ratio for the effective reversal potential V^S, which represents the intracellular voltage, to be unimodal, as in [113].

For individual model neurons, we obtain their steady-state tuning curves by averaging their responses under drifting-grating stimuli presented at different orientation angles. We study these curves for the firing rates, membrane potentials, and conductances, which include that arising from the geniculate excitation, as well as those resulting from excitatory and inhibitory cortico-cortical connections. Sample neurons, presented in Fig. 11, show well-tuned firing rates for both simple and complex cells, regardless of their positions in the orientation columns, as in [12, 107, 114].

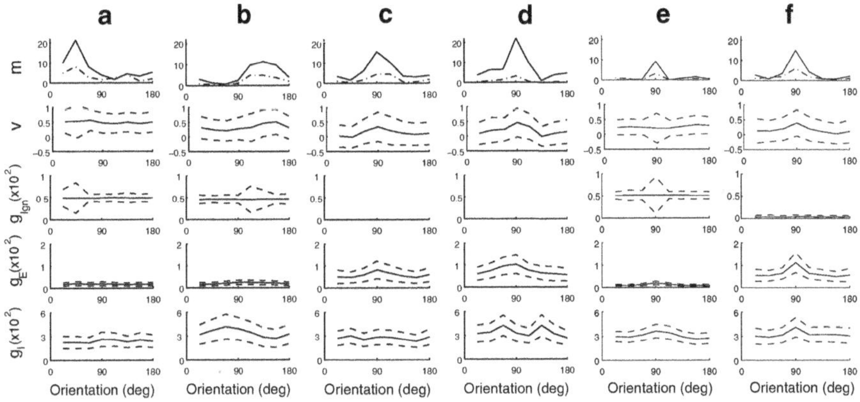

Fig. 11 Tuning curves for neurons near the pinwheel center (**a–d**) and in iso-orientation domains (**e, and f**); (**a, b, e**) are simple and (**c, d, f**) complex cells. Firing rate, membrane potential, geniculate excitation, excitatory and inhibitory cortico-cortical conductances are plotted. *Solid lines* represent the mean values at medium contrast. *Dash-dotted* lines represent the mean values at low contrast. *Dashed lines* represent the mean plus/minus one standard deviation. There is little or no LGN input to complex cells in (**c, d, f**). [Reproduced with permission from [3], http://www.pnas.org (Copyright 2006, National Academy of Sciences, USA).]

On the other hand, the membrane potential and total conductance are tuned more broadly in cells near the pinwheel centers (Fig. 11a–d) than in iso-orientation domains (Fig. 11e, f) [12, 114]. Moreover, in iso-orientation domains, the firing-rate, membrane-potential, and conductance tuning curves for a given neuron are well aligned in orientation angle with one-another (their peaks are at the same angle locations), while near the pinwheel centers the relationship between the conductance and firing-rate tuning curves is in general more varied and complicated (for example, their peak locations can differ) [12].

One quantitative measure of orientation selectivity for drifting grating stimuli is the circular variance (CV), defined as $\mathrm{CV}\,[m] = 1-\left|\int_0^{\pi} m\,(\theta)\,\mathrm{e}^{2\mathrm{i}\theta}\,\mathrm{d}\theta\right| / \int_0^{\pi} m\,(\theta)\,\mathrm{d}\theta$, where $m(\theta)$ is the time-averaged firing rate. CV is near 0 for well-tuned neurons, near 1 for poorly tuned neurons, and in-between otherwise. We display the statistical distribution of the CV for the excitatory neurons in our network in Fig. 12. In particular, Fig. 12a reveals the approximate contrast invariance of orientation selectivity [7], and Fig. 12b shows that orientation selectivity for the firing rates is almost independent of the neuron's location within the orientation column [12, 107, 114].

The mechanism for orientation tuning in iso-orientation domains is relatively simple: all neurons receive spikes only from neighbors with like orientation preference, so all the cortical conductances, membrane potentials, and firing rates are simply sharpened versions of the LGN drive. Near pinwheel singularities, however, sparsity of connections in our model network is needed for conductances and membrane potentials of complex cells to be tuned. Namely, in a densely connected network, they would be untuned as they would be composed of roughly equal contributions from a number of LGN drive dominated simple cells with all possible

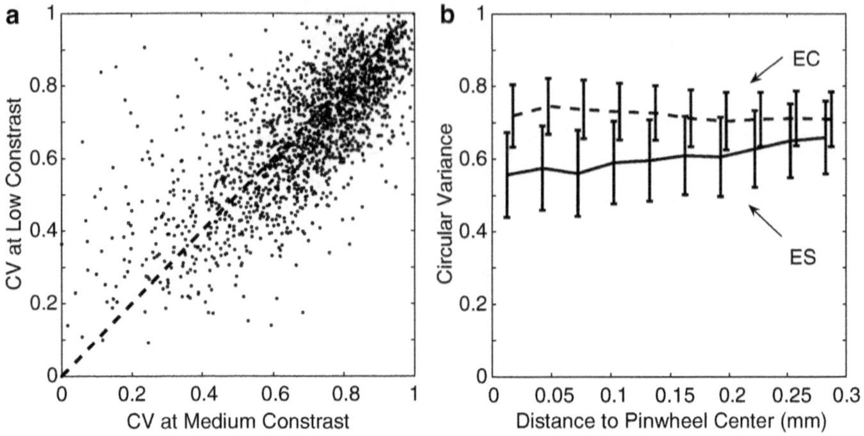

Fig. 12 (**a**) Circular variance at medium vs. low contrasts. (**b**) Dependence of circular variance on the distance from a pinwheel center: EC and ES are excitatory complex and simple cells, respectively. [Reproduced with permission from [3], http://www.pnas.org (Copyright 2006, National Academy of Sciences, USA).]

orientation preferences. Sparsity is thus also needed to achieve tuned firing rates for complex cells near pinwheel centers, since untuned conductances could not confer any tuning upon them. (See [3] for details.) In addition, strong cortical amplification is needed to sharpen the complex cell tuning, and strong gain for contrast invariance. Synaptic fluctuations in the network, again induced by its sparsity, give it stability. Our modeling work further reveals that there is a bifurcation mechanism, the "fluctuation-controlled criticality" [3], underlying the orientation tuning dynamics of simple and complex cells.

We arrived at the bifurcation structure associated with the "fluctuation-controlled criticality" through studying the effects of synaptic fluctuations on the behavior of the model network. By varying the connectivity sparsity, we control the effective network connectivity N_{eff}, which is the average number of pre-synaptic neurons coupled to a given neuron. If N_{eff} is too large, i.e., fluctuations are small, many complex cells become bistable when there is sufficient excitation for them to become tuned. As usual, there is a hysteresis related to this bistability. To bring this hysteretic behavior to the fore, for fixed values of N_{eff}, we slowly ramped up and then down the stimulus contrast. We then recorded the distributions of $\Delta N_{\mathrm{spikes}}$ for both simple and complex cells, where $\Delta N_{\mathrm{spikes}}$ is the difference in the number of spikes during the contrast decrease and increase, respectively. The results for two such effective connectivities are shown in Fig. 13. For a relatively small $N_{\mathrm{eff}} = 96$, i.e., a sparse network with strong fluctuations, both distributions are roughly even in $\Delta N_{\mathrm{spikes}}$. On the other hand, for a relatively large $N_{\mathrm{eff}} = 768$, i.e., a mean-driven network, the distribution of $\Delta N_{\mathrm{spikes}}$ is centered at $\Delta N_{\mathrm{spikes}} > 0$, indicative of hysteresis.

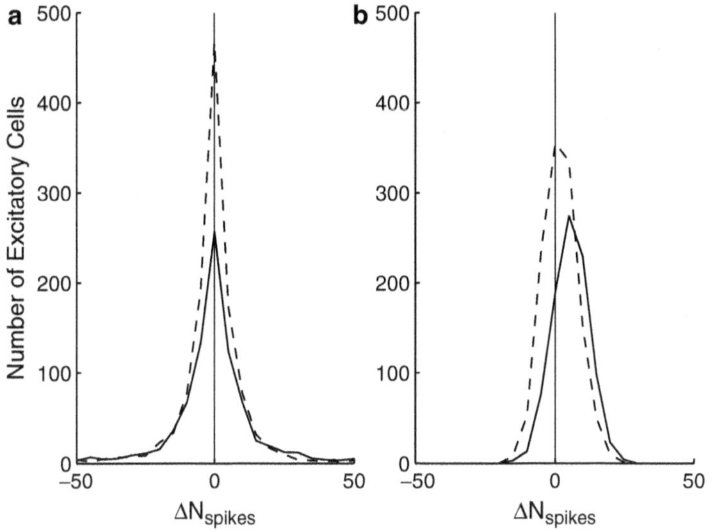

Fig. 13 Spike-rate hysteresis in the V1 model as characterized by $\Delta N_{\mathrm{spikes}}$. $\Delta N_{\mathrm{spikes}} > 0$ if the cell spiked more during the contrast decrease than increase. Simple cells are depicted by *broken line* and complex by *solid line*. (**a**) $N_{\mathrm{eff}} = 96$. (**b**) $N_{\mathrm{eff}} = 768$. [Reproduced with permission from [3], http://www.pnas.org (Copyright 2006, National Academy of Sciences, USA).]

This "fluctuation-controlled criticality" can be further illustrated using a highly idealized, minimal network model. In this model, we let one half of the neurons receive feedforward drive in the form of Poisson spike trains with identical rates ν_0 and spike strengths f (simple cells), and the other half only strong intracortical excitation (complex cells). Both receive strong intracortical inhibition. We ignore any spatial structure of the cortico-cortical coupling, but we do include sparsity in the network connections. We include both fast AMPA and slow NMDA components in the excitatory conductances. In the absence of any NMDA component in the excitatory conductances, our numerical analysis reveals a two-parameter bifurcation diagram of the firing rate per neuron vs. the driving strength $f\nu_0$ of the simple cells and the effective network connectivity N_{eff}, which controls the synaptic fluctuations, as shown in Fig. 14. Bistability and hysteresis arise for sufficiently large N_{eff}. We find a similar bifurcation structure when we add an NMDA component to the cortico-cortical excitation, with less NMDA component increasing the amount of fluctuations, thus smoothing the gain curves and taking the network out of the bistable range.

In the model network, the IDS state operates just below this "fluctuation-driven criticality" in order to have strong cortical amplification and gain, yet not be in the bistable regime, so that our complex cells exhibit experimentally observed tuning properties.

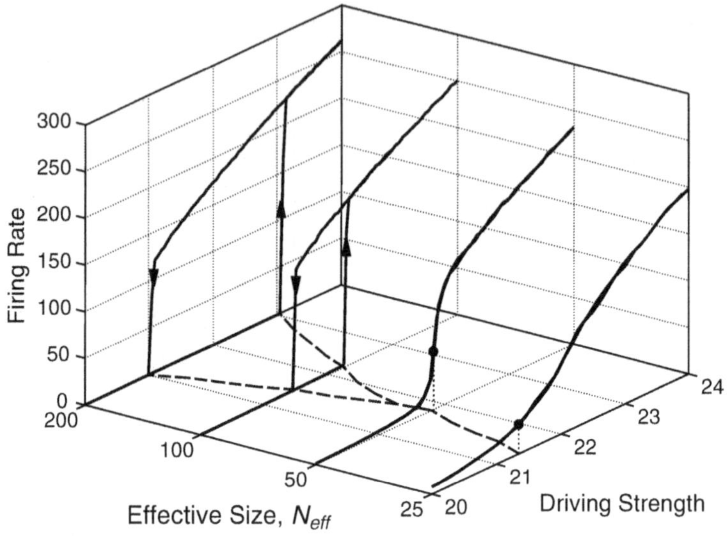

Fig. 14 Bifurcation diagram near the "fluctuation-controlled criticality" in an idealized network. The firing rate is for complex cells. There is a transition to bistability as N_{eff} is increased. [Reproduced with permission from [3], http://www.pnas.org (Copyright 2006, National Academy of Sciences, USA).]

Discussion

In our modeling, we have not only successfully reproduced experimentally observed phenomena quantitatively and qualitatively, but also investigated the network mechanisms underlying these phenomena. In addition, we have carefully verified the structural robustness of dynamical regimes that exhibit the desired physiological behavior types. We have described some of these verifications in the previous sections. They include the studies of the bistability and hysteresis responsible for neuronal orientation tuning in V1, as well as the investigations of how the corresponding bifurcation point depends on both the driving strength and the percentage Λ of the NMDA conductances, described in section "Orientation Tuning." Our structural robustness verifications further include the studies of the alternative cortical operating states in sections "Patterns of Spontaneous Cortical Activity" and "Line-Motion Illusion," various scenarios depicted in Fig. 9c–l that may be potentially related to pharmacological manipulation in the real experimental setting, and the variety of stimuli that induce similar spatiotemporal V1 activity associated with the line-motion illusion shown in Fig. 10.

Furthermore, we have incorporated the consequences of additional physiological properties including synaptic failure, spike frequency adaptation, synaptic depression, and axonal delays, and have verified that they do not significantly alter the mechanisms underlying the cortical activities in our model V1.

Structural robustness also manifests itself with respect to certain parameter changes and small changes in cortical architecture. In particular, by readjusting the cortical coupling strengths, we can sustain the neuronal dynamics in the IDS regime over broad network parameter and architectural ranges, which allows for the same network mechanisms to underlie the physiological phenomena we have modeled. We have observed that the IDS state can remain the cortical operating state for (1) the percentage Λ of G^{NMDA} in the total conductance of the long-range connections ranging from $\sim$5% to 100%; (2) the ratio of major to the minor axis, i.e., the eccentricity, of the long-range elliptical coupling kernel [11] ranging from $\sim$1 to 2, as well as arbitrary orientations of the ellipse; (3) the orientation spread projected by the long-range couplings, $\pm\Delta\theta_{\mathrm{LR}}$ ranging from $\pm 5°$ to $\pm 11°$; (4) the extent of the local excitatory or inhibitory interaction length-scales ranging from 100 μm to 300 μm, and their ratio from $\sim$0.5 to $\sim$2; (5) the background firing rate R_{B} ranging from 2 to $\sim$20 spikes/s; and (6) the NMDA decay time ranging from $\sigma^{\mathrm{NMDA}} = 40$ ms to 80 ms.

Finally, we should emphasize that our results are robust even to the individual neuronal model we used. In particular, by retuning the cortical parameters, we can reproduce our results with the same underlying cortical mechanisms regardless of whether we use linear or exponential [115] I&F point neurons.

Acknowledgment A.V.R. and D.C. were partly supported by the NSF grant DMS-0506396 and by the Schwartz foundation. L.T. was supported by the NSF grant DMS-0506257. G.K. was supported by NSF grants IGMS-0308943 and DMS-0506287, and gratefully acknowledges the hospitality of the Courant Institute of Mathematical Sciences and Center for Neural Science during his visits at New York University in 2003/04 and 2008.

References

1. D. Cai, A. V. Rangan, and D. W. McLaughlin. Architectural and synaptic mechanisms underlying coherent spontaneous activity in V1. *Proc. Natl. Acad. Sci. USA*, 102:5868–5873, 2005.
2. A. V. Rangan, D. Cai, and D. W. McLaughlin. Modeling the spatiotemporal cortical activity associated with the line-motion illusion in primary visual cortex. *Proc. Natl. Acad. Sci. USA*, 102:18793–18800, 2005.
3. L. Tao, D. Cai, D. W. McLaughlin, M. J. Shelley, and R. Shapley. Orientation selectivity in visual cortex by fluctuation-controlled criticality. *Proc. Natl. Acad. Sci. USA*, 103:12911–12916, 2006.
4. A. Arieli, D. Shoham, R. Hildesheim, and A. Grinvald. Coherent spatiotemporal patterns of ongoing activity revealed by real-time optical imaging coupled with single-unit recording in the cat visual cortex. *J. Neurophysiol.*, 73:2072–2093, 1995.
5. D. Fitzpatrick. Cortical imaging: capturing the moment. *Curr. Biol.*, 10:R187–R190, 2000.
6. R. C. Kelly, M. A. Smith, J. M. Samonds, A. Kohn, A. B. Bonds, J. A. Movshon, and T.-S. Lee. Comparison of recordings from microelectrode arrays and single electrodes in the visual cortex. *J. Neurosci.*, 27:261–264, 2007.
7. J. Anderson, I. Lampl, D. Gillespie, and D. Ferster. The contribution of noise to contrast invariance of orientation tuning in cat visual cortex. *Science*, 290:1968–1972, 2000.

8. M. J. Shelley, D. W. McLaughlin, R. Shapley, and J. Wielaard. States of high conductance in a large-scale model of the visual cortex. *J. Comp. Neurosci.*, 13:93–109, 2002.
9. D. Pare, E. Shink, H. Gaudreau, A. Destexhe, and E. J. Lang. Impact of spontaneous synaptic activity on the resting properties of cat neocortical pyramidal neurons In vivo. *J. Neurophysiol.*, 79:1450–1460, Mar 1998.
10. R. Dingledine, K. Borges, D. Bowie, and S. F. Traynelis. The glutamate receptor ion channels. *Pharmacol. Rev.*, 51:7–61, 1999.
11. A. Angelucci, J. B. Levitt, E. J. Walton, J. M. Hupe, J. Bullier, and J. S. Lund. Circuits for local and global signal integration in primary visual cortex. *J. Neurosci.*, 22:8633–8646, 2002.
12. J Marino, J. Schummers, D. C. Lyon, L. Schwabe, O. Beck, P. Wiesing, K. Obermayer, and M. Sur. Invariant computations in local cortical networks with balanced excitation and inhibition. *Nat. Neurosci.*, 8:194–201, 2005.
13. M. Tsodyks, T. Kenet, A. Grinvald, and A. Arieli. Linking spontaneous activity of single cortical neurons and the underlying functional architecture. *Science*, 286:1943–1946, 1999.
14. T. Kenet, D. Bibitchkov, M. Tsodyks, A. Grinvald, and A. Arieli. Spontaneously emerging cortical representations of visual attributes. *Nature*, 425:954–956, 2003.
15. A. Grinvald and R. Heildesheim. VSDI: a new era in functional imaging of cortical dynamics. *Nat. Rev. Neurosci.*, 5:874–885, 2004.
16. M. N. Shadlen and W. T. Newsome. The variable discharge of cortical neurons: implications for connectivity, computation and information coding. *J. Neurosci.*, 18:3870–3896, 1998.
17. O. Hikosaka, S. Miyauchi, and S. Shimojo. Focal visual attention produces illusory temporal order and motion sensation. *Vis. Res.*, 33:1219–1240, 1993.
18. D. Jancke, F. Chavance, S. Naaman, and A. Grinvald. Imaging cortical correlates of illusion in early visual cortex. *Nature*, 428:423–426, 2004.
19. I. Lampl, I. Reichova, and D. Ferster. Synchronous membrane potential fluctuations in neurons of the cat visual cortex. *Neuron*, 22:361–374, 1999.
20. L. Borg-Graham, C. Monier, and Y. Fregnac. Voltage-clamp measurement of visually-evoked conductances with whole-cell patch recordings in primary visual cortex. *J. Physiol. (Paris)*, 90:185–188, 1996.
21. L. J. Borg-Graham, C. Monier, and Y. Fregnac. Visual input evokes transient and strong shunting inhibition in visual cortical neurons. *Nature*, 393:369–373, 1998.
22. M. N. Shadlen and W. T. Newsome. The variable discharge of cortical neurons: implications for connectivity, computation, and information coding. *J. Neurosci.*, 18:3870–3896, 1998.
23. A. Destexhe, M. Rudolph, and D. Pare. The high-conductance state of neocortical neurons in vivo. *Nat. Rev. Neurosci.*, 4:739–751, 2003.
24. J. Anderson, I. Lampl, I. Reichova, M. Carandini, and D. Ferster. Stimulus dependence of two-state fluctuations of membrane potential in cat visual cortex. *Nat. Neurosci.*, 3:617–621, 2000.
25. M. Volgushev, J. Pernberg, and U. T. Eysel. A novel mechanism of response selectivity of neurons in cat visual cortex. *J. Physiol.*, 540:307–320, 2002.
26. M. Volgushev, J. Pernberg, and U. T. Eysel. Gamma-frequency fluctuations of the membrane potential and response selectivity in visual cortical neurons. *Eur. J. Neurosci.*, 17:1768–1776, 2003.
27. C. Rivadulla, J. Sharma, and M. Sur. Specific roles of NMDA and AMPA receptors in direction-selective and spatial phase-selective response in visual cortex. *J. Neurosci.*, 21:1710–1719, 2001.
28. R. Ben-Yishai, R. Bar-Or, and H. Sompolinsky. Theory of orientation tuning in the visual cortex. *Proc. Natl. Acad. Sci. USA*, 92:3844–3848, 1995.
29. J.A. Goldberg, U. Rokni, and H. Sompolinsky. Patterns of ongoing activity and the functional architecture of the primary visual cortex. *Neuron*, 13:489–500, 2004.
30. A. V. Rangan and D. Cai. Fast numerical methods for simulating large-scale integrate-and-fire neuronal networks. *J. Comput. Neurosci.*, 22:81–100, 2007.
31. D. Hubel and T. Wiesel. Receptive fields, binocular interaction and functional architecture of the cat's visual cortex. *J. Physiol. (Lond.)*, 160:106–154, 1962.

32. J. A. Movshon, I. D. Thompson, and D. J. Tolhurst. Spatial summation in the receptive fields of simple cells in the cat's striate cortex. *J. Physiol. (Lond.)*, 283:53–77, 1978.
33. J. A. Movshon, I. D. Thompson, and D. J. Tolhurst. Receptive field organization of complex cells in the cat's striate cortex. *J. Physiol. (Lond.)*, 283:79–99, 1978.
34. K. Toyama, M. Kimura, and K. Tanaka. Organization of cat visual cortex as investigated by cross-correlation technique. *J. Neurophysiol.*, 46:202–214, 1981.
35. D. Ringach, R. Shapley, and M. Hawken. Orientation selectivity in macaque V1: Diversity and laminar dependence. *J. Neurosci.*, 22:5639–5651, 2002.
36. K. P. Hoffman and J. Stone. Conduction velocity of afferents to cat visual cortex: a correlation with cortical receptive field properties. *Brain Res.*, 32:460–466, 1971.
37. W. Singer, F. Tretter, and M. Cynader. Organization of cat striate cortex: a correlation of receptive-field properties with afferent and efferent connections. *J. Neurophysiol.*, 38:1080–1098, 1975.
38. D. Ferster and S. Lindstrom. An intracellular analysis of geniculo-cortical connectivity in area 17 of the cat. *J. Physiol.*, 342:181–215, 1983.
39. J. A. Movshon. The velocity tuning of single units in cat striate cortex. *J. Physiol.*, 249:445–468, 1975.
40. P. Hammond and D. M. MacKay. Differential responsiveness of simple and complex cells in cat striate cortex to visual texture. *Exp. Brain. Res.*, 30:275–296, 1977.
41. J. G. Malpeli. Activity of cells in area 17 of the cat in absence of input from layer a of lateral geniculate nucleus. *J. Neurophysiol.*, 49:595–610, 1983.
42. J. G. Malpeli, C. Lee, H. D. Schwark, and T. G. Weyand. Cat area 17. I. pattern of thalamic control of cortical layers. *J. Neurophysiol.*, 56:1062–1073, 1986.
43. M. Mignard and J. G. Malpeli. Paths of information flow through visual cortex. *Science*, 251:1249–1251, 1991.
44. K. Tanaka. Organization of geniculate inputs to visual cortical cells in the cat. *Vis. Res.*, 25:357–364, 1985.
45. J.-M. Alonso, W. M. Usrey, and R. Reid. Rules of connectivity between geniculate cells and simple cells in cat primary visual cortex. *J. Neurosci.*, 21:4002–4015, 2001.
46. D. Ringach. Spatial structure and symmetry of simple-cell receptive fields in macaque primary visual cortex. *J. Neurophysiol.*, 88:455–463, 2002.
47. L. Tao, M. J. Shelley, D. W. McLaughlin, and R. Shapley. An egalitarian network model for the emergence of simple and complex cells in visual cortex. *Proc. Natl. Acad. Sci. USA*, 101:366–371, 2004.
48. K. Miller and D. MacKay. The role of constraints in hebbian learning. *Neural Comput.*, 6:100–126, 1994.
49. K. Miller. Synaptic economics: Competition and cooperation in synaptic plasticity. *Neuron*, 17:371–374, 1996.
50. S. Royer and D. Pare. Bidirectional synaptic plasticity in intercalated amygdala neurons and the extinction of conditioned fear responses. *Neuroscience*, 115:455–462, 2002.
51. S. Royer and D. Pare. Conservation of total synaptic weight through balanced synaptic depression and potentiation. *Nature*, 422:518–522, 2003.
52. J. Wielaard, M. J. Shelley, R. Shapley, and D. W. McLaughlin. How simple cells are made in a nonlinear network model of the visual cortex. *J. Neurosci.*, 21:5203–5211, 2001.
53. T. Bonhoeffer and A. Grinvald. Iso-orientation domains in cat visual cortex are arranged in pinwheel like patterns. *Nature*, 353:429–431, 1991.
54. G. Blasdel. Differential imaging of ocular dominance and orientation selectivity in monkey striate cortex. *J. Neurosci.*, 12:3115–3138, 1992.
55. G. Blasdel. Orientation selectivity, preference, and continuity in the monkey striate cortex. *J. Neurosci.*, 12:3139–3161, 1992.
56. G. DeAngelis, R. Ghose, I. Ohzawa, and R. Freeman. Functional micro-organization of primary visual cortex: Receptive field analysis of nearby neurons. *J. Neurosci.*, 19:4046–4064, 1999.
57. M. Hubener, D. Shoham, A. Grinvald, and T. Bonhoeffer. Spatial relationships among three columnar systems in cat area 17. *J. Neurosci.*, 17:9270–9284, 1997.

58. R. Everson, A. Prashanth, M. Gabbay, B. Knight, L. Sirovich, and E. Kaplan. Representation of spatial frequency and orientation in the visual cortex. *Proc. Natl. Acad. Sci. USA*, 95:8334–8338, 1998.
59. N. P. Issa, C. Trepel, and M. P. Stryker. Spatial frequency maps in cat visual cortex. *J. Neurosci.*, 20:8504–8514, 2000.
60. L. Sirovich and R. Uglesich. The organization of orientation and spatial frequency in primary visual cortex. *Proc. Natl. Acad. Sci. USA*, 101:16941–16946, 2004.
61. S. Molotchnikoff, P.-C. Gillet, S. Shumikhina, and M. Bouchard. Spatial frequency characteristics of nearby neurons in cats' visual cortex. *Neurosci. Lett.*, 418:242–247, 2007.
62. D. Fitzpatrick, J. Lund, and G. Blasdel. Intrinsic connections of macaque striate cortex Afferent and efferent connections of lamina 4C. *J. Neurosci.*, 5:3329–3349, 1985.
63. J. S. Lund. Local circuit neurons of macaque monkey striate cortex: Neurons of laminae 4C and 5A. *J. Comp. Neurology*, 257:60–92, 1987.
64. E. Callaway and A. Wiser. Contributions of individual layer 2 to 5 spiny neurons to local circuits in macaque primary visual cortex. *Vis. Neurosci.*, 13:907–922, 1996.
65. E. Callaway. Local circuits in primary visual cortex of the macaque monkey. *Ann. Rev. Neurosci.*, 21:47–74, 1998.
66. C. Koch. *Biophysics of Computation*. Oxford University Press, Oxford, 1999.
67. W. H. Bosking, Y. Zhang, B. Schofield, and D. Fitzpatrick. Orientation selectivity and the arrangement of horizontal connections in tree shrew striate cortex. *J. Neurosci.*, 17:2112–2127, 1997.
68. L. Sincich and G. Blasdel. Oriented axon projections in primary visual cortex of the monkey. *J. Neurosci.*, 21:4416–4426, 2001.
69. A. Angelucci, J. B. Levitt, E. J. S. Walton, J. Hupe, J. Bullier, and J. S. Lund. Circuits for local and global signal integration in primary visual cortex. *J. Neurosci.*, 22:8633–8646, 2002.
70. A. Angelucci and J. Bullier. Reaching beyond the classical receptive field of V1 neurons: horizontal or feedback axons? *J. Physiol. (Paris)*, 97(2-3):141–154, 2003.
71. J. S. Lund, A. Angelucci, and P. C. Bressloff. Anatomical substrates for functional columns in macaque monkey primary visual cortex. *Cereb. Cortex*, 12:15–24, 2003.
72. A. Angelucci, J. B. Levitt, P. Adorjan, Y. Zheng, L. C. Sincich, N. P. McLoughlin, G. P. Blasdel, and J. S. Lund. Bar-like patterns of lateral connectivity in layers 4B and upper 4Cα of macaque primary visual cortex, area V1. preprint.
73. Z. F. Kisvárday, K. A. C. Martin, T. F. Freund, Z. Magloczky, D. Whitteridge, and P. Somogy. Synaptic targets of HRP-filled layer III pyramidal cells in the cat striate cortex. *Exp. Brain Res.*, 64:541–552, 1986.
74. K. A. C. Martin and D. Whitteridge. Form, function and intracortical projections of spiny neurons in the striate cortex of the cat. *J. Physiol. (Lond.)*, 353:463–504, 1984.
75. B. A. McGuire, C. D. Gilbert, P. K. Rivlin, and T. N. Wiesel. Targets of horizontal connections in macaque primary visual cortex. *J. Comp. Neurol.*, 305:370–392, 1991.
76. J. A. Hirsch and C. D. Gilbert. Synaptic physiology of horizontal connnections in the cat's visual cortex. *J. Neurosci.*, 11:1800–1809, 1991.
77. Y. Yoshimura, H. Sato, K. Imamura, and Y. Watanabe. Properties of horizontal and vertical inputs to pyramidal cells in the superficial layers of the cat visual cortex. *J. Neurosci.*, 20:1931–1940, 2000.
78. K. S. Rockland and T. Knutson. Axon collaterals of Meynert cells diverge over large portions of area V1 in the macaque monkey. *J. Comp. Neurol.*, 441:134–147, 2001.
79. H. J. Chisum, F. Mooser, and D. Fitzpatrick. Emergent properties of layer 2/3 neurons reflect the collinear arrangement of horizontal connections in tree shrew visual cortex. *J. Neurosci.*, 23:2947–2960, 2003.
80. D. Shoham, D. E. Glaser, A. Arieli, T. Kenet, C. Wijnbergen, Y. Toledo, R. Hildesheim, and A. Grinvald. Imaging cortical dynamics at high spatial and temporal resolution with novel blue voltage-sensitive dyes. *Neuron*, 24:791–802, 1999.
81. Z. Kisvarday, E. Toth, M. Rausch, and U. Eysel. Orientation-specific relationship between populations of excitatory and inhibitory lateral connections in the visual cortex of the cat. *Cereb. Cortex*, 7:605–618, 1997.

82. R. Malach, Y. Amir, M. Harel, and A. Grinvald. Relationship between intrinsic connections and functional architecture revealed by optical imaging and in vivo targeted biocytin injections in primate striate cortex. *Proc. Natl. Acad. Sci. USA*, 90:10469–10473, 1993.
83. T. Yoshioka, G. Blasdel, J. Levitt, and J. Lund. Relation between patterns of intrinsic lateral connectivity, ocular dominance, and cytochrome oxidase-reactive regions in macaque monkey striate cortex. *Cereb. Cortex*, 6:297–310, 1996.
84. B. Roerig and J. P. Kao. Organization of intracortical circuits in relation to direction preference maps in ferret visual cortex. *J. Neurosci.*, 19:RC44:1–5, 1999.
85. N. W. Daw, P. G. S. Stein, and K. Fox. The role of NMDA receptors in information transmission. *Annu. Rev. Neurosci.*, 16:207–222, 1993.
86. H. Sato, Y. Hata, and T. Tsumoto. Effects of blocking non-N-methyl-D-aspartate receptors on visual responses of neurons in the cat visual cortex. *Neuroscience*, 94:697–703, 1999.
87. C. E. Schroeder, D. C. Javitt, M. Steinschneider, A. D. Mehta, S. J. Givre, H. G. Vaughan, Jr., and J. C. Arezzo. N-methyl-D-aspartate enhancement of phasic responses in primate neocortex. *Exp. Brain Res.*, 114:271–278, 1997.
88. P. Seriès, J. Lorenceau, and Y. Frégnac. The "silent" surround of V1 receptive fields: theory and experiments. *J. Physiol. (Paris)*, 97:453–474, 2003.
89. S. Friedman-Hill, P. E. Maldonado, and C. M. Gray. Dynamics of striate cortical activity in the alert macaque: I. Incidence and stimulus-dependence of gamma-band neuronal oscillations. *Cereb. Cortex*, 10:1105–1116, 2000.
90. A. Kohn and M. A. Smith. Stimulus dependence of neuronal correlation in primary visual cortex of the macaque. *J. Neurosci.*, 25:3661–3673, 2005.
91. W. Singer and C. M. Gray. Visual feature integration and the temporal correlation hypothesis. *Annu. Rev. Neurosci.*, 18:555–586, 1995.
92. P. E. Maldonado, S. Friedman-Hill, and C. M. Gray. Dynamics of striate cortical activity in the alert macaque: II. Fast time scale synchronization. *Cereb. Cortex*, 10:1117–1131, 2000.
93. D. Ringach, M. Hawken, and R. Shapley. Dynamics of orientation tuning in macaque primary visual cortex. *Nature*, 387:281–284, 1997.
94. D. Xing, R. Shapley, M. Hawken, and D. Ringach. The effect of stimulus size on the dynamics of orientation selectivity in macaque V1. *J. Neurophysiol.*, 94:799–812, 2005.
95. D. C. Somers, E. V. Todorov, A. G. Siapas, L. J. Toth, D. S. Kim, and M. Sur. A local circuit approach to understanding integration of long-range inputs in primary visual cortex. *Cereb. Cortex*, 8:204–217, 1998.
96. P. C. Bressloff, J. D. Cowan, M. Golubitsky, P. J. Thomas, and M. C. Wiener. Geometric visual hallucinations, euclideansymmetry and the functional architecture of striate cortex. *Phil. Trans. R. Soc. Lond. B*, 356:299–330, 2001.
97. P. C. Bressloff. Spatially periodic modulation of cortical patterns by long-range horizontal connections. *Physica D*, 185:131–157, 2002.
98. L. Schwabe, K. Obermayer, A. Angelucci, and P. C. Bressloff. The role of feedback in shaping the extra-classical receptive field of cortical neurons: A recurrent network model. *J. Neurosci.*, 26:9117–9129, 2006.
99. F. Chance, S. Nelson, and L. F. Abbott. Complex cells as cortically amplified simple cells. *Nature Neurosci.*, 2:277–282, 1999.
100. T. Troyer, A. Krukowski, N. Priebe, and K. Miller. Contrast invariant orientation tuning in cat visual cortex with feedforward tuning and correlation based intracortical connectivity. *J. Neurosci.*, 18:5908–5927, 1998.
101. D. W. McLaughlin, R. Shapley, M. J. Shelley, and J. Wielaard. A neuronal network model of macaque primary visual cortex (V1): Orientation selectivity and dynamics in the input layer $4C\alpha$. *Proc. Natl. Acad. Sci. USA*, 97:8087–8092, 2000.
102. C. Myme, K. Sugino, G. Turrigiano, and S. B. Nelson. The nmda-to-ampa ratio at synapses onto layer 2/3 pyramidal neurons is conserved across prefrontal and visual cortices. *J. Neurophysiol.*, 90:771–779, 2003.
103. G. W. Huntley, J. C. Vickers, N. Brose, S. F. Heinemann, and J. H. Morrison. Distribution and synaptic localization of immunocytochemically identified nmda receptor subunit proteins in sensory motor and visual cortices of monkey and human. *J. Neurosci.*, 14:3603–3619, 1994.

104. R. C. Reid and J.-M. Alonso. Specificity of monosynaptic connections from thalamus to visual cortex. *Nature*, 378:281–284, 1995.
105. G. DeAngelis, I. Ohzawa, and R. Freeman. Receptive-field dynamics in the central visual pathways. *Trends Neurosci.*, 18:451–458, 1995.
106. C. D. Gilbert. Horizontal integration and cortical dynamics. *Neuron*, 9:1–13, 1992.
107. P. Maldonado, I. Godecke, C. Gray, and T. Bonhoeffer. Orientation selectivity in pinwheel centers in cat striate cortex. *Science*, 276:1551–1555, 1997.
108. U. Eysel. Turning a corner in vision research. *Nature*, 399:641–644, 1999.
109. D. Hansel, G. Mato, C. Meunier, and L. Neltner. Numerical simulations of integrate-and-fire neural networks. *Neural Comp.*, 10:467–483, 1998.
110. M. J. Shelley and L. Tao. Efficient and accurate time-stepping schemes for integrate-and-fire neuronal networks. *J. Comput. Neurosci.*, 11:111–119, 2001.
111. K. Fox, H. Sato, and N. Daw. The effect of varying stimulus intensity on NMDA-receptor activity in cat visual cortex. *J Neurophysiol.*, 64:1413–1428, 1990.
112. F. Mechler and D. Ringach. On the classification of simple and complex cells. *Vis. Res.*, 42:1017–1033, 2002.
113. N. Priebe, F. Mechler, M. Carandini, and D. Ferster. The contribution of spike threshold to the dichotomy of cortical simple and complex cells. *Nat. Neurosci.*, 7:1113–1122, 2004.
114. J. Schummers, J. Marino, and M. Sur. Synaptic integration by v1 neurons depends on location within the orientation map. *Neuron*, 36:969–978, 2002.
115. N. Fourcaud-Trocmé, D. Hansel, C. van Vreeswijk, and N. Brunel. How spike generation mechanisms determine the neuronal response to fluctuating inputs. *J. Neurosci.*, 23:11628–11640, 2003.

Index

A
Aaron, G., 47
Abbott, L.F., 129, 211
Abouzeid, A., 238
Absolute integrate-and-fire (IF) model, gap junctions
  aif model, 79–80
  phase response curve (PRC), 83–84
  spike adaptation
    parameters variation, 82
    periodic orbit, 81
    tonic and bursting behavior, 80–81
  voltage dynamics and firing times, 79
Akaike information criterion (AIC), 176–177
AMPA conductances, 274, 279, 284
Andersen, R.A., 123
Arbitrary sensory-motor remapping
  context-dependent classification task, 132
  contextual switch (*see* Contextual switch)
  neurophysiological recordings, 132
Asynchronous synaptic transmission
  calcium elevation, 69, 70
  experimental evidence, 69
  polysynaptic currents, 68
  silent periods, 67
  vs. synchronous phase, 69
Attentional modulation
  firing rate *vs.* bar orientation, 125
  visual responses, 131
Autoregressive (AR) model. *See* Oscillatory cortical networks, granger causality
Averbeck, B.B., 207

B
Baigent, S., 92
Baker, S.N., 10
Battacharyya distance (BD), noise correlations, 219–220
Bayesian information criterion (BIC), 177
Behavioral aspects, interceptive action, 192–193
Behavioral flexibility, gain modulation, 121–122
Bentley, N.M., 121
Bi, G.-Q., 61
Bogoliubov, N.N., 97
Bollimunta, A., 169
Brain dynamics, perception
  Chapman–Kolmogorov equation, 249
  Fokker–Planck equation, 250
  leaky integrate-and-fire (LIF) model, 248–249
  mean-field technique, 250–251
  population transfer function, 250
  RC-circuit, 249
  true dynamics generation, 251
Broer, H.W., 100
Bryant, H.L., 239
Buzsáki, G., 79, 80

C
Cai, D., 263
Cerebral cortex network, excitation and inhibition
  end of up states
    EPSPs/IPSPs, population analysis, 29, 33
    in vitro, 28–30
    in vivo, 31–32
  experimental procedures and detection
    amplitudes and time-course, 24–25
    comparative statistics, 26
    data analysis, 22, 43
    intracellular and extracellular recordings, 21–22, 42
    IPSPs and EPSPs detection, 23

method, 22–24
V_m deflections, 25
timing
conductances, 40
down states, 40
fast spiking neurons, 39
parameters, 41
up and down states
activated periods, 19
auditory cortex, slow oscillations, 19–20
in computo, 35–38
functional balance, 19–21
inhibitory events, 37, 38
model architecture and function, 34
network model, 33–35
slow rhythmic activity, 17–18
up states risetime
normalized synaptic events, 28, 32
synaptic event rates, 27, 31
in vitro, 26–28
in vivo, 30–31
Chapman–Kolmogorov equation, perception, 249
Chen, Y., 169
Chorev, E., 91
Circular variance (CV), 287, 288
Competing Yes-No-Neurons (CYNN), 259
Complex stimuli, flexible responses
curse of dimensionality, 138
invariant object recognition, 138
model network, 139
under natural conditions, 139
neural mechanisms, need for, 139
neurophysiological recordings, 140
optimizing detection of target, 140
target-dependent modulation, 139
Compte, A., 17
Computational neuronal network modeling. *See* Primary visual cortex (V1)
Contextual switch
change in motor responses, of second layer, 135
coefficients $u_j(1)$, 134
gain-modulated responses
activity of downstream neuron, 133–134
response of sensory neuron, 133
model network, 134
sensory activity, changes, 135
stimulus-context combinations, 135
Coombes, S., 77
Cortical neurons, synaptic inputs dynamics
excitation and inhibition
balance of, 5
correlation of, 7
lag analysis, 8
shape and amplitude, 6–7
intracellular recordings, 2–4
membrane potential fluctuations repeats
comparison, 12
high resolution index (HRI) scores, 11
physiological data, 11
spike motifs, 9–10
surrogate data generation, 10–11
synfire chain model, 8–9
time domain interval shuffling, 10
subthreshold spontaneous activity, 3
synchrony, 4–5
up-down dynamics, 2
Cortico-cortical connections, 265–267, 274, 275, 280, 286
Coupling coefficients, 273

D
Dayan, P., 211
Decay time scales σ^{ϱ}, 272
Deco, G., 247, 259
de Lafuente, V., 252, 255, 259, 260
de Vries, G., 91
Ding, M., 169, 170
Divisive normalization models, 126

E
Effective network connectivity N_{eff}, 288
Eguíluz, V.M., 97, 111
Encoding and decoding target parameters. *See* Interceptive behavior, parietofrontal system
Ermentrout, B., 91, 229

F
Feed-forward chain, 95
band-pass filters, 98
cochlear modeling
basilar membrane (BM), 117
Hopf models, 117–118
two-frequency forcing, 118–119
differential equations, 96
Duffing equation, 97
flow-invariant subspace, 96
periodic forcing system
experiments, 104–105

Hopf bifurcation, 105–116
network motif, 100–101
simulations, 101–104
synchrony-breaking Hopf bifurcations
asymptotic dynamics, 100
coupled-cell system, 99
Jacobian matrix, 98–99
periodic solution, 100
quasiperiodic solution, 100–101
synchronous equilibrium, 99
Filter-amplifier motif. *See* Feed-forward chain
Fisher, I., 143
Fokker–Planck equation, perception, 250
Fusi, S., 249

G
Gabor spatial kernel $K^{\theta,\phi,\omega}(\rho)$, 273
Gain modulation
arbitrary sensory-motor remapping
context-dependent classification task, 132
Contextual switch (*see* Contextual switch)
neurophysiological recordings, 132
attentional modulation, 125
behavioral flexibility, 121–122
context- and task-dependent
color based tasks, 126–127
inferior-temporal cortex (IT) neurons, 127
prefrontal cortex, 126
coordinate transformations
downstream neurons, 129
eye position, 131
gain fields relationship, 128
hypothetical responses, 129, 130
mathematical description, 129
motor error vector, 130
target location and gaze angle, 128–129
visual information, 131
experimental evidence for, 124–127
eye position, 123
gain field, 123
hypothetical responses, 123
modulation strength, quantification, 123–124
nonlinear interactions, multiple stimuli, 124–125
proprioceptive information modulation, 124–125
reference points, 122
Galán, R.F., 229, 240
Gap junctions and emergent rhythms
absolute integrate-and-fire model
aif model, 79–80
phase response curve, 83–84
spike adaptation, 80–83
voltage dynamics and firing times, 79
direct gap-junction coupling model
adjoint calculation, 88
asynchronous state existence, 85–86
bifurcation point, 89–90
conductance, 84
differential equation, 87
parameters, 89
perturbation mean field, 86–87
phase reduction techniques, 86
spectrum, 89
splay solution, 88
neural signatures, 92
neuronal firing, 78
olivary neurons, 91
Gaze angle, 124, 128
Geweke, J., 170, 173, 175
Gollo, L.L., 143
Golubitsky, M., 95
Granger causality analysis. *See* Oscillatory cortical networks, Granger causality
Granger, C.W.J., 171

H
Hebb, D.O., 61, 62, 65, 66, 71, 72
High resolution index (HRI), 51
Hikosaka line-motion illusion, 265, 266, 275, 281
Hodgkin and Huxley (HH) cells, dynamical relaying
firing phases, 149
postsynaptic potential (PSP), 150
reciprocal connection, 150
zero-lag synchronization dependence, 150–151
Hodgkin–Huxley-type equations, 265
Hopf bifurcations
asymptotic dynamics, 100
coupled-cell system, 99
Jacobian matrix, 98–99
periodic solution, 100
Q-factor, 115–116
quasiperiodic solution, 100–101
response curve
multiplicity solution, 107
singularity theory, 108
theorem, 108–111
two stable solutions, 106–107

simulations, 105–106
solution amplitudes
cube root growth, 111–112
log–log plots, 111
theorem, 113–115
synchronous equilibrium, 99
Hysteresis, 274, 288, 289

I
Ikegaya, Y., 10, 53, 54
Inferior-temporal cortex (IT), 127
Instantaneous neuronal membrane potentials, 280
Integrate-and-fire (I&F) neuron model, 265, 267
Interceptive behavior, parietofrontal system
decoding, angular position and tau
Bayesian reconstruction method, 200–201
spikes number and Poisson distribution, 200
TTC target (TTC_{tar}) information, 202–203
variance and bias, 201, 204
encoding, angular position and time-to-contact (TTC)
independent linear regression model, 196–197
time shifts and log-ratio values, 198–199
motor cortex and area 7a, 196, 202, 205
psychophysics of, 192
sensorimotor processing, circularly moving targets, 194, 196
task and strategy, 194, 195
types, 191
visual motion processing, 193, 204
Intermittent de-suppressed (IDS) operating state
characterization, 267
correlation between G^{NMDA} and G^{I} in, 279
cortical activity patterns in, 276, 278
as cortical operating state for, 291
fluctuations in, 274
high conductance state of network in, 267
in model V1, analysis of, 278
role of long range connections in, 280
significance of, 275
stimulus in, 278
Intracellular membrane potentials, 271
Itti, L., 140
Izhikevich, E.M., 91

K
Karbowski, J., 79
Katz, B., 69
Katz, Y., 1
Kazanci, F.G., 91
Kern, A., 97, 103, 111
Kopell, N., 79
Kovačič, G., 263

L
Lampl, I., 1
Large-scale computational model
computational method, 274
LGN input and background noise, 273–274
long-range connections, 272–273
short-range connections, 272
Large-scale neuronal networks
coupled oscillatory activity, 154
direct interaction dynamics, 156
dynamical relay dynamics, 155
firing cycles, 154
V-shaped network, reciprocal coupling, 157
Lateral geniculate nucleus (LGN), 268, 272, 274, 275
dominated behavior, 285
drive-dominated simple cells, 287
model, 266, 274
cortical input from, 281
Lau, P.-M., 61
Lauwereyns, J., 126
Leakage conductance, 271
Leaky integrate-and-fire (LIF) model, perception, 248–249
Lee, D, 127
Lemon, R.N., 10
Levinson, Wiggins, Robinson (LWR) algorithm, 176
Local field potentials (LFP), 187–188
Long-range kernel $K^{LR,Q}(\mathbf{x}i, \theta i, \mathbf{x}j, \theta j\ )$, 273
Low resolution index (LRI), 49
Luczak, A., 13
Lyapunov exponents
noise color and reliability, 240
and phase reduction
linear adjoint equation, 236
oscillators, 235
phase resetting curve (PRC), 236–237

M
Mainen, Z.F., 239
Marella, S., 241, 242
Matveev, V., 92

McCullen, N.J., 95, 97, 104
McCulloch, W.S., 62
Merchant, H., 191
Mirasso, C.R., 143
Mitral cells
gamma oscillations, 231
noise-induced oscillatory synchrony, 232–233
oscillators frequency, 231–232
recurrent and lateral inhibition, 230–231
stochastic synchrony, 233–235
Mitropolsky, Y.A., 97
Modulation ratio F1/F0, 286
Mokeichev, A., 1
Montgomery, K.A., 97, 117, 118
Multiple stimuli, nonlinear interactions, 125–126

N
Nakao, H., 241
Navalpakkam, V., 140
Neocortical neuron
improved repeat detector
PHRI (PSP-based detection), 54, 56, 57
PSPs, 54–55
shuffle surrogate technique, 56, 57
in vivo cat recording, 56, 57
pattern generator, 47–48
recording conditions and effects, 57–59
repeat detection
correlation coefficients, 52
cross-covariance function, 48–49
extracted motif-repeats alignment, 50
high resolution index (HRI), 51
intracellular recordings, 48
low resolution index (LRI), 49
Matlab software, 49–50
putative repeats, 49, 51
stages, 48
voltage-clamp recordings, 52–53
significance testing
implanted, artificial repeats, 54
PSCs detection, 53
shuffling procedure limitations, 53–54
surrogate recordings, 52–53
spontaneous up states, 47
Neurotransmitter, 269
NMDA conductance, 267, 279, 283
NMDA-induced priming, in cortical dynamics, 283
NMDA receptor, 265, 270, 273, 280
Noise correlations, information encoding and decoding
definition, 208–210
effects on, 221–223
empirical studies
diagonal and shuffled covariance matrix, 214–216
effects, 215–216
Gaussian distribution, 213
Shannon and Fisher information, 213
signal to noise ratio (d'), 213–214
empirical validation
assumption, 217–219
Battacharyya distance (BD), 219–220
classifier, 220–221
population effects
Gaussian estimator, 226
linear and nonlinear information, 224
quadratic estimator, 226–227
signal independent and signal dependent, 208
theoretical studies
data processing inequality, 211–212
information loss, 212
tuning curve model, 210–211
variance scaling, 211
Non-Competing Yes-Neurons (NCYN) model, 256–259
Normalized spatial kernel $K^{SR,Q}(\mathbf{r})$, 272

O
Okun, M., 1
Olfactory bulb, stochastic synchrony
circuits, 230
gamma oscillations, 231
50 Hodgkin–Huxley oscillators, 233–234
input/output correlations, 241–242
lateral inhibition, slow kinetics, 230–231
mitral cells
gamma oscillations, 231
noise-induced oscillatory synchrony, 232–233
oscillators frequency, 231–232
recurrent and lateral inhibition, 230–231
stochastic synchrony, 233–235
noise color and reliability
HH model, 239
Lyapunov exponent, 240
Monte Carlo simulation, 240–241
neuron spikes, 238–239
noise-induced oscillatory synchrony, 232–233
phase reduction and Lyapunov exponents, 238

linear adjoint equation, 236
oscillators, 235
phase resetting curve (PRC), 236–237
recurrent and lateral inhibition, 230
Optical imaging, 269
Oram, M.W., 10
Orientation tuning
containing $O(10^4)$ model neurons, 286
fluctuation-controlled criticality, 289–290
in input layer 4$C\alpha$ *of macaque V1, 271*
model V1 cortical area, 276
time scales of, 275
Ornstein–Uhlenbeck process, perception, 250
Oscillatory cortical networks, granger causality
autoregressive (AR) model
Akaike information criterion, 176–177
Bayesian information criterion (BIC), 177
causal influence, 171, 174
coherence, 173
coupling schemes, three time series, 175
LWR algorithm, 176
spectral matrix, 172–173, 177
spectral representation, 172
classification and rhythmicity, 169–170
cortical alpha rhythm, laminar organization
AR model, 183–185
current source density (CSD), 181–183
Fourier based power spectra, 184
infragranular (IG) and supragranular (SG) layers, 182, 186
multiunit activity (MUA) coherence, 182–183
phase index contact, 182
spectral analysis and power data, 185
spontaneous neural activity, 181
neuronal interaction analysis
LFP and MUA, 187–188
single-trial CSDs, 186
spectral analysis, 188
numerical simulations
pairwise analysis, 180
simple neural model, 178–180
three coupled AR models, 180
two variable model, 178

P

Panzeri, S., 212
Parietofrontal system. *See* Interceptive behavior, parietofrontal system
Perception, stochastic neural dynamics
brain dynamics
Chapman–Kolmogorov equation, 249
Fokker–Planck equation, 250
leaky integrate-and-fire (LIF) model, 248–249
mean-field technique, 250–251
population transfer function, 250
RC-circuit, 249
true dynamics generation, 251
computational model, probabilistic detection
Competing Yes-No-Neurons (CYNN), 259
detection curve plot, 260
minimal network model, 256–258
spiking simulations analysis, 257–259
neurophysiology
detection task trial, 251–253
medial premotor cortex (MPC) response, 254–255
near-threshold stimulus, 251
neurometric and psychometric curves, 252–254
stimulus and S1 neuron activity, 254
spiking noise, 247–248
Pérez, O., 191
Periodic forcing system
experiments, 104–105
Hopf bifurcation
asymmetry and multiplicity, 106–111
Q-factor, 115–116
simulations, 105–106
solution amplitudes, 111–115
network motif, 100–101
oscillator experiment, 105
simulations
amplitude ε, 103
cell dynamics, 102
differential equations, 101–102
dotted, dashed and solid curves, 103
Fitzhugh–Nagumo equations, 102–103
harmonic and spiking forcing, 104
Phase resetting curve, 236–237
Phase response curve, 83–84
Phase sensitivity, 268
Pipa, G., 143
Pitts,W., 62
Postlethwaite, C., 95
Postsynaptic currents (PSCs), 53
Postsynaptic potential (PSP), 54–55
Preferred cortical state V_P, 277
Prefrontal neurons, 126

Primary visual cortex (V1), 263
alternative Hikosaka stimuli, 286
bistability, 274, 288–290
computational
approach, 267–268
model (*see* Large-scale computational model)
iso-orientation domains, 275, 280
line-motion illusion, 281–286
optical imaging, anatomical, 269
orientation tuning (*See* Orientation tuning)
parsimonious modeling strategy, 265–267
physiological background, 268–271
similarity index, 276–278
drifting grating stimuli, 266
temporal dynamics, 279
sparsity, 288
spontaneous cortical activity patterns, 275–280
Proprioceptive signals, 124
PSP-based detection, high resolution index (PHRI), 54, 56, 57

R
Rangan, A.V., 263
Reig, R., 17
Reverberatory activity, neuronal networks
asynchronous synaptic transmission
calcium elevation, 69, 70
experimental evidence, 69
polysynaptic currents, 68
silent periods, 67
vs. synchronous phase, 69
cell assembly, 62
cultured neuronal networks
dynamic features, 66
hippocampal neurons, 64–65
primary cultures, 63–64
history, 61–63
intrinsic bistability *vs.* recurrent excitation, 66–67
mechanism, 71
native circuits, 72
short-term synaptic dynamics, 70–71
Reversal potentials ε^Q
Ricciardi, L., 250
Richmond, B.J., 212
Robinson, H.P., 238
Romo, R., 247, 252, 255, 259, 260
Rubin, J.E., 91

S
Salinas, E., 121, 129
Sanchez-Vives, M.V., 17
Schroeder, C.E., 169
Schultz, S.R., 212
Segundo, J.P., 239
Sejnowski, T.J., 239
Sensorimotor processing, circularly moving targets, 194–196
Shamir, M., 212, 225
Sherman, A., 91
Shiau, L., 95
Shu, Y., 40
Sohn, J.W., 127
Sompolinsky, H., 212, 224, 225
Spatiotemporal pattern, 265, 266, 282, 283
Spatiotemporal V1 neuronal activity, 266
Spike-rate hysteresis, in V1 model, 289
Spike–spike interactions, 274
Spike-triggered, spontaneous activity pattern, 275, 277
Spontaneous cortical activity, 275, 279, 280
Stoop, R., 97, 103, 111
Synaptic couplings, 265
Synaptic events, cerebral cortex network
amplitudes and time-course, 24–25
comparative statistics, 26
data analysis, 22, 43
intracellular and extracellular recordings, 21–22, 42
IPSPs and EPSPs detection, 23
method, 22–24
V_m deflections, 25
Synaptic fluctuations, 267, 288
Synaptic repeats, neocortical neuron, 57–59

T
Tao, L., 263
Task-dependent modulation, 126
Tateno, T., 238
Traub, R., 147
Treves, A., 212
Tuning curves, 287

U
Urban, N., 229

V
van Vreeswijk, C., 78, 86
Vegter, G., 100
Vicente, R., 143

Visual cortex (V1). *See* Primary visual cortex (V1)
Visually evoked activity, cortical neurons, 124
Visual motion processing, interceptive behavior, 193
Visual pathway, 268
Visual stimulation, 267, 270, 271

W
Wang, Q., 105
Wang, X.J., 79, 80
Weiner, N., 171
Wu, S., 212

Y
Young, L.S., 105

Z
Zachariou, M., 77
Zador, A.M., 2
Zero-lag long-range synchronization
  and anatomical structure, 159
  conduction delays
    action potential, 145
    axons, 146
    dendritic propagation, 146
    electrical synapses and gap junction, 146–147
    interneuron doublet, 147
    neuronal plasticity, 147
    synaptic potential, 145
  cortico-thalamic interactions, 159
  data analysis, 163–164
  dynamical relaying
    conduction delay and broad distribution effect, 151–153
    Hodgkin and Huxley (HH) cells module, 148–151
    large-scale neuronal networks, 153–157
  models
    excitatory and inhibitory transmissions, 161–162
    Poissonian distribution, 162
    temporal evolution, 160
    voltage-dependent transition rates, 160–161
    voltage dynamics, 162
  multielectrode technique, 143–144
  neocortex connectivity, 158
  simulations, 163
  temporal coding strategy, 159
Zhang, Y., 95
Zohary, E., 210, 211, 224

GPSR Compliance

The European Union's (EU) General Product Safety Regulation (GPSR) is a set of rules that requires consumer products to be safe and our obligations to ensure this.

If you have any concerns about our products, you can contact us on ProductSafety@springernature.com

In case Publisher is established outside the EU, the EU authorized representative is:

Springer Nature Customer Service Center GmbH
Europaplatz 3
69115 Heidelberg, Germany

Batch number: 10761599

Printed by Printforce, the Netherlands